A Special Issue of
Visual Cognition

Real-world scene perception

Guest Editor

John M. Henderson

Routledge
Taylor & Francis Group
LONDON AND NEW YORK

First published 2005 by Psychology Press Ltd

Published 2018 by Routledge
2 Park Square, Milton Park, Abingdon, Oxon OX14 4RN
52 Vanderbilt Avenue, New York, NY 10017

First issued in paperback 2018

Routledge is an imprint of the Taylor & Francis Group, an informa business

British Library Cataloguing in Publication Data
A catalogue record for this book is available from the British Library

ISSN 1350-6285

Cover design by Tom Hybert
Typeset in the UK by DP Photosetting, Aylesbury

ISBN 13: 978-1-138-87327-8 (pbk)
ISBN 13: 978-1-84169-993-6 (hbk)

Contents*

* This book is also a special issue of the journal *Visual Cognition* and forms Issue 6 of Volume 12 (2005). The page numbers used here are taken from the journal and so begin on p. 849.

VISUAL COGNITION, 2005, 12 (6), 849–851

Introduction to real-world scene perception

John M. Henderson

Michigan State University, East Lansing, MI, USA

Real-world scene perception involves sensory and cognitive processing of visual input that in some important sense is like that typically encountered during the natural course of everyday human activity. Unlike objects, which are viewed from a vantage point outside of themselves, scenes are specific views of the environment within which the viewer is embedded. Described in this way, real-world scene perception differs from perception of the visual stimuli that are often used in psychophysical and visual cognition experiments. For example, real-world scenes fill the entire visual field and produce specific wavelength, intensity, and spatial frequency profiles that are constrained by properties of the world itself. These regularities in image statistics arise in part because real-world scenes are composed of objects and backgrounds that have specific semantic and spatial constraints that in turn give rise to the semantics of the scene as a whole.

Scene can be defined as a semantically coherent (and often nameable) human-scaled view of a real-world environment comprising background elements and multiple discrete objects arranged in a spatially licensed manner (Henderson & Ferreira, 2004; Henderson & Hollingworth, 1999). Background elements are larger scale, immovable surfaces and structures, and objects are smaller scale discrete entities that move or can be moved within the scene. Real-world scenes have a hierarchical spatial structure. For example, a kitchen viewed from the vantage point of the doorway would likely include floor, ceiling, and walls as background elements, and a stove, refrigerator, and dishwasher as a few prominent objects among others, spatially arranged according to the laws of physics and the constraints associated with the functions of a kitchen. But, from the vantage point of standing in front of the counter, the countertop might be considered a scene, with its surface forming most of the background and a loaf of bread, knife, peanut butter jar, and plate serving as individuated objects, themselves spatially arranged according to the same physical laws and the semantics of sandwich preparation. One way to bound the

Please address all correspondence to: John M. Henderson, 285B Psychology Building, Michigan State University, East Lansing, MI 48824-1116, USA. Email: john@eyelab.msu.edu

DOI:10.1080/13506280444000544

concept of *scene* is to limit scale to that affording human locomotion and motor interaction (Henderson & Ferreira, 2004).

Spatial licensing includes adherence to the physical constraints of the universe, including gravity, space, and time, and to the semantic constraints imposed by object identity and function (Biederman, Mezzanotte, & Rabinowitz, 1982). Examples of the former constraint include the fact that objects generally cannot float unsupported in the air and that two objects cannot occupy the same region of space at the same time. Examples of the latter constraint include the fact that a fire hydrant does not belong on top of a mailbox and that couches are not typically found outdoors (Biederman et al., 1982). Some objects of course can float without visible support, and some objects do allow other objects to be seen through them (i.e., violate standard rules of interposition), so determining whether a physical constraint has been violated often requires knowledge of the identity and semantic attributes of the objects (Henderson, 1992). Nevertheless, a well-formed scene must ultimately conform to a set of basic physical and semantic constraints that impose structure on the environment.

Henderson and Ferreira (2004) noted two orthogonal dimensions along which the stimuli used to study real-world scene perception can differ. First, the stimuli in scene experiments can either be the real environment itself, or some form of depiction of it. Second, stimuli can be complete or can be degraded in some way. Henderson and Ferreira (2004) referred to the former as true real-world scenes, and the latter as ersatz scenes. Examples of ersatz scenes used in visual cognition experiments include simple arbitrary arrays of objects (either in the environment or in depictions) with no natural structure or semantic interpretation. To the extent that such exhortations are persuasive, we have urged investigators to reserve the term *scene* for true scenes, because there are likely to be important differences in how true scenes versus (for example) object arrays are processed visually and cognitively. For example, as argued in several of the papers in this issue, true real-world scenes are likely identified as coherent meaningful entities using global image properties that transcend the specific objects in them. Arrays, in contrast, have no route to semantic interpretation beyond that which can be inferred from the identities of the objects they contain, and so cannot benefit from this additional global level of visual analysis. In addition, perceptual factors such as the functional field of view are likely to differ for objects in simple arrays and in scenes because scenes are semantically denser and also because objects in scenes tend to obscure each other via lateral masking. Scene depictions capture some of the important properties of real-world scene perception by including what are thought to be the important properties of scenes such as visual complexity, structural and semantic constraint, and meaning, while allowing for control over factors that would be difficult if not impossible to hold constant in the actual environment. At the same time, scene depictions such as drawings or photographs necessarily reduce the amount of information normally available to the visual system.

The current volume brings together an eclectic group of investigators, all of whom study critical issues in the perception of true real-world scenes. Topics include the rapid acquisition of scene gist, scene recognition, spatial layout and spatial scale, distance perception in scenes, updating of scene views over time, visual search for meaningful objects in scenes, scene context effects on object perception, scene representation in memory, the allocation of attention including eye fixations during scene viewing, and the neural implementation of these representations and processes in the brain. Because the study of real-world scene perception benefits from an interdisciplinary approach, contributors to this special issue use a variety of research methods including psychophysical and behavioural techniques, eyetracking, functional neuroimaging (including fMRI and ERP), and mathematical and computational modelling. While much has been learned from studying simplified visual stimuli, many of the papers in this special issue make the important point that understanding the functional and neural architectures of the visual system requires studying how that system operates when faced with the types of real-world stimuli that evolution crafted it to handle.

REFERENCES

Biederman, I., Mezzanotte, R. J., & Rabinowitz, J. C. (1982). Scene perception: Detecting and judging objects undergoing relational violations. *Cognitive Psychology*, *14*, 143–177.

Henderson, J. M. (1992). Object identification in context: The visual processing of natural scenes. *Canadian Journal of Psychology*, *46*, 319–341.

Henderson, J. M., & Ferreira, F. (2004). Scene perception for psycholinguists. In J. M. Henderson & F. Ferreira (Eds.), *The interface of language, vision, and action: Eye movements and the visual world*. New York: Psychology Press.

Henderson, J. M., & Hollingworth, A. (1999). High-level scene perception. *Annual Review of Psychology*, *50*, 243–271.

VISUAL COGNITION, 2005, 12 (6), 852–877

How long to get to the "gist" of real-world natural scenes?

Guillaume A. Rousselet, Olivier R. Joubert, and Michèle Fabre-Thorpe

Centre de Recherche Cerveau et Cognition, UMR 5549 (CNRS-UPS), Toulouse, France

This study aimed at assessing the processing time of a natural scene in a fast categorization task of its context or "gist". In Experiment 1, human subjects performed 4 go/no-go categorization tasks in succession with colour pictures of real-world scenes belonging to 2 natural categories: "Sea" and "mountain", and 2 artificial categories: "Indoor" and "urban". Experiment 2 used colour and grey-level scenes in the same tasks to assess the role of colour cues on performance. Pictures were flashed for 26 ms. Both experiments showed that the gist of real-world scenes can be extracted with high accuracy (>90%), short median RT (400–460 ms) and early responses triggered with latencies as short as 260–300 ms. Natural scenes were processed faster than artificial scenes. Categories for which colour could have a diagnostic value were processed faster in colour than in grey. Finally, processing speed is compared for scene and object categorization tasks.

Natural scenes are more than a simple collection of objects. However, much of the research on scene processing has been devoted to the understanding of object processing in scenes, leaving aside the question of how we process the whole scene itself.

This issue is important given that we do not only process objects but we also analyse the context in which they appear. Global coarse information about a scene (mainly its category, or *gist*, and its spatial structure, or *layout*) is relatively crucial in memory-free models of scene perception in which little information is integrated across saccades. According to this idea, perception is constructed by integrating abstract scene representations with volatile object

Please address all correspondence to: Guillaume A. Rousselet, McMaster University, Department of Psychology, Hamilton, L8S4K1, ON, Canada. Email: rousseg@mcmaster.ca

This work was supported by the CNRS, the ACI "Integrative and Computational Neurosciences", and the Cognitique grant no. IC2. Financial support was provided to G. A. Rousselet by a PhD grant from the French government. We thank Anne-Sophie Paroissien for her very valuable help running the experimental sessions in Experiment 1.

DOI:10.1080/13506280444000553

representations available at the locus of attention (O'Regan, 1992; Rensink, 2000, 2002; Wolfe, 1999; but see Henderson & Hollingworth, 2003; Hollingworth, 2003; Hollingworth & Henderson, 2002; Simons, Chabris, Schnur, & Levin, 2002). More generally, visual context has been shown to guide attention toward potential target objects (for a review see Chun, 2000). The influence of the background of a scene on object identification is still controversial (see reviews in Henderson, 1992; Henderson & Hollingworth, 1999), with evidence both in favour and against such a view (see among many others Biederman, Mezzanotte, & Rabinowitz, 1982; Boyce, Pollatsek, & Rayner, 1989; but see Ganis & Kutas, 2003; Hollingworth & Henderson, 1998, 1999). But our knowledge about the global processing of a scene is severely limited compared to the knowledge accumulated about object processing. In particular, for scene context to influence object identification, one fundamental constraint is the speed at which scene context can be extracted.

What do we know exactly about scene processing? Only recently this topic has received more attention from cognitive neuroscience researchers, revealing a set of cortical areas involved in different aspects of scene processing, that include the parahippocampal and parietal cortices (Nakamura et al., 2000; Sato et al., 1999). More specifically, the parahippocampal area has been attributed functions such as the processing of the spatial layout of the scene (Epstein, Graham, & Downing, 2003; Epstein & Kanwisher, 1998) and the learning and recognition of buildings and landscapes (Maguire, Frith, & Cipolotti, 2001). It is also thought to mediate, in conjunction with the retrosplenial cortex, both spatial and nonspatial contextual processing (Bar & Aminoff, 2003). Finally, the right lingual sulcus has been implicated in the perception of buildings (Aguirre, Zarahn, & D'Esposito, 1998). This distributed system, largely separated from the object system, might lead to the hypothesis that scenes are processed very efficiently, perhaps fast enough to be able to influence object processing.

Scene categorization is often regarded as the ultimate representation generated along the ventral pathway in which a scene would be reconstructed progressively by integrating local contrasts (Marr, 1982). Following such a view, objects would be processed almost systematically before the scene (Biederman, 1987; Riesenhuber & Poggio, 2000).

Alternatively, many studies have suggested that scene categorization can be performed very efficiently from very brief visual presentations (Biederman et al., 1982; Intraub, 1997; Oliva & Schyns, 1997, 2000; Potter, 1975, 1976; Potter & Levy, 1969; Schyns & Oliva, 1994). The fact that scene categorization is possible with very brief presentations is often taken as an evidence for fast underlying mechanisms. Thus, scene categorization could be performed simultaneously or even precede object identification. Indeed, although some theories have suggested that scene categorization might result from the identification of the component objects (e.g., Friedman, 1979), other theories have emphasized that scenes might also be identified from scene-specific cues. For instance,

Biederman (1988) suggested that his original structural model of object recognition using ''geons'' (3-D primitives) might be extended to scene recognition. He proposed that primitives with a larger spatial scale than those used to represent objects could represent scene specific information independently of object information. Although this proposal has not been tested empirically, Henderson and Hollingworth (1999) have suggested that, given the lack of strong constraints on their structural organization, scenes are not likely to be represented as large objects. However, it remains possible that the spatial organization of a scene (even at a coarse level) might mediate its identification. Sanocki and Epstein (1997, p. 378) suggested that the representations of the spatial layout ''may include information about the extent and location of the ground plane and other reference objects and surfaces, as well as size and distance relations within the scene''. This information about the spatial structure of the scene might be used to extract its meaning. Indeed, the gist of a scene could be extracted from low spatial frequency versions of photographs preserving coarse spatial layouts, or spatially arranged colour blobs, but in which information to categorize component objects was not preserved (Oliva & Schyns, 2000; Schyns & Oliva, 1994). This strengthens the idea that object and scene categorization might be mediated by distinct visual cues. Finally, computational evidence suggests that scene-based visual filters derived from the combination of a restricted set of low-level filters might be sufficient to perform most of the visual discrimination needed to categorize the context of our environment (Oliva & Torralba, 2001; Torralba & Oliva, 2003).

Given such a reduced dictionary of scene-based physical properties and the known capacity of the visual system to dynamically adjust its strategies as a function of task constraints to pick the most adequate image features (Schyns, 1998), it is plausible that scenes might be identified very fast, probably as fast or even faster than objects. So far, the evidence for fast scene processing comes from experiments using brief visual presentations (see above). Unfortunately, brief visual presentations and particularly RSVP sequences (Rapid Sequential Visual Presentations; e.g., Intraub, 1997; Potter, 1975, 1976; Potter & Levy, 1969) provide a rate of visual processing rather than an absolute evaluation of the visual processing time. Furthermore, experiments like those performed by Oliva and Schyns (1997, 2000) used vocal responses or involved a matching task between a written name and a visual scene with a two-choice response. Using 16 different categories in the matching task, they showed that the averaged mean reaction times (RT) were largely distributed from 476 ms (*city* category) to 631 ms (*valley* category).

This mean RT can be compared to the 400 ms mean RT that has been commonly found for object categorization in various studies from our group. These studies employed a go/no-go animal categorization task in which human subjects were required to respond as fast and as accurately as possible each time a natural photograph, that was flashed for the first time and for only 20–40 ms, contained an animal (Delorme, Richard, & Fabre-Thorpe, 2000; Fabre-Thorpe, Delorme, Marlot, & Thorpe, 2001; Thorpe, Fize, & Marlot, 1996). This finding has been extended to other object categories such as means of transport (VanRullen & Thorpe, 2001), human faces and animal faces (Rousselet, Macé, & Fabre-Thorpe, 2003), although food objects might take up to 30 ms longer (Delorme et al., 2000). Compared with Schyns and Oliva's studies, objects might thus be identified before extraction of scene context. But the difference in processing time might also originate in the motor response required in both tasks as our go/no-go task relies only on a single motor output, whereas their matching task requires a choice of response. In keeping with this idea, Goffaux, Jacques, Mouraux, Oliva, Schyns, and Rossion (2005) report an averaged median reaction time of 428 ms in a go/no-go verification task.

In this study, we have assessed the time course of the categorization of the scene context, or its ''gist'', with the same go/no-go visual categorization task previously used to study object categorization. We selected four categories of colour pictures, two of them were natural, ''sea'' and ''mountain'', and the two others were artificial, ''indoor and urban'' scenes. In a first experiment, subjects were asked to perform four go/no-go categorization tasks, one per category. A second experiment was designed to assess the effect of colour on gist processing speed.

EXPERIMENT 1

The aim was to provide an estimate of the temporal constraints in the visual processing of the gist of a natural scene, for four scene categories representative of our environment (sea, mountain, urban, and indoor). These categories were relatively coarsely defined in order to present subjects with pictures that were as varied as possible (see Stimuli and Figure 1).

Method

Participants. Twenty-four adults (12 women and 12 men, mean age 30, ranging from 19 to 51, three of them left handed), volunteered in this study and gave their informed written consent. All participants had normal or corrected to normal vision.

Stimuli. We used 24-bit (16 million colours) photographs of natural scenes (768 × 512 pixels, sustaining a visual angle of about 15.6° × 10.5°) taken from a large commercial CD-ROM library (Corel Stock Photo Libraries). From this databank, we selected 384 images for each of the four environmental categories. For each category, half of them were horizontal photographs, the other half were vertical photographs. They were all chosen to be as varied as possible, representing the four types of scenes from a large range of viewpoints and

Figure 1. Tasks and stimuli in Experiments 1 and 2. A. Tasks: While performing one of the four scene categorization tasks (sea, mountain, indoor, urban), nontargets belonged equally to the three other categories. Note the variety of stimuli used in this experiment. B. Pixel by pixel average picture for each stimulus category presented separately for horizontal and vertical images. For each category, the top two images represent the raw average pictures and the two bottom images are the equalized versions obtained using the "equalize" function in Photoshop 5.5. For each colour channel and the luminance channel, the function attributes a "black" value to the darkest pixel and a "white" value to the brightest one. It then redistributes regularly the intermediate pixel values of the distribution between these two extremes. C.. For each category, the nine target pictures associated with the fastest reaction times of individual subjects in Experiment 1 are presented. In Experiment 1, all stimuli were in colour while in Experiment 2 half were in colour and half in grey levels. To view this figure in colour please see the online issue.

perspectives (Figure 1). Each image was seen only once by a given subject to prevent learning.

Sea pictures were composed of various coast scenes (including beach scenes, cliff scenes, or showing various rocks, icebergs, etc.) as well as ''full sea'' pictures with boats, sailboards, and surfboards. In all cases the sea was clearly visible on the pictures. The mountain category contained pictures that showed large mountain backgrounds at different distances in all seasons as well as various photographs taken from the point of view of mountain hikers. Urban pictures were almost exclusively taken from the point of view of someone walking in towns ranging from small villages to large cities. Photographs depicted streets, buildings, houses, public squares, etc., from many places around the world. Indoor scenes were photographs taken from inside various urban constructions like houses, apartments, churches, museums, and stores, and so on.

There was no overlap in the pictures from the four target categories: Sea scenes did not contain harbours or mountains in the background; mountain scenes did not contain villages; street scenes did not include streets constructed along the sea, etc.

Procedure. Image presentation and behavioural response measurement were carried out using the software Presentation (NeuroBehavioral Systems; http://nbs.neuro-bs.com/). Subjects sat in a dimly lit room at 100 cm from a computer screen (horizontal resolution = 1024 pixels, vertical resolution = 768 pixels, vertical refresh rate: 75 Hz) piloted by a PC computer. To start a block of trials, they had to place their finger on an infrared response pad for 1 s. A trial was organized as following: A fixation cross (0.1° of visual angle) appeared for 300–900 ms and was immediately followed by the stimulus presented for two frames, i.e., about 26 ms, in the middle of the screen. These brief presentations prevented any exploratory eye movements. Participants had to lift their finger as quickly and as accurately as possible (go response) each time a target scene was presented and to withhold their response (no-go response) when the photograph did not belong to the target category. Responses were detected using infrared diodes. Subjects had 1000 ms to respond; longer reaction times were considered as no-go responses. This maximum response time delay was followed by a 300 ms black screen, before the fixation point was presented again for a variable duration, resulting in a random 1600–2200 ms intertrial interval.

Subjects were tested in two experimental sessions on two different days. A given picture category was the target category for four consecutive series. In each session they performed two categorization tasks for a total of eight blocks of 96 trials with target and nontarget trials being equiprobable. This led to a total of 3072 trials per subject. The order in which the subjects performed the four category tasks was counterbalanced across subjects. In a given task, the 48 nontarget images belonged equally to the three other environmental categories.

Thus, when performing the sea categorization task, a 96 trial series contained 48 target sea pictures, 16 nontarget mountain scenes, 16 nontarget urban scenes, and 16 nontarget indoor scenes. To avoid any bias, the design was organized so that across subjects each of the 384 pictures of a given category was seen 12 times as a target and 12 times as a distractor. Furthermore, when seen as a distractor, each image appeared the same number of times in the three different categorization tasks. Subjects had one training block of 48 images before starting the four series of a given categorization task. Training pictures were not used during testing.

Performance was evaluated by determining the percentage of correct trials and the latency (computed between stimuli onset and finger lift) at which subjects triggered their finger movement response. When repeated measures ANOVA were used, a Greenhouse-Geisser correction for nonsphericity was applied.

Results

Performance in the four tasks was evaluated by analysing separately accuracy and reaction times (RT). A summary of the results can be found in Table 1.

Accuracy. Subjects performed remarkably well in the four tasks. Mean accuracy was 96.2% correct in the sea categorization task; 95.6% in the mountain task; 95.5% in the indoor task; and 95.1% in the urban task. A one-way analysis on ranks was performed on these data and showed no significant effect, Friedman test: $\chi^2(3\ df) = 6.8$, n.s. We then analysed separately go and no-go responses.

In all four tasks subjects were better at responding on target scenes (97.4% correct) than at withholding their response on nontarget trials (93.8% correct). Correct go responses reached 98.1% in the sea task; 97.5% in the mountain task; 97.0% in the indoor task; and 97.1% in the urban task. These results were not homogenous, Friedman test: $\chi^2(3\ df) = 10.4$, $p = .016$, subjects scored better with the sea targets. Planned post hoc Wilcoxon tests showed that this higher accuracy with sea scenes reached significance when compared to indoor and mountain targets (both $Z < -2.3$, both $p < .02$). All other comparisons failed to reach significance.

Correct no-go responses reached 94.5% in the sea task; 93.6% in the mountain task; 93.9% in the indoor task; and 93.4% in the urban task. Sea and mountain scenes both belonged to natural categories, whereas indoor and urban scenes belonged to artificial categories. This means that, in each categorization task, one third of the distractors had a very strong relationship with the target category. Thus, the performance on distractors was studied separately depending on whether distractors were "related" or "unrelated" to the target category. Data were analysed using repeated measures ANOVA with category (four

TABLE 1

Experiment 1: Summary of results. Correct no-go related (or unrelated) accuracy refers to correctly categorized distractor images that belonged (or did not belong) to the same high-level category (natural vs. artificial scenes) as the target images

	Sea	*Mountain*	*Indoor*	*Urban*
Accuracy (%)				
Mean	96.2 (1.9) [91.1–99.7]	95.6 (2.6) [87.8–98.4]	95.5 (2.6) [88.0–98.4]	95.1 (2.7) [88.0–99.2]
Correct go	98.1 (2.4) [88.5–100]	97.5 (1.8) [92.2–99.5]	97.0 (2.6) [91.1–100]	97.1 (3.1) [86.0–100]
Correct no-go	94.5 (2.7) [88.5–99.5]	93.6 (4.5) [81.8–98.4]	93.9 (3.7) [84.9–99.0]	93.4 (4.8) [80.7–99.0]
Correct no-go related	85.9 (7.9)	83.5 (10)	84.3 (9.7)	83.8 (13)
Correct no-go unrelated	98.8 (1.2)	98.7 (2.0)	98.7 (1.2)	98.2 (1.7)
d'	3.9 (0.6) [2.7–5.7]	3.7 (0.4) [2.4–4.3]	3.7 (0.6) [2.4–4.6]	3.6 (0.6) [2.5–5.3]
RT (ms)				
Mean	422 (37) [348–494]	444 (46) [358–535]	466 (50) [373–555]	482 (45) [403–569]
Median	405 (37) [332–477]	425 (45) [334–499]	448 (47) [359–523]	463 (42) [384–538]
Minimal RT (ms)				
Overall data	260	290	300	300
Individual data	331 (28) [280–370]	346 (35) [290–420]	363 (32) [310–420]	372 (36) [310-460]

Standard deviation is indicated in brackets. Range of individual responses is indicated in square brackets [min-max].

levels) and related/unrelated (two levels) as within-subject factors. The analysis showed that subjects performed equally well at ignoring distractors regardless of the task, category factor: $F(2.6, 23) = 1.0$, n.s. However, correct no-go responses were strongly modulated by the categorical relationship between distractors and targets. Accuracy was significantly worse with distractors that belonged to the related category (84.4%) compared to the two others (98.6%), related/unrelated effect: $F(1, 23) = 82.5, p < .0001$. This was true for all categories (no interaction with the category factor; further confirmed by separate Wilcoxon tests on each category), all $Z < -4.1$, all $p < .0001$.

Reaction times. Although the analysis of accuracy did not reveal major differences between tasks, speed of processing measured by mean and median reaction times (RT) differed strongly between the four categorization tasks, Friedman tests: Both $\chi^2(3\ df) > 42$, both $p < .0001$. Mean and median RT were respectively 422/405 ms in the sea task; 444/425 ms in the mountain task; 466/448 ms in the indoor task; and 482/463 ms in the urban task. All two by two comparisons on mean and median RT were significant, Wilcoxon tests: All $Z < -2.6$, all $p < .01$, except the comparisons between the urban task and the indoor task (Figure 2A). Thus the four tasks were ranked according to processing speed as follows: (1) Sea, (2) mountain, (3) indoor = urban. These differences in processing speed can be observed in the RT distributions of Figure 3. Speed of processing was thus faster for the sea context and was also less variable (Figure 3, A, B, C, and D) as shown by a narrower RT distribution in the sea task compared to the mountain task (standard deviation: Sea = 37 ms, mountain = 46 ms, indoor = 50 ms, urban = 45 ms). Moreover this faster processing speed for sea pictures and, to a lesser extent, for mountain pictures, could be seen even on the fastest responses produced by the subjects. Thus, a complete shift of the RT distributions towards shorter latencies could be seen for sea and mountain pictures (Figure 3E). Expressing performance as cumulative d' curves as a function of time revealed that discriminative information was available earlier in the sea task than in the three other tasks and accumulated faster to reach a higher value (Figure 3F).

We then assessed more directly whether this average processing speed ranking of the four tasks was also true for the earliest responses that could be triggered. As target and distractor trials were equally likely, a random behaviour should equalize hits and false alarms; hence the minimal behavioural processing time was determined by the latency at which correct go responses started to significantly outnumber incorrect go responses, $\chi^2(1\ df)$, $p < .001$, using a noncumulated RT histogram with 10 ms time bins. Such early correct go responses cannot be considered as anticipations. The analyses were performed both on the overall data (pooling together all trials from all subjects), and for each subject separately. With the overall data set, the minimal processing time was 260 ms in the sea task, 290 ms in the mountain task, and 300 ms in both the

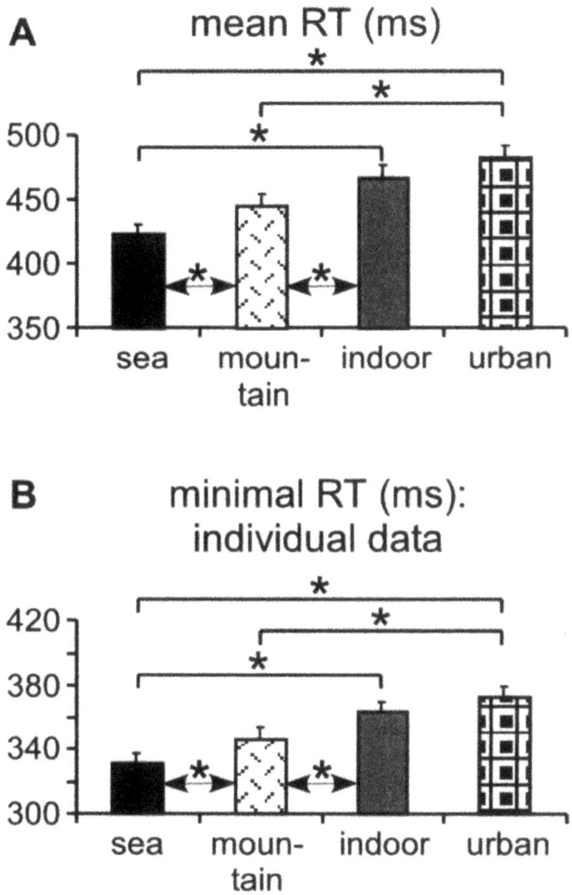

Figure 2. Mean and minimal reaction time with associated standard errors obtained for each of the four scene categorization tasks in Experiment 1. Asterisks indicate statistically significant differences (see text for details). In B, asterisks correspond to Wilcoxon tests where all $Z < -2.4$ and all $p < .02$.

indoor and the street task. Individual data (computed using cumulated RT histograms with 10 ms time bins) confirmed this tendency with a mean individual minimal processing speed of 331, 346, 363, and 372 ms respectively for sea, mountain, indoor, and urban target photographs (Figure 2B).

In conclusion, natural environments could be classified faster than artificial environments and this was true for the whole range of responses produced, from the earliest to the latest responses. Furthermore, among natural environments, sea scenes presented a clear processing speed advantage over mountain scenes. Although the accuracy performance was not very different between the four

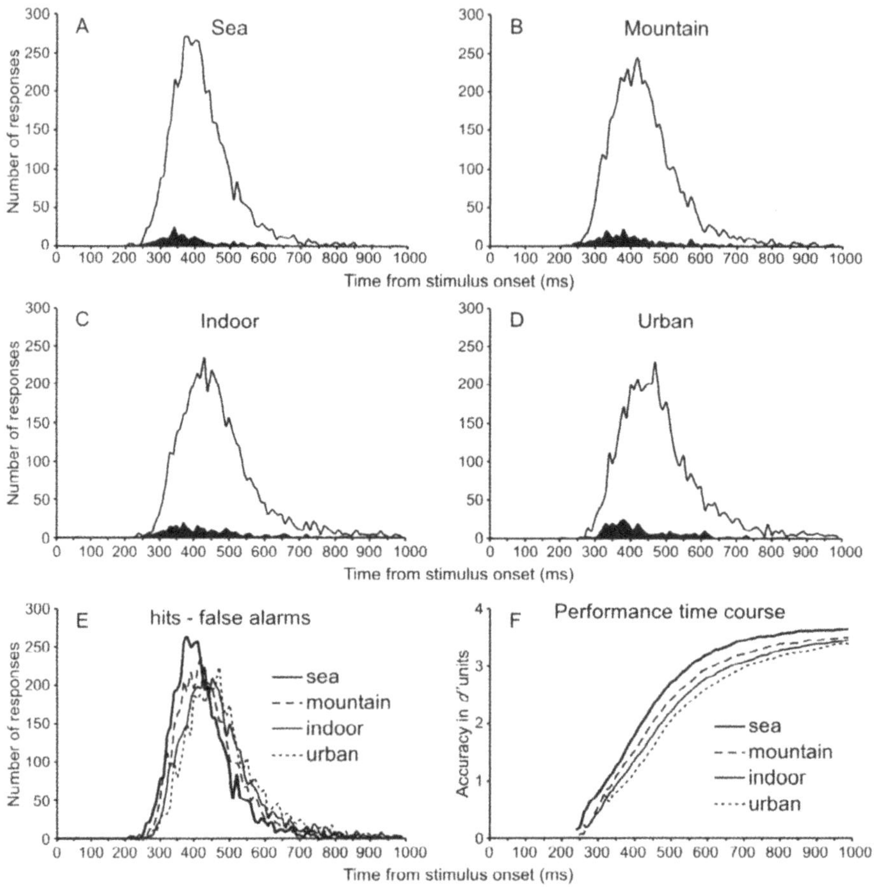

Figure 3. Time course of visual processing in Experiment 1. From A to D, RT distributions on correct (upper curve) and incorrect go responses (black histogram) are presented with the number of responses expressed over time, with 10 ms time bins. In E, false alarm distributions have been subtracted from hit distributions to allow a more direct comparison of the four tasks. In F, average performance accuracy (in *d'* units) is plotted as a function of processing time with 10 ms time bins. Cumulative numbers of responses were used. The *d'* was calculated from the formula $d' = z_n - z_s$, where zn is chosen such that the area of the normal distribution above that value is equal to the false-alarm rate, and where zs is chosen to match the hit rate. Note that the *d'* calculated here is not presumed to represent the actual distributions of signal and noise that underlie performance in the response time task. By taking into account the hit and FA rates in a single value at each time point, this time course of performance gives an estimation of the processing dynamics for the entire subject population. The plateau values correspond to the *d'* calculated from the overall accuracy results.

tasks, the rate of information processing was higher in the natural scenes (especially the sea scenes) compared to the artificial scenes.

Comparison with an object categorization task. Results presented above show that the gist of a natural scene flashed for 26 ms can be extracted both very efficiently and very quickly. But how fast is that processing compared to the categorization of objects in natural scenes? Previous studies from our laboratory have extensively assessed the performance of human subjects with the same go/no-go categorization task using animal as target category. A recent study (Rousselet et al., 2003) is particularly appropriate to compare the present human performance on global scene categorization with animal categorization because the same number of subjects were tested ($n = 24$) with the same number of trials per category, an identical set-up, the same image data bank, the same number of trials per category, and the same behavioural procedures (subjects had to alternate between the animal categorization task and a human face categorization task). A similar level of accuracy was also reached in the animal task (96.3%, n.s.) but the speed of processing was faster than with scenes. This faster processing was seen when using the median RT, which was significantly shorter in the animal task (371 ms) than in any of the four scene categorization tasks used here; two by two comparisons using Mann-Whitney tests: All $U < -2.8$, all $p < .005$. The fastest discriminative responses were found at the same latency than in the sea categorization task; thus earlier than for any other scene context, Mann-Whitney tests: All $U < -2.6$, all $p < .01$, an effect that was true for both the overall data set (260 ms in both tasks) and the individual data. On the other hand, performance accuracy increased more rapidly in the animal task than in the sea task. This can be seen on the RT distribution and even more clearly when accuracy performance is expressed in function of time by a d' curve (Figure 4).

Discussion

This first experiment confirmed that the general meaning of a visual scene can be extracted both very rapidly and highly efficiently with only brief visual presentations (e.g., Biederman, 1972; Potter, 1975, 1976). It also sets a minimal and an average processing time, respectively in the range of 260–300 ms and 400–460 ms, to extract this meaning from natural photographs. These figures are not very different from those obtained for object categorization in a variety of studies performed in our group. However a difference clearly emerged between natural scenes (sea, mountain) and artificial scenes (indoor and urban), natural scenes being categorized faster.

Among many properties that might be used to categorize natural scenes, the most obvious one, and probably the easiest to test, is colour (e.g., Torralba & Oliva, 2003). For instance, it is very plausible that colour was used as a

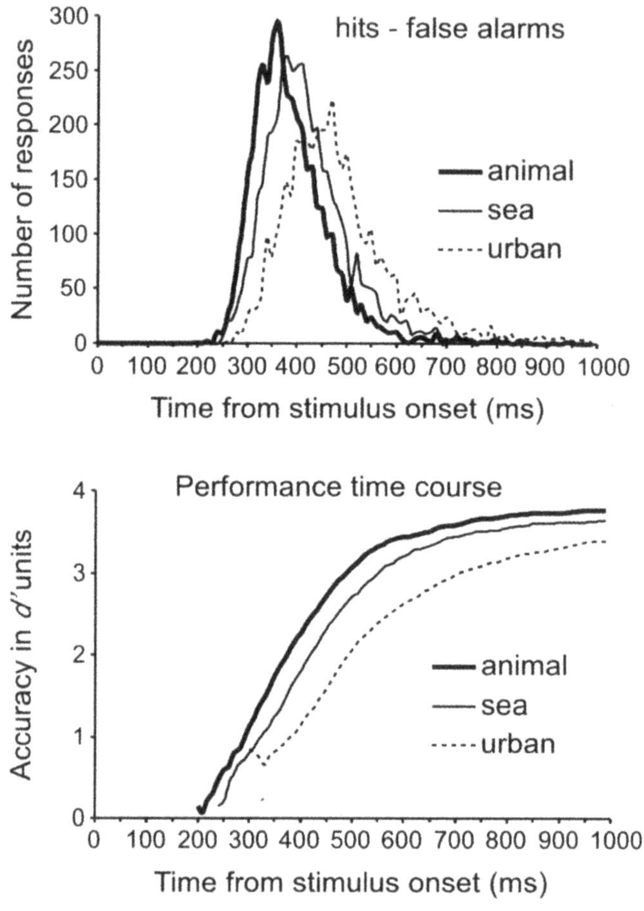

Figure 4. Comparison between scene and object categorization. Top panel: The RT distribution obtained in a preceding study (Rousselet et al., 2003) on an object categorization task (target = animal) is compared to the RT distributions obtained with the fastest (sea task) and the slowest (urban task) gist categorization tasks (false-alarm distributions have been subtracted from hit distributions to allow a more direct comparison). Bottom panel: The *d'* curves show that signal discrimination started earlier in the animal task compared to the sea task, a processing speed advantage that was maintained over the entire range of response latencies. For more details see text and caption of Figure 3.

diagnostic cue to categorize sea scenes. The influence of colour would be maximal in the sea task, because of large blue textured surfaces in the lower part of the picture and might explained the faster speed of processing in this task. The importance of colour cues depends on whether they constitute diagnostic features of the target category (Oliva & Schyns, 2000; Tanaka & Presnell, 1999). Colour contrasts may also be used to improve image segmentation, accelerating image analysis (Gegenfurtner, 2003). For example, the importance of colour has already been demonstrated in a recognition test using natural scenes (Gegenfurtner & Rieger, 2000). Indeed, such diagnostic cues could allow the presetting of specific groups of "diagnostic" neurons and speed up the processing of the expected visual information (Delorme, Rousselet, Macé, & Fabre-Thorpe, 2004).

The use of colour cues in the rapid categorization of objects has been investigated and it was unexpectedly shown that removing chromatic information had little effect on average accuracy and speed of processing as well as on minimal processing speed (Delorme et al., 2000). But colour might not be such an efficient diagnostic cue in animal categorization as in global scene categorization. Experiment 2 was thus designed to test the effect of removing colour cues on the categorization of the gist of natural scenes.

EXPERIMENT 2

In this second experiment, we wanted to assess more specifically how the removing of colour information from natural scenes would impair performance and whether colour could be considered as one of the cues mediating the fast processing of natural scenes revealed in the first experiment. Thus, we tested another group of subjects using the same paradigm and the same set of images employed in Experiment 1. The difference between the two experiments was that half of the images were presented in colour and half in black and white (BW pictures; 256 grey levels). Colour and BW images were mixed at random in the series of stimulation in order to prevent subjects from relying on different strategies when categorizing colour and BW pictures and to allow a more direct comparison of human performance in these two conditions.

Method

Participants. Twenty-four adults (12 women and 12 men, mean age 30, ranging from 20 to 52, three of them left handed) volunteered in this study and gave their informed written consent. All participants had normal or corrected to normal vision. None of them had been tested in the first experiment.

Stimuli. The same set of natural scenes photographs used in Experiment 1 served as stimuli in Experiment 2. For each 24-bit (16 million colours) photograph, an 8-bit version (256 grey levels) was generated using Photoshop 5.5.

Procedure. The design of Experiment 2 was identical to the one of Experiment 1 except on two points. First, subjects were tested in a single session. Second, subjects were presented with 50% colour and 50% BW photographs. Thus, each experimental condition was subdivided into two colour conditions. The design was counterbalanced so that across subjects each image was seen the same number of times in colour and in black and white.

Results

In Experiment 2, the mean accuracy for colour images was not significantly different from the one reached in the first experiment (95.1% and 95.6% correct respectively; between-subject analysis on ranks), Mann-Whitney test: $U = 281$, $Z = -0.4$, n.s. There was also a nonreliable tendency in favour of faster responses in the first experiment compared to the second (mean RT were 454 ms and 476 ms respectively), $U = 216$, $Z = -1.5$, n.s. As in Experiment 1, we analysed separately accuracy and reaction time data. A summary of the results can be found in Table 2.

Accuracy. Global accuracy was very good in Experiment 2. Data were entered in a repeated measure ANOVA with category (four levels) and colour (two levels) as within-subject factors. This analysis showed that the levels of accuracy reached with colour and BW images were not significantly different (colour = 95.1%; BW = 94.7%), $F(1, 23) = 3.8$, n.s. This was true for all four categories of natural scenes as there was no significant interaction between category and colour factors.

Contrary to Experiment 1, accuracy on go and no-go responses did not differ significantly from one another (go = 95.4%; no-go = 94.5%), $F(1, 23) = 1.2$, n.s. Separate analysis showed that the removal of colour cues had no significant effect on either go responses (colour = 95.5%; BW = 95.2%), $F(1, 23) = 0.7$, n.s., and no-go responses (colour = 94.8%, BW = 94.1%), $F(1, 23) = 3.3$, n.s. In addition, there was no difference in accuracy between the four tasks for both go responses, $F(2.2, 51) = 0.2$, n.s., and no-go responses, $F(2.4, 55) = 2.2$, n.s. There was no significant interaction between colour and category factors.

As previously found in Experiment 1, no-go responses were made more frequently toward distractors that belonged to the same higher level category as the targets (natural versus artificial). In other words, subjects proved much better at categorizing distractors that were unrelated (98.6%) than related (86.0%) to the target category, $F(1, 23) = 92.6$, $p < .0001$. The only effect induced by the removal of colour cue was seen on related distractors in the urban task: Indoor pictures were correctly ignored with a higher accuracy when presented in colour (88.2%) than in BW (82.9%), $Z = 2.3$, $p = .02$.

Reaction times. An ANOVA analysis showed that reaction times were affected both by the category of the target scene (category effect on both median and mean RT, both $F > 26$, $p < .0001$) and by the availability of colour cues (mean RT): $F(1, 23) = 9.1$, $p = .006$; median RT: $F(1, 23) = 5.2$, $p = .03$, so that

TABLE 2

Experiment 2: Summary of results. For each condition, the results are indicated separately for colour pictures (colour) and for grey-level pictures (BW)

	Sea		*Mountain*		*Indoor*		*Urban*	
	Colour	*BW*	*Colour*	*BW*	*Colour*	*BW*	*Colour*	*BW*
Accuracy (%)								
Mean	95.1 (2.6)	95.6 (2.8)	95.1 (3.4)	94.5 (2.8)	95.2 (4.3)	94.5 (3.5)	95.0 (3.5)	94.2 (3.1)
	[89–99]	[86–99]	[83–99]	[84–98]	[77–98]	[82–99]	[84–99]	[87–98]
Correct go	95.0 (3.2)	95.6 (3.4)	95.1 (5.2)	95.1 (4.5)	96.1 (7.1)	95.1 (5.3)	95.7 (4.8)	95.1 (4.5)
	[86–100]	[88–100]	[74–100]	[78–100]	[64–100]	[75–100]	[81–100]	[82–100]
Correct no-go	95.2 (3.8)	95.5 (3.4)	95.1 (3.3)	93.9 (3.2)	94.4 (3.7)	93.9 (4.0)	94.4 (4.6)	93.2 (4.7)
	[85–99]	[85–100]	[86–100]	[88–99]	[86–99]	[86–100]	[80–100]	[83–100]
Correct no-go related	87.2 (10.8)	88.8 (9.2)	87.1 (8.8)	83.7 (8.1)	86.2 (8.7)	84.1 (10.3)	88.2 (7.4)	82.9 (11.3)
Correct no-go unrelated	99.2 (1.5)	98.8 (1.5)	99.0 (1.9)	99.0 (1.6)	98.5 (1.9)	98.8 (2.0)	97.5 (3.8)	98.4 (1.9)
d'	3.5 (0.6)	3.6 (0.6)	3.5 (0.6)	3.5 (0.6)	3.7 (0.6)	3.5 (0.8)	3.6 (0.6)	3.4 (0.6)
	[2.4–5.2]	[2.2–4.5]	[2.1–4.9]	[2.0–4.9]	[1.7–4.3]	[1.9–5.0]	[2.0–5.2]	[2.3–4.6]
RT (ms)								
Mean	443 (56)	461 (60)	471 (58)	462 (63)	493 (70)	503 (65)	498 (55)	499 (58)
	[300–566]	[323–586]	[359–603]	[332–595]	[344–640]	[352–632]	[347–599]	[360–636]
Median	429 (56)	444 (61)	452 (59)	443 (61)	475 (70)	485 (67)	479 (57)	478 (61)
	[288–549]	[310–589]	[336–572]	[320–572]	[314–619]	[324–625]	[318–580]	[322–615]
Minimal RT (ms)								
Overall data	290	300	310	310	320	340	330	310
Individual data	388 (45)	398 (50)	401 (50)	400 (54)	425 (64)	434 (55)	429 (48)	432 (53)
	[280–490]	[280–520]	[300–500]	[290–510]	[300–600]	[320–560]	[310–520]	[310–580]

For other details see Table 1 caption.

speed of performance was analysed separately on colour and BW pictures for each categorization task (Figure 5).

However the results were not consistent from one category to another. Whereas sea and indoor pictures were categorized faster in colour than in BW—about 15 ms and 10 ms faster respectively; Wilcoxon tests: both $Z < -3.2$, both

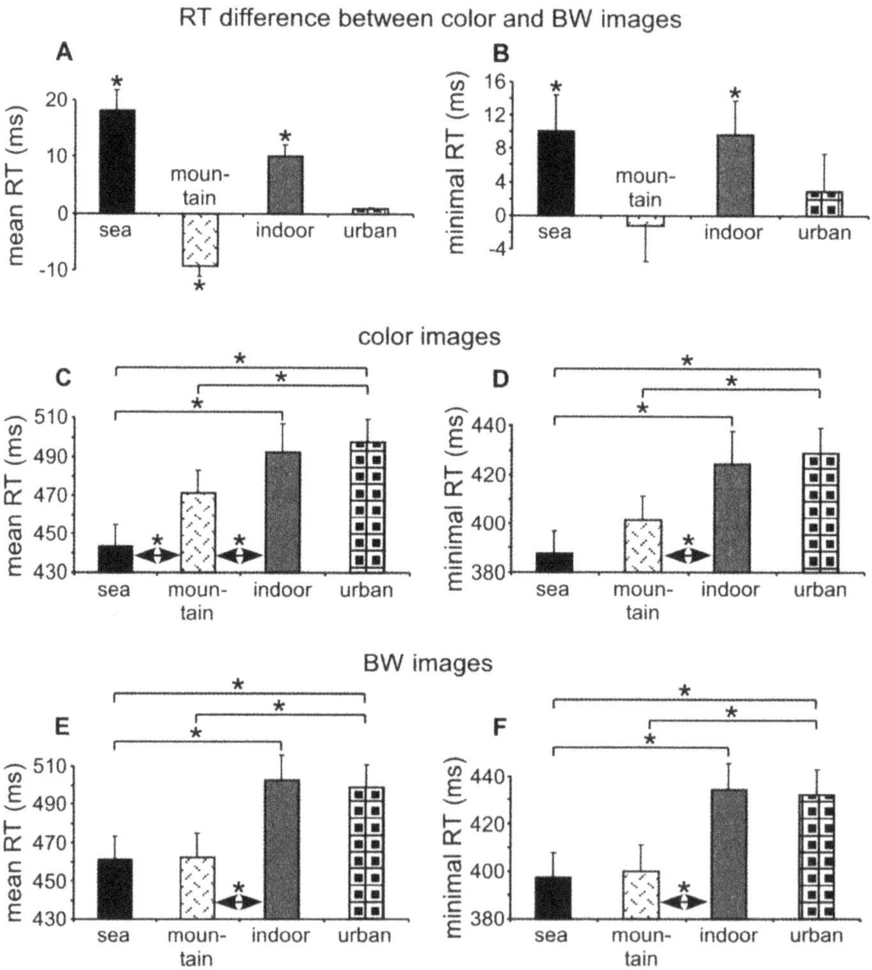

Figure 5. Speed of processing in Experiment 2: Mean RT (left column) and minimal RT (right column). The top two graphs illustrate the colour processing speed advantage by subtracting, for each of the four categorization tasks, the value obtained with colour images from the value obtained with BW images. Mean reaction times for each categorization task and associated standard errors are shown for colour images (middle panel) and for BW images (bottom panel). An asterisk shows statistically significant between-category effects.

$p < .001$, for both mean and median RT—urban scenes showed no effect of colour cue removal—mean and median RT: Both $Z > -0.7$, both n.s.—and, surprisingly, mountain pictures showed a tendency to be categorized with a slower speed in colour (9 ms slower) than in BW. This tendency reached significance for mean RT, $Z = -2.2$, $p = .03$, but not for median RT which presented only a borderline effect, $Z = -2.0$, $p = .05$.

Regarding differences in processing speed between the four categories, the results obtained with colour images tested separately showed the robustness of the results obtained in Experiment 1. Indeed, as in this first experiment, all two by two comparisons were significant for both mean and median RT, all $Z < -5$, all $p < .02$, except between indoor and urban scenes. Thus the two artificial categories were categorized at about the same speed.

When BW pictures were analysed separately, the same pattern of results appeared again for both mean and median RT, all $Z < -3.5$, all $p < .0001$, with the exception that the two natural categories (mountain and sea pictures) were categorized at the same speed.

The limited effect on performance speed linked to the removal of colour information when extracting the gist of natural scenes can be seen in RT distributions (Figure 6). Colour and BW RT distributions were virtually superimposed in the case of urban pictures and the amplitude of the shift towards shorter RT (for sea and indoor scenes) or towards longer RT (for mountain scenes) was indeed limited. The time course of performance (Figure 6, insets) again shows a small effect of colour removal on the subjects' capacity to discriminate between targets and distractors. Cumulated d' curves were virtually superimposed from the earliest responses to the plateau, with very similar slopes, indicating that information accumulated at a similar speed for colour and BW pictures. Regarding the small differences in plateau value, two by two one-way analyses on ranks revealed only one significant difference, namely that signal detection was slightly higher with colour images compared to BW images in the urban task (colour = 3.4; BW = 3.3; $Z = -2.1$, $p = .03$; other comparisons were not significant).

Discussion

Two main points have been stressed by the results obtained in these two studies. First, human subjects are both very accurate and very fast at categorizing the gist of real-world scenes with a processing speed advantage in favour of natural scenes compared to artificial scenes. Second, colour cues do not appear to be essential to trigger fast and accurate behavioural responses (some subjects were even unable to notice that half of the scenes were presented in BW); however, for certain categories where colour has a diagnostic value, there can be a small speed-up that even affect the fastest responses. Moreover, comparing the present

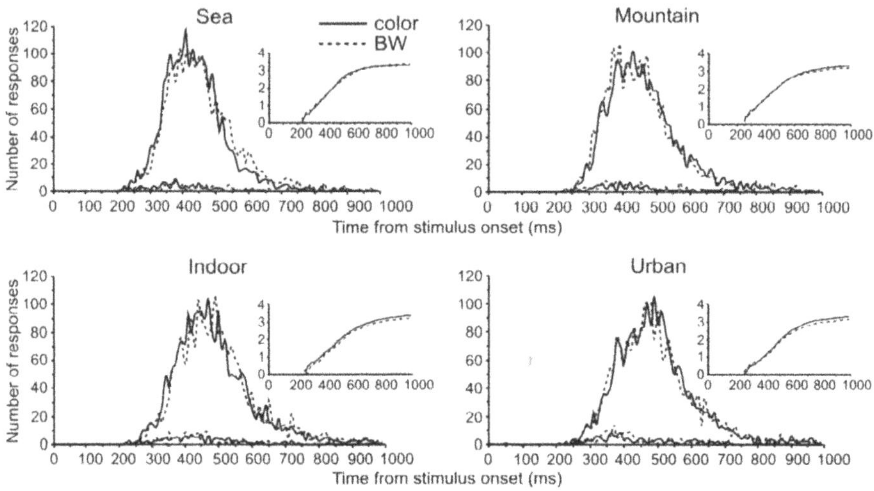

Figure 6. Reaction time distributions for correct hits (top two lines) and for false alarms (bottom two lines) for the four scene categorization tasks in Experiment 2. The graphs compare for each target category the reaction time distributions associated with colour (solid line) and BW images (dotted line). Each time, the insets show the *d'* computed from the cumulative number of responses in the RT histograms in both conditions. For details see caption of Figure 3.

results with those obtained for object categorization using the same task and the same set-up provides information on the possible interactions between the processing of objects and of the context in which they are presented.

Scene categorization is accurate and fast, but faster for natural scenes. Experiment 2 confirmed that subjects could extract very rapidly the global context of a natural scene. It also strengthened the finding that natural scenes can be processed faster than artificial scenes, and confirmed a slight processing speed advantage of sea images over mountain images.

Overall, Experiments 1 and 2 showed that human subjects can differentiate complex scene categories with median RT ranging from about 400 to 460 ms but with early responses observed at latencies that can be as short as 260–300 ms. This processing time is remarkably short compared to RT observed when subjects are simply asked to detect the appearance of a natural scene on the screen. Indeed, human subjects are able to detect that a scene has appeared, whatever its category, with a mean RT of about 230 ms (VanRullen & Thorpe, 2001) or even shorter (Rousselet, unpublished data: 20 subjects, mean RT = 211 ms). Thus, using a mean RT of 230 ms in such detection task as a reference, the additional cost to realize a complex gist categorization task is on average 170–230 ms, but can be as low as 30–70 ms for early responses. These strong temporal constraints are in favour of models of visual processing relying essentially on coarse, feedforward, and massively parallel mechanisms to achieve scene recognition (Oliva & Torralba, 2001; Torralba & Oliva, 2003; VanRullen & Thorpe, 2002).

Both experiments also revealed a clear processing speed advantage in favour of natural scenes compared to artificial scenes. This advantage was not mediated by colour cues, given that the same was true with BW pictures. Aside from colour, several combinations of low-level scene-based properties could potentially explain how subjects classified pictures in these two experiments (Oliva & Torralba, 2001; Torralba & Oliva, 2003). But the bias towards natural scenes might find its explanation in a stronger variability in the physical properties of artificial environments compared to the natural environments used in these experiments. Indeed, although we used very varied natural scenes, they were still limited to sea and mountain pictures. On the other hand, indoor and outdoor artificial scenes covered almost the whole collection of imaginable artificial scenes. This resulted in a more constrained dictionary of physical properties for natural compared to artificial scenes, which could recruit more specifically preset groups of ''diagnostic'' neurons responding to target properties in the natural scene images. The stronger variability in artificial scenes compared to natural scenes can be appreciated from the average scenes and the examples in Figure 1 (B and C respectively). This in turn might allow faster decision making, on the basis of limited need for evidence, in the case of natural scenes compared to artificial scenes. Future experiments will be needed to evaluate more carefully how gist processing speed is affected by target diagnosticity, for instance by varying the physical similarities among targets and between target and distractor scenes.

Influence of colour cues in scene categorization. The second experiment provided evidence about the influence of colour information in task performance for the four scene contexts studied here. In the first place, it is surprising to note that colour had virtually no effect on subjects' capacity to detect targets. Thus, colour is not necessary to categorize real-world pictures in this fast go/no-go paradigm. Concerning the speed of processing of urban scenes, the lack of effect of removing the colour cues is in agreement with one previous report from Oliva and Schyns (2000). However, in the present study we found that colour helped the subjects at correctly ignoring nontarget indoor pictures in the urban task (artificial category-related distractors) as they scored better when such scenes were presented in colour than in BW. Colour also speeded up the processing of indoor scenes as subjects were also faster to make a decision on coloured than on BW indoor target pictures. This effect was not found by Oliva and Schyns (2000), and might have been strengthened by our protocol. The two artificial scene categories used in this experiment might share many low-level properties, forcing subjects to rely on distinctive features to discriminate between those two categories. One of these features might have been the diagnostic yellowish/

brownish colour present in many indoor scenes because of artificial lighting (see the colour version of Figure 1B online). Note that this small colour advantage was also seen on the earliest responses produced. Thus, in this task, colour information was used very early during the course of visual processing, in keeping with other reports (Delorme et al., 2004; Edwards, Xiao, Keysers, Foldiak, & Perrett, 2003; Gegenfurtner & Rieger, 2000; Goffaux et al., 2005).

Processing speed was also faster (15 ms on average) when colour was available in the sea task. This is not surprising, given that the blue of the sea was largely predictive of the category. However, colour alone does not appear to be able to explain the good performance in this task. First, BW sea pictures were also categorized very efficiently. Second, the blue feature was not a specific attribute of the sea category; it was present in large proportions in the skies of the mountain task and some urban pictures. Thus, it seems that the use of a single strategy relying on the diagnostic use of blue would not be sufficient to perform the task. But subjects could also rely on a slightly more sophisticated strategy based on the detection of a blue surface situated in the lower part of the scene (Delorme et al., 2004).

The use of blue textured surfaces as a cue in the sea task might also explain the unexpected negative effect of colour in the mountain task in which large surfaces of blue sky were often present and could induce ambiguity as a common feature with some of the distractors. Anyway, colour effects in this task (about 9 ms), contrary to the indoor and the sea task, did not affect the earliest responses, but essentially behavioural responses triggered with average RT larger than the median RT (>450ms).

The magnitude of the colour effects reported here is relatively small compared to the average 50 ms advantage in naming RT reported for colour pictures over BW pictures by Oliva and Schyns (2000). The category that is the more related to our *sea* category was their *coastline* category, which presented a colour effect of about 50 ms. This effect is much larger than the one reported here (about 15 ms). A very plausible interpretation for this discrepancy relies on the use of two different tasks, a go/no-go task in the present study and a naming task in Oliva and Schyns. It is possible that the necessity to name pictures, which was also associated with an additional 200–300 ms RT, forced subjects to rely on different representations, more sensitive to the colour factor. Another more plausible explanations stems in the fact that in the Oliva and Schyns experiment, subjects had to name a picture belonging to eight possible categories on every trial, while in our experiment subjects had a unique target for 384 consecutive trials, a protocol used to allow subjects to respond as fast as they could. This, in addition with the intermixed appearance of BW and colour pictures, might have biased subjects' strategy towards the use of noncolour properties like spatial frequencies, textures, depth of field, and other properties that have been shown to constitute valid cues for scene categorization (Torralba & Oliva, 2003). On the other hand, Goffaux et al. (2005) report a colour advantage of about 30 ms using a task very similar to the one we used here. Two main differences might explain the discrepancy between the two studies. First, subjects in the Goffaux et al. study were presented with colour, grey, as well as abnormally coloured images, which might have affected their response strategy. It would be interesting to see if the 30 ms colour effect can be replicated when subjects are only presented with normal colour and grey-level images. Second, Goffaux et al. used four categories known to be colour diagnostic (deserts, forests, canyons, and coastlines), while our stimuli were not as well controlled in this respect. Thus, both targets and distractors could be categorized efficiently by relying on colour cues in their task, while only two of our categories were potentially colour diagnostic. The different results between the two studies are compatible with the idea that diagnostic cues are used whenever they are available and that diagnosticity in itself cannot be determined a priori but depends on such factors as the physical differences between targets and distractors (Schyns, 1998). In keeping with this idea, Delorme et al. (2000) showed virtually no effect of colour on the fast categorization of animals in natural scenes whereas a significant effect on accuracy was found with food objects.

Overall, there is now sufficient evidence showing that colour can be extracted very rapidly during the course of visual processing and can be used to improve performance when it has diagnostic value for the target scene category (Edwards et al., 2003; Gegenfurtner & Rieger, 2000; Goffaux et al., 2005; and the present study). However, our study also demonstrates that at least in some circumstances, colour is not a crucial feature for processing speed in a fast go/no-go categorization task of real-world scenes. More general conclusions about the importance of colour cues on scene categorization will need a more systematic investigation than the one presented in this paper.

Object categorization versus scene categorization. In Experiment 1, we performed a tentative comparison between object and scene categorization performances. This analysis revealed that although subjects were fast in categorizing a whole scene, this fast processing of scene gist was actually on average 30–90 ms slower than the processing of objects in natural scenes, like animals, but also like humans, human and animal faces, and means of transport (Delorme et al., 2000; Rousselet et al., 2003; VanRullen & Thorpe, 2001). This object advantage might very plausibly find its origin in the weaker structural constraints found in natural scenes. Indeed, the same gist can be assigned to scenes with relatively different low-level features and spatial arrangements. It is probably this relatively loosely defined structure of the scenes compared to component objects (like animals, vehicles, faces, etc.) that can explain their slower processing speed. However, this does not mean that scene categorization relies on higher level representations than object categorization. First, the fact that subjects made systematically more errors on distractors that belonged to the same higher level category as the target of the task (''natural'' vs. ''artificial'' categories) might be taken as an evidence for the use of relatively low-level cues in these tasks. Second, as we have argued recently (Rousselet et al., 2003), the rapid categorization of objects like faces and animals in natural scenes might depend on coarsely defined features of intermediate complexity rather than on high-level complete descriptions (see also Ullman, Vidal-Naquet, & Sali, 2002). Thus, the slower processing of scenes compared to objects might find its explanation in the need to integrate in parallel a larger conjunction of relatively low-level features in order to reach a decision level in the processing of a natural context. The rapid categorization of objects in natural scenes would rely on the conjunction of a more limited number of features than the categorization of gist. Hence, because a given object category like animals can often be defined on the basis of a local combination of features, neurons coding for target objects in the ventral pathway would benefit from a finer task related top-down presetting in the animal task compared to the scene tasks. It is thus predicted that in a task where scene categories can be discriminated on the basis of a very limited set of basic features (e.g., natural vs. artificial scenes), processing speed could be even faster for scenes than objects, a hypothesis that we are currently testing. In addition, the difference in processing speed between objects and scenes could reflect the limitations of our visual system to process natural scenes in parallel. Indeed, although we recently demonstrated that two scenes can be processed in parallel (Rousselet, Fabre-Thorpe, & Thorpe, 2002), we have now evidence that this capacity is limited to certain conditions (Rousselet, Thorpe, & Fabre-Thorpe, 2004; VanRullen, Reddy, & Koch, 2004). Constraints on parallel processing would be even stronger on gist categorization because, to make a decision, it requires integrating low-level features over a wider spatial scale than for object detection.

However, these results should not be considered as an argument in favour of models postulating that there is no early interaction between scene and object representations (e.g., Henderson & Hollingworth, 1999). Indeed, in everyday situations, the gist of a scene is not changing abruptly from one image to the next but is much stable over time, probably allowing predictive hypotheses about possible objects to build up, as proposed by modern interactive frameworks (Bullier, 2001; Rao & Ballard, 1999; Ullman, 1995). But even here, using very short presentations, a considerable overlap was observed between the RT distributions of object and scene categorization. This overlap shows that the processing of an object might benefit from the simultaneous processing of the context in which it appears. The activation of congruent populations of neurons would probably allow a faster identification of a cow in a field than in a church. This issue clearly deserves further investigation.

CONCLUSIONS

Confirming earlier studies that have used brief visual presentations, data from the two experiments reported here showed that the gist of real-world scenes could be extracted with a high accuracy and with short RT in a fast go/no-go visual categorization task. Furthermore, it was shown that natural scenes used in these experiments were processed faster than artificial scenes, probably because features of natural scenes might be more diagnostic than those of artificial scenes, allowing a stronger top-down presetting. In addition, we showed that colour information does not appear the most crucial feature used by human subjects to perform the fast go/no-go categorization tasks of real-world scenes studied here. Even if under certain conditions where colour is diagnostic of a specific scene category it can be used to produce a small performance increase, the fact that subjects can do so well with grey scale images shows that form-based information is typically self-sufficient.

REFERENCES

Aguirre, G. K., Zarahn, E., & D'Esposito, M. (1998). An area within human ventral cortex sensitive to "building" stimuli: Evidence and implications. *Neuron*, *21*(2), 373–383.

Bar, M., & Aminoff, E. (2003). Cortical analysis of visual context. *Neuron*, *38*(2), 347–358.

Biederman, I. (1972). Perceiving real-world scenes. *Science*, *177*(43), 77–80.

Biederman, I. (1987). Recognition-by-components: A theory of human image understanding. *Psychological Review*, *94*(2), 115–147.

Biederman, I. (1988). Aspects and extensions of a theory of human image understanding. In Z. W. Pylyshyn (Ed.), *Computational processes in human vision: An interdisciplinary perspective* (pp. 370–428). Norwood, NJ: Ablex.

Biederman, I., Mezzanotte, R. J., & Rabinowitz, J. C. (1982). Scene perception: Detecting and judging objects undergoing relational violations. *Cognitive Psychology*, *14*(2), 143–177.

Boyce, S. J., Pollatsek, A., & Rayner, K. (1989). Effect of background information on object identification. *Journal of Experimental Psychology: Human Perception and Performance*, *15*(3), 556–566.

Bullier, J. (2001). Integrated model of visual processing. *Brain Research Reviews*, *36*(2–3), 96–107.

Chun, M. M. (2000). Contextual cueing of visual attention. *Trends in Cognitive Sciences*, *4*(5), 170–178.

Delorme, A., Richard, G., & Fabre-Thorpe, M. (2000). Ultra-rapid categorisation of natural scenes does not rely on colour cues: A study in monkeys and humans. *Vision Research*, *40*(16), 2187–2200.

Delorme, A., Rousselet, G. A., Macé, M. J.-M., & Fabre-Thorpe, M. (2004). Interaction of top-down and bottom-up processing in the fast visual analysis of natural scenes. *Cognitive Brain Research*, *19*(2), 103–113.

Edwards, R., Xiao, D., Keysers, C., Foldiak, P., & Perrett, D. (2003). Color sensitivity of cells responsive to complex stimuli in the temporal cortex. *Journal of Neurophysiology*, *90*(2), 1245–1256.

Epstein, R., Graham, K. S., & Downing, P. E. (2003). Viewpoint-specific scene representations in human parahippocampal cortex. *Neuron*, *37*(5), 865–876.

Epstein, R., & Kanwisher, N. (1998). A cortical representation of the local visual environment. *Nature*, *392*(6676), 598–601.

Fabre-Thorpe, M., Delorme, A., Marlot, C., & Thorpe, S. (2001). A limit to the speed of processing in ultra-rapid visual categorization of novel natural scenes. *Journal of Cognitive Neuroscience*, *13*(2), 171–180.

Friedman, A. (1979). Framing pictures: The role of knowledge in automatized encoding and memory for gist. *Journal of Experimental Psychology: General*, *108*(3), 316–355.

Ganis, G., & Kutas, M. (2003). An electrophysiological study of scene effects on object identification. *Cognitive Brain Research*, *16*(2), 123–144.
Gegenfurtner, K. R. (2003). Cortical mechanisms of colour vision. *Nature Reviews Neuroscience*, *4*(7), 563–572.
Gegenfurtner, K. R., & Rieger, J. (2000). Sensory and cognitive contributions of color to the recognition of natural scenes. *Current Biology*, *10*(13), 805–808.
Goffaux, V., Jacques, C., Mouraux, A., Oliva, A., Schyns, P. G., & Rossion, B. (2005). Diagnostic colours contribute to the early stages of scene categorization: Behavioural and neurophysiological evidence. *Visual Cognition*, *12*, 878–892.
Henderson, J. M. (1992). Object identification in context: The visual processing of natural scenes. *Canadian Journal of Psychology*, *46*(3), 319–341.
Henderson, J. M., & Hollingworth, A. (1999). High-level scene perception. *Annual Review of Psychology*, *50*, 243–271.
Henderson, J. M., & Hollingworth, A. (2003). Eye movements and visual memory: Detecting changes to saccade targets in scenes. *Perception and Psychophysics*, *65*(1), 58–71.
Hollingworth, A. (2003). Failures of retrieval and comparison constrain change detection in natural scenes. *Journal of Experimental Psychology: Human Perception and Performance*, *29*(2), 388–403.
Hollingworth, A., & Henderson, J. M. (1998). Does consistent scene context facilitate object perception? *Journal of Experimental Psychology: General*, *127*(4), 398–415.
Hollingworth, A., & Henderson, J. M. (1999). Object identification is isolated from scene semantic constraint: Evidence from object type and token discrimination. *Acta Psychologica*, *102*(2–3), 319–343.
Hollingworth, A., & Henderson, J. M. (2002). Accurate visual memory for previously attended objects in natural scenes. *Journal of Experimental Psychology: Human Perception and Performance*, *28*(1), 113–136.
Intraub, H. (1997). The representation of visual scenes. *Trends in Cognitive Sciences*, *1*(6), 217–222.
Maguire, E. A., Frith, C. D., & Cipolotti, L. (2001). Distinct neural systems for the encoding and recognition of topography and faces. *Neuroimage*, *13*(4), 743–750.
Marr, D. (1982). *Vision*. San Francisco, CA: Freeman.
Nakamura, K., Kawashima, R., Sato, N., Nakamura, A., Sugiura, M., Kato, T., Hatano, K., Ito, K., Fukuda, H., Schormann, T., & Zilles, K. (2000). Functional delineation of the human occipito-temporal areas related to face and scene processing: A PET study. *Brain*, *123*(9), 1903–1912.
Oliva, A., & Schyns, P. G. (1997). Coarse blobs or fine edges? Evidence that information diagnosticity changes the perception of complex visual stimuli. *Cognitive Psychology*, *34*(1), 72–107.
Oliva, A., & Schyns, P. G. (2000). Diagnostic colors mediate scene recognition. *Cognitive Psychology*, *41*(2), 176–210.
Oliva, A., & Torralba, A. (2001). Modeling the shape of the scene: A holistic representation of the spatial envelope. *International Journal of Computer Vision*, *42*(3), 145–175.
O'Regan, J. K. (1992). Solving the "real" mysteries of visual perception: The world as an outside memory. *Canadian Journal of Psychology*, *46*(3), 461–488.
Potter, M. C. (1975). Meaning in visual search. *Science*, *187*(4180), 965–966.
Potter, M. C. (1976). Short-term conceptual memory for pictures. *Journal of Experimental Psychology: Human Learning and Memory*, *2*(5), 509–522.
Potter, M. C., & Levy, E. I. (1969). Recognition memory for a rapid sequence of pictures. *Journal of Experimental Psychology*, *81*(1), 10–15.
Rao, R. P., & Ballard, D. H. (1999). Predictive coding in the visual cortex: A functional interpretation of some extra-classical receptive-field effects. *Nature Neuroscience*, *2*(1), 79–87.
Rensink, R. A. (2000). The dynamic representation of scenes. *Visual Cognition*, *7*(1–3), 17–42.
Rensink, R. A. (2002). Change detection. *Annual Review of Psychology*, *53*, 245–277.

Riesenhuber, M., & Poggio, T. (2000). Models of object recognition. *Nature Neuroscience*, *3*(Suppl), 1199–1204.

Rousselet, G. A., Fabre-Thorpe, M., & Thorpe, S. J. (2002). Parallel processing in high-level categorization of natural images. *Nature Neuroscience*, *5*(7), 629–630.

Rousselet, G. A., Macé, M. J.-M., & Fabre-Thorpe, M. (2003). Is it an animal? Is it a human face? Fast processing in upright and inverted natural scenes. *Journal of Vision*, *3*(6), 440–455.

Rousselet, G. A., Thorpe, S. J., & Fabre-Thorpe, M. (2004). Processing of one, two or four natural scenes in humans: The limits of parallelism. *Vision Research*, *44*(9), 877–894.

Sanocki, T., & Epstein, W. (1997). Priming spatial layout of scenes. *Psychological Science*, *8*(5), 374–378.

Sato, N., Nakamura, K., Nakamura, A., Sugiura, M., Ito, K., Fukuda, H., & Kawashima, R. (1999). Different time course between scene processing and face processing: A MEG study. *Neuroreport*, *10*(17), 3633–3637.

Schyns, P. G. (1998). Diagnostic recognition: Task constraints, object information, and their interactions. In M. J. Tarr & H. H. Bülthoff (Eds.), *Object recognition in man, monkey, and machine* (pp. 147–179). Amsterdam: Elsevier Science Publishers.

Schyns, P. G., & Oliva, A. (1994). From blobs to boundary edges: Evidence for time and spatial scale dependant scene recognition. *Psychological Science*, *5*, 195–200.

Schyns, P. G., & Oliva, A. (1997). Flexible, diagnosticity-driven, rather than fixed, perceptually determined scale selection in scene and face recognition. *Perception*, *26*(8), 1027–1038.

Simons, D. J., Chabris, C. F., Schnur, T., & Levin, D. T. (2002). Evidence for preserved representations in change blindness. *Consciousness and Cognition*, *11*(1), 78–97.

Tanaka, J. W., & Presnell, L. M. (1999). Color diagnosticity in object recognition. *Perception and Psychophysics*, *61*(6), 1140–1153.

Thorpe, S., Fize, D., & Marlot, C. (1996). Speed of processing in the human visual system. *Nature*, *381*(6582), 520–522.

Torralba, A., & Oliva, A. (2003). Statistics of natural image categories. *Network: Computation in Neural Systems*, *14*, 391–412.

Ullman, S. (1995). Sequence seeking and counter streams: A computational model for bidirectional information flow in the visual cortex. *Cerebral Cortex*, *5*(1), 1–11.

Ullman, S., Vidal-Naquet, M., & Sali, E. (2002). Visual features of intermediate complexity and their use in classification. *Nature Neuroscience*, *5*(7), 682–687.

VanRullen, R., Reddy, L., & Koch, C. (2004). Visual search and dual-tasks reveal two distinct attentional resources. *Journal of Cognitive Neuroscience*, *16*(1), 4–14.

VanRullen, R., & Thorpe, S. J. (2001). Is it a bird? Is it a plane? Ultra-rapid visual categorisation of natural and artifactual objects. *Perception*, *30*(6), 655–668.

VanRullen, R., & Thorpe, S. J. (2002). Surfing a spike wave down the ventral stream. *Vision Research*, *42*(23), 2593–2615.

Wolfe, J. M. (1999). Inattentional amnesia. In V. Coltheart (Ed.), *Fleeting memories* (pp. 71–94). Cambridge, MA: MIT Press.

VISUAL COGNITION, 2005, 12 (6), 878–892

Diagnostic colours contribute to the early stages of scene categorization: Behavioural and neurophysiological evidence

Valérie Goffaux and Corentin Jacques

Face Categorization Lab, UCL, Louvain-la-Neuve, and Laboratoire de Neurophysiologie, UCL, Bruxelles, Belgium

André Mouraux

Laboratoire de Neurophysiologie, UCL, Bruxelles, Belgium

Aude Oliva

Department of Brain and Cognitive Sciences, MIT, Cambridge, MA, USA

Philippe G. Schyns

Department of Psychology, University of Glasgow, UK

Bruno Rossion

Face Categorization Lab, UCL, Louvain-la-Neuve, and Laboratoire de Neurophysiologie, UCL, Bruxelles, Belgium

We examined the effects of colour cues on the express categorization of natural scenes. Using a go/no-go paradigm sensitive to fast recognition processes, we measured early event-related potential (ERP) correlates of scene categorization to elucidate the processing stage at which colour contributes to scene recognition. Observers were presented with scenes belonging to four colour-diagnostic categories (desert, forest, canyon and coastline). Scenes were presented in one of three forms: Diagnostically coloured, nondiagnostically coloured, or greyscale images. In a verification task, observers were instructed to respond whenever the presented stimulus matched a previously presented category name. Reaction times and accuracy were optimal when the stimuli were presented as their original

Please address all correspondence to: Valérie Goffaux, Unité de Neurosciences Cognitives, UCL, Place Cardinal Mercier, 10, 1348 Louvain-la-Neuve, Belgium.
Email: Valerie.goffaux@psp.ucl.ac.be

This work was supported by grant ARC 01/06-267 (Communauté Française de Belgique—Actions de Recherche Concertées). VG, CJ, and BR are supported by the Belgian National Fund for National Research (FNRS). The authors would like to thank Quoc Vuong for helpful comments on the manuscript.

DOI:10.1080/13506280444000562

diagnostically coloured version, followed by their greyscale version, and lastly by their nondiagostically coloured version. These effects were mirrored in the early (i.e., 150 ms following stimulus onset) ERP frontal correlates. Their onset was delayed for greyscale scenes compared to diagnostically coloured scenes, and for nondiagnostically coloured scenes compared to the other two conditions. Frontal ERP amplitudes also decreased for greyscale and nondiagnostically coloured scenes. Together, the results suggest that diagnostic colours are part of the scene gist responsible for express scene categorization.

Our ability to recognize complex visual scenes is an outstanding property of the human visual system. Under natural viewing conditions, a scene is typically composed of a large number of objects, which can vary in pose, illumination, and occlusions arising from other objects or from cast shadows. To a scene recognition system, the resulting image variations are mostly noise from which diagnostic scene signals must be extracted. To date, this task has eluded all recognition algorithms, except that implemented by the human brain: We recognize complex visual scenes at a glance.

Recently, researchers have turned their attention to the visual information responsible for such effective scene categorization. Several studies have revealed that natural scenes are indeed identified very quickly (e.g., Potter, 1975), as fast as individual objects (Biederman, Mezzanotte, & Rabinowitz, 1982), and from information that can be extracted from a single fixation (Henderson & Hollingworth, 2003). This information, called the ''scene gist'' (or *scene schema*), is hypothesized to act as a representational skeleton that guides subsequent eye movements to further selectively flesh out important scene components (Antes, Penland, & Metzger, 1981; Biederman, 1981; De Graef, de Troy, & d'Ydewalle, 1992; Friedman, 1979; Oliva, Torralba, Castelhano, & Henderson, 2003; Schyns & Oliva, 1994).

The visual information making up the scene gist

Schyns and Oliva (1994) hypothesized that the regularity of spatial organization of scene categories might provide the necessary scene gist for a mechanism to bootstrap scene recognition. Research in early vision suggests that the bases for recognition are low-level visual cues such as luminance, chromaticity, movement, and depth (Livingstone & Hubel, 1987). In line with this suggestion, Oliva and Schyns (1997) and Schyns and Oliva (1994; see also Parker, Lishman, & Hughes, 1992) showed that luminance variations at coarse scales (i.e., ''blobs'') were sufficient to mediate scene recognition without prior recognition of component objects (see Oliva & Torralba, 2001, for a formal analysis). Hence, luminance variations at a coarse scale provide one component of the scene gist.

Another potential component of the scene gist is chromatic variations. However, the role of colour cues for scene and object recognition remains controversial (see Oliva & Schyns, 2000; Tanaka, Weiskopf, & Williams, 2001;

Wichmann, Sharpe, & Gegenfurtner, 2002 for reviews). For example, Delorme, Richard, and Fabre-Thorpe (2000) reported little effect of colour when observers categorized briefly presented (20–30 ms) coloured or greyscale pictures of animals and food. The effects of colour only manifested when the slowest responses and observers were examined. In contrast to luminance cues that are used at an early stage, these results suggest that colour influences a comparatively later stage of scene processing. According to this ‘‘colouring-book’’ interpretation, the scene gist would first include luminance variations as a skeleton that is then filled in with appropriate colours.

To establish whether colour contributed to scene recognition, Oliva and Schyns (2000) compared naming and verification performance of natural (e.g., forest, desert) and artificial (e.g., city, motorway) scenes that were presented as normally coloured, abnormally coloured, or greyscale images (see Figure 1). The authors controlled the colour composition of the stimuli using a 3-D colour space (L*a*b space) that separated luminance and chromaticity. Importantly, this separation isolated variations in scene colours from variations in luminance. The colour diagnosticity of scene categories—i.e., the degree of overlap between the colours making up scenes within a category—was also controlled. When colour was diagnostic (i.e., for natural scenes like deserts, forests, coastlines, and canyons), reaction times were faster with normally coloured than with greyscale images, and reaction times in both of these conditions were faster than with abnormally coloured images. Oliva and Schyns (2000) therefore hypothesized that chromatic variations could also be part of the scene gist that indexes scene memory.

There are at least two main reasons why chromatic variations, in addition to luminance variations, would provide an advantage for speeded scene recogni-

Figure 1. Pictures of natural scenes were used (desert, forest, coastline, and canyon). This figure illustrates the variations of a forest and a coastline picture across the three coloured versions: Normal, grey, and abnormal (from left to right).

tion. The first reason is a simple problem of image segmentation: Chromatic variations provide two additional dimensions (the red–green and yellow–blue colour axes) to segment a scene into its main parts, compared to greyscale versions of the same image. A second advantage is that the additional colour dimension provides supplementary information to index scene memory, speeding up categorization processes when colours are diagnostic of scene categories (Oliva & Schyns, 2000; Tanaka et al., 2001).

Our experiment used diagnostic versus nondiagnostic colouring of the same scenes to tease apart these two possibilities for express scene categorization (segmentation versus memory indexing). We also used greyscale versions of the same scenes to measure baseline performance. In a go/no-go task, we expected to replicate Oliva and Schyns' (2000, Exp. 1) finding obtained in a naming categorization task that recognition of normally coloured scenes was fastest, followed by greyscale scenes, and lastly by abnormally coloured scenes.

In addition to behavioural data (reaction time and accuracy), we also recorded neurophysiological event-related potentials (ERPs) to obtain a more direct measure of the neural processes related to express scene categorization. While typical reaction times in scene recognition tasks are in the range of 400–600 ms (Oliva & Schyns, 2000; Rousselet, Fabre-Thorpe, & Thorpe, 2002; Schyns & Oliva, 1994; Thorpe, Fize, & Marlot, 1996), ERPs reveal, at a much finer temporal resolution (millisecond range), the time course of neural processes underlying categorization. For example, Thorpe and his colleagues (Rousselet et al., 2002; Thorpe et al., 1996; VanRullen & Thorpe, 2001) have shown that express scene categorization was indexed by early ERP activities. In a go/no-go paradigm, observers were instructed to decide whether briefly flashed scenes contained an animal. No-go trials were selectively associated with a frontal negative activity around 150 ms following stimulus onset, suggesting that, at this time, sufficient processing was achieved to perform the categorization task. The ERP technique could thus provide earlier measures of the effects of colour information on scene categorization.

In the present experiment, we investigated whether colour information modulates the time course of scene categorization as early as the fast go/no-go ERP difference reported by Thorpe and his colleagues, thought to reflect the categorization decision. If colour influences the early stages of scene categorization—i.e., the extraction of the scene gist—this response should be modulated by the scene colour content.

METHOD

Participants

Eighteen students from Louvain University (four males, mean age = 24, two left handed), with normal or corrected to normal vision, were paid to participate in the experiment.

Stimuli

Stimuli were presented using Eprime® software on a PC computer monitor. The stimulus set comprised 80 pictures of natural scenes selected from the Corel CD Photo Library (Oliva & Schyns, 2000; see Figure 1). They belonged to four categories of colour-diagnostic scene categories (canyon, forest, coastline, and desert). The computation of colour diagnosticity was based on the L*a*b method detailed in Oliva and Schyns (2000). Twenty exemplars per category represented the scenes from a variety of viewpoints and perspectives. Three different versions of each scene were used in the experiment: Normally coloured ("normal"), greyscale ("grey"), and abnormally coloured ("abnormal"). Abnormally coloured scenes were obtained by first transposing red–green pixels with blue–yellow pixels and vice versa. These transposed images were then inverted, such that red pixels became green, and blue pixels became yellow and vice versa (Figure 1; see Oliva & Schyns, 2000 for technical details on stimulus generation). Thus, 60 images were available for each scene category for a total of 240 stimuli. Luminance—i.e., grey-level means and standard deviations—was kept constant across categories and across these different versions. Viewed at 30 cm from screen, the stimuli (472 $\times$ 325 pixels) subtended 15.2° $\times$ 10.4° of visual angle.

Procedure

In a go/no-go paradigm, we instructed participants to press a button, as quickly and as accurately as possible, whenever the presented stimulus belonged to a target scene category. The experiment was divided into blocks of 60 trials, and the scene category to respond to varied for each block. A block began with the label of the target scene category (either "desert", "forest", "coastline", or "canyon"). The order of blocks (i.e., target scene category) was randomized and counterbalanced across participants. Blocks comprised 30 targets randomly interleaved with 30 distractors. We repeated each stimulus four times, twice as targets and twice as distractors. Low-level image properties were thus balanced across go and no-go trials. The experiment comprised 960 trials in total. Trials began with a 100 ms presentation of a scene at the centre of the screen against a black background. Stimulus onset asynchrony (SOA) varied randomly between 1500 ms and 1800 ms.

EEG recordings and analyses

Continuous EEG was recorded from 54 tin electrodes, mounted in an electrode cap (Quick cap). Electrode positions included the standard 10–20 system locations and additional intermediate positions. During EEG recordings, all electrodes were referenced to the left mastoid. Impedance was kept below 10 kΩ. EEG signal was sampled at a rate of 1024 Hz. EEG signal was rereferenced offline relative to linked mastoids and linearly detrended in order to eliminate baseline drifts on frontal and central electrodes. The epochs were excluded from averaging if the standard deviation of the EEG within a sliding 200 ms time window exceeded 35 µV. Incorrect-response trials were also excluded. For 11 participants, ERPs were corrected for eye blinks by the subtraction of PCA-transformed EOG components for each electrode, weighted according to EOGV propagation factors (computed via linear regression; Nowagk & Pfeifer, 1996). This procedure resulted in a mean number of 147 ($\pm$ 2.5) trials per participant and condition. Electrodes were grouped into frontal (left: FP1, F1; right: FP2, F2; midline: FPZ, FZ, according to 10–20 system nomenclature) and parietal (left: P1, P3, PO3; right: P2, P4, PO4; midline: PZ, POZ) groups according to position. We computed four ERP difference waveforms for each participant: (1) Go minus no-go trials, independently of colour conditions; (2) go minus no-go trials in the normal condition, (3) go minus no-go trials in the abnormal condition, and (4) go minus no-go trials in the grey condition.

Statistical analyses

Repeated-measures analyses of variance (ANOVAs) were performed on behavioural go responses (RT and accuracy) with colour (normal versus abnormal versus grey) as a within-subjects factor. ERP data were analysed according to previous studies that used the go/no-go response paradigm (e.g., Thorpe et al., 1996; VanRullen & Thorpe, 2001). Intrasubject *t*-tests ($df = 17$) were performed at the $p < .05$ level between -100 and 600 ms on the four differential waveforms and for the nine averaged electrodes. This analysis enabled the precise tracking of categorization effects on onset latencies across the colour conditions. We further characterized the go/no-go differential activities across colour conditions in terms of their amplitude, by computing pairwise intrasubject *t*-tests between normal, grey, and abnormal differential waveforms. Overall, an effect was considered to be significant if 30 consecutive *t*-values (29.2 ms here) were below the $p < .05$ level (Rugg, Doyle, & Wells, 1995).

RESULTS

Behavioural results

The behavioural responses were fast and accurate (Figure 2), indicating that our categorization task effectively tapped express scene perception. Despite the ease with which participants categorized scenes, task performance was influenced by the colour content of the scenes. Response times significantly differed across colour conditions, $F(2, 34) = 141.7$, $p < .0001$. They increased from the normal condition (median: 427.5 ms; mean: 428.4 ms $\pm$ 15 ms, following stimulus onset) to the grey condition (median: 459 ms; mean: 461.2 ms $\pm$ 15 ms),

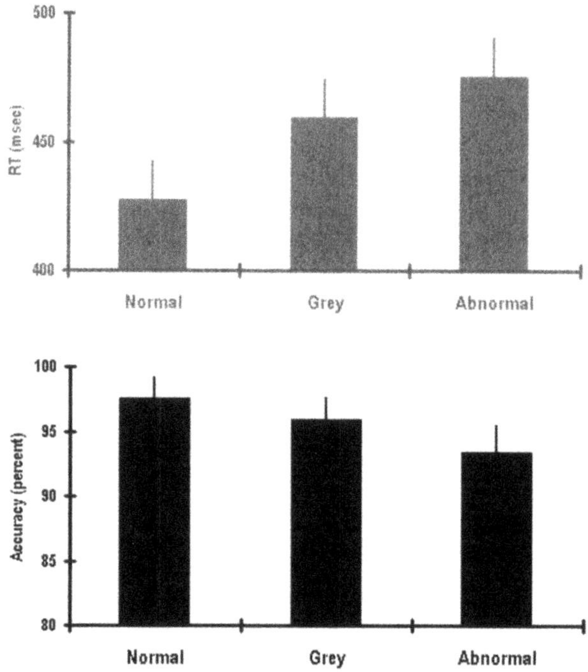

Figure 2. These histograms illustrate the influence of colour on go mean response times (RTs, in ms, top) and accuracy (bottom); $n = 18$. The bars indicate the standard errors (in ms).

$p < .0001$, and they were longest in the abnormal condition (median: 475.5 ms; mean: 484 ms $\pm$ 15 ms); grey–abnormal contrast, $p < .0001$.

Although participants performed at ceiling, accuracy also revealed systematic variations across colour conditions, $F(2, 34) = 24.1$, $p < .0001$. Accuracy decreased from the normal (97.5% $\pm$ 1.7%) to the grey conditions (96% $\pm$ 1.8%), $p < .003$, and it reached its lowest level in the abnormal condition (93% $\pm$ 2.2 %); grey–abnormal contrast: $p < .001$. Overall, the false alarm rate was very low (3.2% $\pm$ 0.6 %), and was not influenced by picture colour, $F(2, 34) = 1$, $p > .37$.

In summary, the behavioural results indicated that the colour content, rather than the mere presence or absence of chromatic cues, affected the speed and accuracy of express scene categorization. Although the absence of colour decreased performance relative to a normally coloured scene, an incongruent colour content, such as in the abnormal condition, produced an even larger interference. These results replicate Oliva and Schyns' (2000) findings and confirm the important role of diagnostic colours when observers categorize natural scenes at a glance.

Neurophysiological results

The comparison of go and no-go trials, with colour conditions collapsed, led to strong ERP differential activities (Figure 3). In agreement with Thorpe et al.'s results (Rousselet et al., 2002; Thorpe et al., 1996; VanRullen & Thorpe, 2001), no-go trials led to larger negative activities at approximately 150–200 ms following stimulus onset (Figure 3). The go/no-go difference is first evident on frontal electrodes (midline frontal), as soon as 147 ms following stimulus onset, and continued 420 ms on average poststimulus onset. The next consistent differential activity occurred at parietal sites approximately 300 ms after stimulus onset and persisted approximately until 500 ms post onset; 400 ms following stimulus onset, a negative differential component was observed on frontocentral electrodes (see Figure 3). However, several aspects strongly suggest that this negative difference did not constitute an early correlate of scene categorization, and was not analysed further. First, it was related to the go and no-go waveforms returning back to baseline after their early categorization-related differentiation (see for example Fz box in Figure 3). Second, its late latency actually overlapped with manual response execution. Finally, this negative difference displayed a highly restricted scalp distribution since it was only significant on FC1 and FCz electrodes.

We examined the effect of colour content on the frontal and parietal differential responses by tracking the onset latency of significant go/no-go differences for each colour condition separately (Figures 4 and 5). In the normal colour condition, the frontal difference occurred as soon as 137 ms (averaged onset latencies across left, right, and midline locations). The parietal difference occurred around 140 ms later, at 276 ms. Overall, the onset latencies of go/no-go differences were consistently delayed in the grey and abnormal colour conditions (Figures 4 and 5). The frontal ERP difference started at 184 ms on average in the grey condition and at 226 ms on average in the abnormal condition. On frontal sites, the grey condition thus revealed intermediate onset latencies between the normal and abnormal conditions. Parietal sites revealed large delays between the normal condition, on the one hand, and the abnormal and grey conditions, on the other hand. The first parietal go/no-go difference occurred at 327 ms in the grey condition, and at 323 ms in the abnormal condition (on right parietal electrodes; Figure 4).

As can be seen in Figures 4 and 5, not only did scene colour contribute to onset latency modulation of the experimental effects, it also influenced the magnitude of these effects. When comparing the differential waveforms observed for the three colour conditions separately (Figure 4), normally coloured images led to the largest differential activities on frontal and mostly on parietal sites. It differed from greyscale or abnormally coloured images at around the same latencies (200 ms, on average, on frontal sites and 302 ms, on average, on parietal sites). The abnormal and grey conditions differed later, at

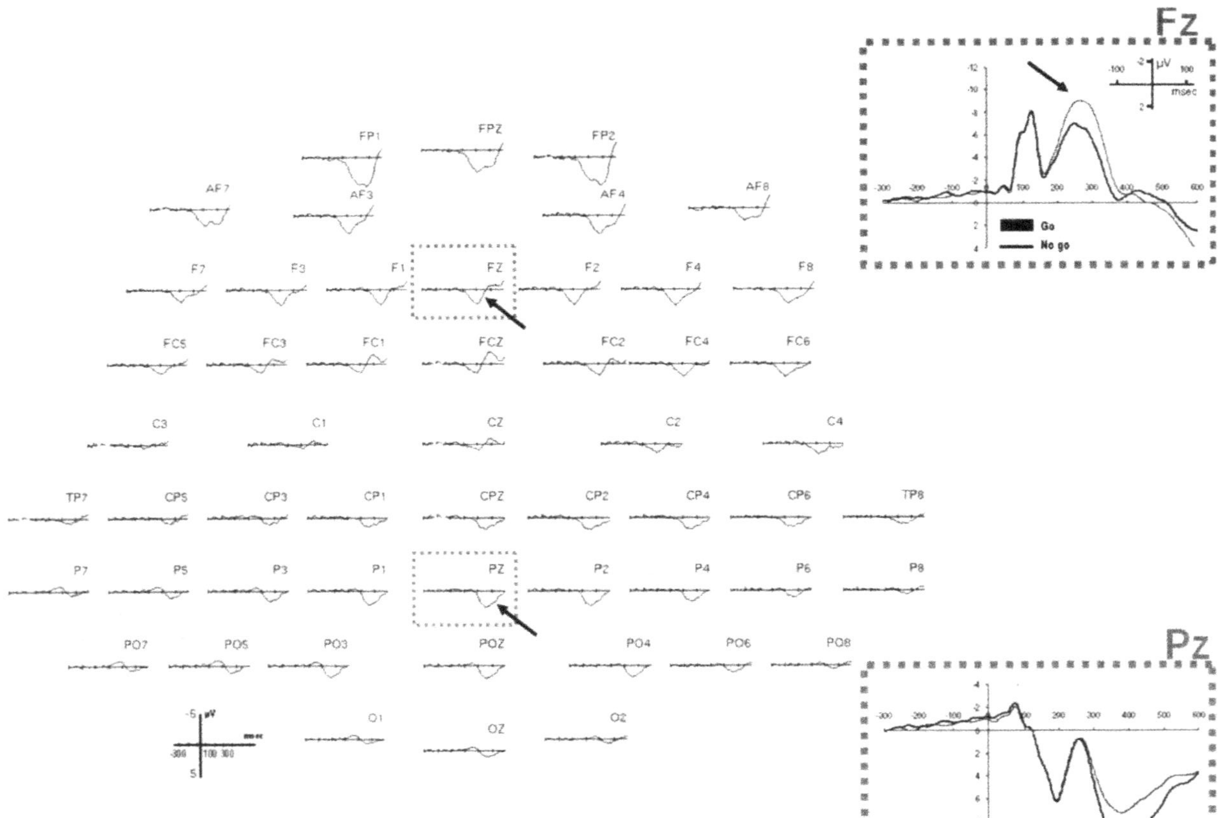

Figure 3. Grand-averaged subtraction ERP (n = 18) resulting from go minus no-go contrast, with colour conditions collapsed. The boxes detail go (thick line) and no-go (thin line) waveforms for FZ and PZ.

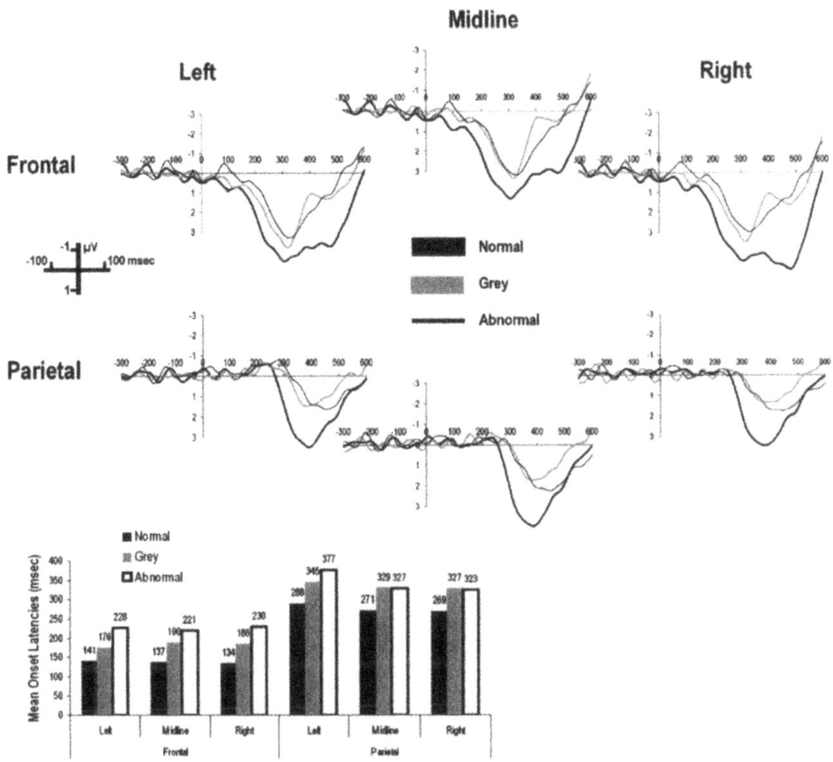

Figure 4. Top: This figure displays the waveforms resulting from go minus no-go ERPs subtraction (n = 18), computed for each colour condition separately and for each scalp electrode site. A 20 Hz low-pass was used for the display. Bottom: Histograms displaying the onset latencies of significant go–no-go statistical differences for colour conditions separately.

251 ms, on average, on frontal electrodes and 371 ms, on average, on parietal electrodes.

In summary, millisecond-by-millisecond statistical analyses revealed that the onset latencies of scene categorization neurophysiological correlates were consistently delayed for abnormally coloured or greyscale images conditions as compared to normally coloured images. The amplitude of these differential activities was also significantly larger for normally coloured images, as compared to abnormally coloured or greyscale images. The onset latency of go/no-go differences was delayed in the grey and abnormal conditions, but the latter condition revealed the most delayed latencies, at least at frontal sites. Grey and abnormal conditions differed later and to a lesser extent at the level of the amplitude of differential activity. In conjunction with the behavioural results, these neurophysiological results suggest that it is the diagnosticity of the colour

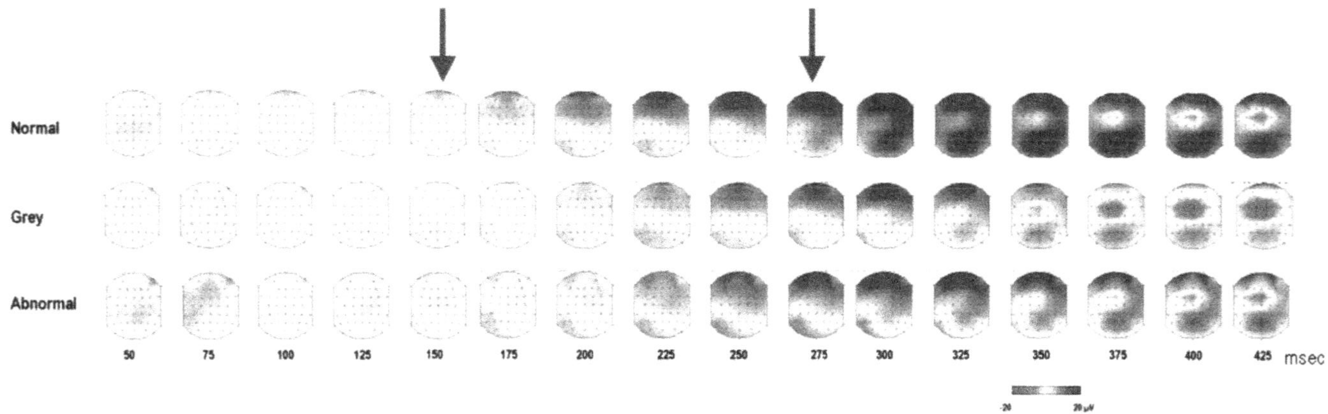

Figure 5. Scalp topographies of differential waveforms, obtained through the go minus no-go trials ERP subtraction (n = 18), are displayed for each colour condition separately. The first arrow indicates the approximate onset of frontal differential activity and the second one indicates the onset of parietal differences for normal condition.

content that contributes to express scene categorization rather than the mere presence of chromatic cues.

DISCUSSION

The aim of the present study was to examine express scene categorization by systematically manipulating the colour content of natural scenes. Behavioural responses were fast and accurate, confirming that our paradigm tapped into express categorization processes. Median RT for scenes presented in their diagnostic colours (427.5 ms in the normal condition) was indeed within the time range reported by Delorme et al. (2000; $\pm$420 ms), Rousselet et al. (2002; 390 ms), Thorpe et al. (1996; 445 ms), and VanRullen and Thorpe (2001; 350 ms), despite dissimilarities in the categorization tasks and measurement apparatus.[1] Consistent effects of colour were observed on categorization with performance being improved (versus worsened) when scenes were shown in their diagnostic (versus nondiagnostic) colours, compared with the baseline greyscale condition. These results obtained in a go/no-go scene verification task replicated those obtained with a naming task by Oliva and Schyns (2000). Thus, we can be reasonably sure that colour helps the speeded categorization of colour-diagnostic scenes.

By measuring the neurophysiological responses underlying categorization at a high temporal resolution, we clarified the stage at which chromatic information influences scene categorization. In line with previous reports (Rousselet et al., 2002; Thorpe et al., 1996; VanRullen & Thorpe, 2001), scene categorization was associated with consistent and early ERP differential activities for go and no-go trials. These differential activities occurred as soon as 150 ms on frontal electrodes, showing that sufficient processing was accomplished at this latency for participants to decide whether or not the scene belonged to a given target category. A second go/no-go difference was initiated on parietal sites at around 300 ms. This latter differential activity can be identified as a P300, a large positive component involved in both stimulus processing and response selection (for a review see Verleger, 1997) and has been described previously in a go/no-go discrimination task (Roberts, Rau, Lutzenberger, & Birbaumer, 1994).

What processing stage may early ERP frontal differential activity reflect? The onset latency of the ERP frontal activity largely preceded the motor output (which occurred at 458 ms on average) and so it is unlikely that this activity

[1] Using a sophisticated infrared diodes touch sensitive apparatus, Thorpe et al.'s experiments measured the onset of button releases, a much faster response than button presses as in the present experiment. Previous go/no-go experiments also used a different type of task where participants had to detect a particular object (an animal, for example) in the scenes. In the present experiment, we instructed participants to decide whether the scene, considered as a whole, corresponded to a target category.

relates to motor activations involved in response execution (see also Thorpe et al., 1996). Low-level visual differences between the conditions cannot account for these activity differences either, given the conditions comprised exactly the same stimulus sets, once presented as target, and once as nontarget (as in VanRullen & Thorpe, 2001). Thus, in line with previous studies, we can attribute frontal differential ERP activities as an index of the lowest processing time limit at which scene categorization is accomplished by the brain.

The colour content of scenes modulated the onset latencies and amplitudes of the frontal ERP differential activities. They were delayed by about 46 ms with greyscale scenes compared with diagnostically coloured scenes, and an 89 ms average shift was observed with nondiagnostically coloured scenes. This finding clearly demonstrates that diagnostic colours do speed up early categorization processes, which then give rise to faster behavioural responses. Note that two further conclusions can be made from the design of the experiment: (1) The colour-related ERP modulations do not stem from disparities in scene luminance gradients, because these were exactly matched across the different colour conditions; and (2) the ERP modulations do not arise simply because coloured images are easier to segment than greyscale images—nondiagnostic colours gave rise to the longest shifts and lowest amplitudes. Briefly stated, our results indicate that diagnostic colours provided faster indexing of scene memory and that this categorization can be accomplished by the brain within the first 150 ms of visual processing (see Gegenfurtner & Rieger, 2000 for supportive evidence). Such colour contribution to express scene categorization should be located at a visual processing stage where the stored knowledge of scene colour properties is represented.

A recent electrophysiological investigation in monkeys might provide more indications about the precise stage of visual processing in which diagnostic colours is involved (Edwards, Xiao, Keysers, Foldiak, & Perrett, 2003). Compared to diagnostically coloured face pictures, a majority of inferotemporal (IT) face selective cells (70%) displayed significantly reduced responses not only for greyscale face images, but also when faces were presented in nondiagnostic colours. In the latter condition, the cells' response amplitudes were lowest. This response difference was evident from stimulus onset (median 91 ms latency), suggesting that information about diagnostic colour is processed at the same time as object shape in monkey IT cortex. These findings support the idea that diagnostic colour information is an integral part of the visual processing of complex shapes (see also Zeki & Marini, 1998).

CONCLUDING REMARKS

Our results demonstrate that the neurophysiological processes underlying scene categorization use diagnostic colours as soon as 150 ms. The reported latency of frontal differential activities, together with the evidence for the involvement of

IT in diagnostic colour coding (Edwards et al., 2003; Zeki & Marini, 1998), suggest that express scene categorization relies on the fast extraction of a scene gist comprising not only the luminance information giving the global scene structure, but also its diagnostic colour content.

REFERENCES

Antes, J. R., Penland, J. G., & Metzger, R. L. (1981). Processing global information in briefly presented pictures. *Psychological Research*, *43*, 277–292.

Biederman, I. (1981). On the semantics of a glance at a scene. In M. Kubovy & J. R. Pomerantz (Eds.), *Perceptual organization* (pp. 213–233). Hillsdale, NJ: Lawrence Erlbaum Associates, Inc.

Biederman, I., Mezzanotte, R. J., & Rabinowitz, J. C. (1982). Scene perception: Detecting and judging objects undergoing relational violations. *Cognitive Psychology*, *14*, 143–177.

De Graef, P., de Troy, A., & d'Ydewalle, G. (1992). Local and global contextual constraints on the identification of objects in scenes. *Canadian Journal of Psychology*, *46*, 489–508.

Delorme, A., Richard, G., & Fabre-Thorpe, M. (2000). Ultra-rapid categorisation of natural scenes does not rely on colour cues: A study in monkeys and humans. *Vision Research*, *40*, 2187–2200.

Edwards, R., Xiao, D., Keysers, C., Foldiak, P., & Perrett, D. (2003). Color sensitivity of cells responsive to complex stimuli in the temporal cortex. *Journal of Neurophysiology*, *90*, 1245–1256.

Friedman, A. (1979). Framing pictures: The role of knowledge in automatized encoding and memory for gist. *Journal of Experimental Psychology: General*, *108*, 316–355.

Gegenfurtner, K. R., & Rieger, J. (2000). Sensory and cognitive contributions of color to the recognition of natural scenes. *Current Biology*, *10*, 805–808.

Henderson, J. M., & Hollingworth, A. (2003). Eye movements, visual memory, and scene representation. In M. A. Peterson & G. Rhodes (Eds.), *Analytic and holistic processes in the perception of faces, objects, and scenes* (pp. 356–383). New York: Oxford University Press.

Livingstone, M. S., & Hubel, D. H. (1987). Psychophysical evidence for separate channels for the perception of form, color movement and depth. *Journal of Neuroscience*, *7*, 3416– 3468.

Nowagk, R., & Pfeifer, E. (1996). Unix implementation of the ERP evaluation package (EEP 3.0). In A. D. Friederici & D. Y. von Cramon (Eds.), *Annual report of Max-Planck-Institute of Cognitive Neuroscience* (p. 124). Leipzig, Germany: Max-Planck Institute.

Oliva, A., & Schyns, P. G. (1997). Coarse blobs or fine edges? Evidence that information diagnosticity changes the perception of complex visual stimuli. *Cognitive Psychology*, *34*, 72–107.

Oliva, A., & Schyns, P. G. (2000). Diagnostic colors mediate scene recognition. *Cognitive Psychology*, *41*, 176–210.

Oliva, A., & Torralba, A. (2001). Modeling the shape of the scene: A holistic representation of the spatial envelope. *International Journal of Computer Vision*, *42*, 145–175.

Oliva, A., Torralba, A., Castelhano, M. S., & Henderson, J. M. (2003). Top-down control of visual attention in object detection. *Proceeding of the IEEE International Conference Image Processing*, *1*, 253–256.

Parker, D. M., Lishman, J. R., & Hughes, J. (1992). Temporal integration of spatially filtered visual images. *Perception*, *21*, 147–160.

Potter, M. C. (1975). Meaning in visual search. *Science*, *187*, 965–966.

Roberts, L. E., Rau, H., Lutzenberger, W., & Birbaumer, N. (1994). Mapping P300 waves onto inhibition: Go/no-go discrimination. *Electroencephalography and Clinical Neurophysiology*, *92*, 44–55.

Rousselet, G. A., Fabre-Thorpe, M., & Thorpe, S. J. (2002). Parallel processing in high-level categorization of natural images. *Nature Neuroscience*, *5*, 629–630.

Rugg, M. D., Doyle, M. C., & Wells, T. (1995). Word and non-word repetition within- and across-modality: An event-related potential study. *Journal of Cognitive Neuroscience, 7*, 209–227.

Schyns, P. G., & Oliva, A. (1994). From blobs to boundary edges: Evidence for time and spatial scale dependent scene recognition. *Psychological Science, 5*, 195–200.

Tanaka, J., Weiskopf, D., & Williams, P. (2001). The role of color in high-level vision. *Trends in Cognitive Sciences, 5*, 211–215.

Thorpe, S., Fize, D., & Marlot, C. (1996). Speed of processing in the human visual system. *Nature, 381*, 520–522.

VanRullen, R., & Thorpe, S. J. (2001). The time course of visual processing: From early perception to decision-making. *Journal of Cognitive Neuroscience, 13*, 454–461.

Verleger, R. (1997). On the utility of P3 latency as an index of mental chronometry. *Psychophysiology, 34*, 131–156.

Wichmann, F. A., Sharpe, L. T., & Gegenfurtner, K. R. (2002). The contributions of color to recognition memory for natural scenes. *Journal of Experimental Psychology: Learning, Memory, and Cognition, 28*, 509–520.

Zeki, S., & Marini, L. (1998). Three cortical stages of colour processing in the human brain. *Brain, 121*, 1669–1685.

VISUAL COGNITION, 2005, 12 (6), 893–924

Why does natural scene categorization require little attention? Exploring attentional requirements for natural and synthetic stimuli

Li Fei-Fei

Department of Electrical Engineering, California Institute of Technology, Pasadena, CA, USA

Rufin VanRullen

Centre de Recherche Cerveau et Cognition, Toulouse, France

Christof Koch

Division of Biology, California Institute of Technology, Pasadena, CA, USA

Pietro Perona

Department of Electrical Engineering, California Institute of Technology, Pasadena, CA, USA

It was recently demonstrated that detecting target objects (e.g., animals) in natural scenes can be done in a dual-task paradigm, in the near absence of spatial attention. Under the same conditions, subjects were unable to perform apparently simpler tasks involving synthetic stimuli (e.g., discriminating randomly rotated Ts and Ls, or a bisected colour disc and its mirror image). Classical theories predict that attention is more critical for the recognition of complex stimuli that cannot be easily separated on a single feature dimension. Therefore, these puzzling results have raised a number of questions. If it is not the complexity of a stimulus, what then determines the recognition task's attentional requirements? How does this differ between natural and artificial stimuli? What can these observations tell us about the mechanism of natural scene processing as well as its relation to attention? Here we show that removing colour information, or doubling the amount of information to be analysed, failed to make the natural scene categorization tasks significantly more "attention demanding". Conversely, increasing discrimin-

Please address all correspondence to: Li Fei-Fei, Department of Electrical Engineering, Mail Code 136-93, California Institute of Technology, Pasadena, CA 91125, USA. Email: lifeifei@cs.bu.edu

This research was supported by grants from the NSF-sponsored Engineering Research Center at Caltech, the National Institutes of Health, the Keck Foundation and McDonnell Foundation. F.F.L. was supported by a Paul and Daisy Soros Fellowship for New Americans and a NSF Graduate Fellowship. R.V. was supported by a Caltech Fellowship.

DOI:10.1080/13506280444000571

ability or predictability did not diminish the need for attention in the case of synthetic stimuli. However, when the familiar letters, such as Ts or Ls, were presented upright, full attention was no longer required for discrimination. This suggests that familiarity and meaningfulness might be among the factors that determine attentional requirements for both natural and synthetic stimuli.

The visual system heavily relies on attentional resources for performing numerous discrimination, categorization, or recognition tasks (Broadbent, 1958; Deutsch & Deutsch, 1963; Kahneman, 1973; Neisser, 1967; Posner, Snyder, & Davidson, 1980). Previous research on attention has revealed that only some simple feature detection tasks seem to be free of attentional requirements, while increasingly complex tasks (e.g., feature conjunction discrimination, pattern recognition) all seem to involve some degree of attention (Braun, 1994; Treisman & Gelade, 1980; Wolfe, 1998). It has even been suggested that there could simply be no visual perception in the absence of attention (Joseph, Chun, & Nakayama, 1997; Mack & Rock, 1998). Most of these studies, however, have been performed using artificial, computer-generated stimuli; yet the visual system might be optimized for processing the stimuli that it most frequently encounters in its natural environment (Vinje & Gallant, 2000).

It was recently demonstrated that natural scene processing does not necessarily follow the same attentional constraints (Li, VanRullen, Koch, & Perona, 2002; see also Rousselet et al., 2002): Detecting the presence of an animal, or a vehicle, in a natural photograph, a genuinely challenging task for today's computer vision algorithms, can be carried out by human observers in the near absence of attention. Subjects could perform this task equally well alone or simultaneously with another attention-demanding task (i.e., deciding whether five randomly rotated Ts and Ls presented at fixation were all identical or whether one of them differed from the others). Under the same dual-task conditions, subjects could not perform apparently simpler discrimination tasks involving synthetic stimuli (discriminating between a single peripherally rotated T or L, or discriminating between a red–green and a green–red bisected disc). While this result implies that natural scenes probably hold a special status for our visual systems, it is unclear exactly what about these stimuli is responsible for this distinctiveness: Is it the mere fact that a picture is natural rather than synthetic, or are there some associated (or confounded) factors that could be responsible for determining attentional requirements? The purpose of the present study is to investigate a variety of such potential factors and establish how they affect the attentional requirements of recognition tasks using natural and artificial stimuli. We approach this question using two complementary lines of attack: On one hand, we attempt to increase the difficulty of the natural scene processing task (animal vs. nonanimal, or vehicle versus nonvehicle) and test whether these manipulations will also increase attentional requirements; on the other hand, we simplify the synthetic stimulus tasks (rotated Ls vs. Ts, or bisected two-colour discs) in various ways and test whether the attentional requirements concurrently diminish.

In Experiment 1, we control that the good dual-task performance observed for natural scene processing is not dependent on previous training on this specific task: Subjects were trained on a particular task (e.g., animal vs. nonanimal categorization) and later tested on a different one (e.g., vehicle vs. nonvehicle categorization). Compared to the large dependency of attention in the synthetic stimuli tasks (Li et al., 2002), much less attention seemed needed for categorizing the unfamiliar natural scene. We observed similar results even when the subjects were trained on a completely different dual-task paradigm without having ever practised natural scene categorization. In Experiment 2, we find that removing colour information from the natural scenes does not increase such attentional requirements. In Experiment 3, we double the amount of information to process by presenting two totally unrelated peripheral scenes instead of one, with one of them at most containing a target (animal) on 50% of the trials. Even though correct performance in this task often required both scenes to be analysed, we found no increase in attentional demands compared to the single scene categorization (Li et al., 2002).

In Experiments 4 and 5, we turn to synthetic stimuli. Our goal is to probe a variety of possibilities that might reduce the high attentional requirement for performing the synthetic stimuli tasks in Li et al. (2002). In Experiment 4 we ask whether increasing the total amount of useful signal can help dual-task performance: Four redundant peripheral stimuli were presented instead of one (i.e., four bisected two-colour discs of similar orientation). But this task remained highly attention demanding. In Experiment 5 we increased stimulus predictability by presenting the letter L or T with a fixed orientation, similar on each trial, either vertical, or diagonal. Although this manipulation greatly simplified the task, only in the case where the letters were systematically presented upright did the task become truly "preattentive". We thus conclude that the difference in the associated attentional requirements previously observed between natural and synthetic stimuli cannot be accounted for by a stronger "signal-to-noise ratio" in the case of natural scenes, but rather by the familiarity and meaningfulness of stimuli. Natural scenes are highly familiar to human observers, and their meaning can be used to determine the presence or absence of a particular object category (animal). Synthetic stimuli of the type used here (bisected discs, randomly rotated letters) are both unfamiliar and meaningless to human observers. Only when the letters were presented upright and could thus be interpreted as text (i.e., both familiar and meaningful) did the attentional load change dramatically.

GENERAL METHOD

Participants

Fifteen highly motivated California Institute of Technology undergraduate and graduate students (from 20 to 26 years' old) served as subjects in all or part of the following experiments. Two authors (LF-F, and RV) were among the subjects for some of the experiments. Each subject enrolled for at least 15 daily sessions of 1 hour and received payment (excluding the two authors). Subjects reported normal colour vision and visual acuity (sometimes with corrective lenses or glasses), but underwent no tests in this respect. All subjects were right handed. Other than the two authors, all subjects were naïve about the purpose of the experiments.

Apparatus

Database. The pictures were complex colour scenes taken from a large commercially available CD-ROM library allowing access to several thousand stimuli. The animal category images included pictures of mammals, birds, fish, insects, and reptiles. The vehicle category images included pictures of cars, trucks, trains, airplanes, ships, and hot-air balloons. There was also a very wide range of distractor images, which included natural landscapes, city scenes, photos of food, fruits, plants, houses, and artificial objects.

Equipment. Subjects were seated in a dark room especially designed for psychophysics experiments. The seat was approximately 100 cm from a computer screen, connected to a Macintosh (OS9) computer. The refresh rate of the monitor was 75 Hz. All experimental software was programmed using the Psychophysics Toolbox (Brainard, 1997; Pelli, 1997) and Matlab™.

Procedure

Experimental paradigm. We use a dual-task paradigm in all of our experiments (Braun, 1994; Sperling & Dosher, 1986). Each experiment consists of three different conditions: The primary task—an attentionally demanding central task (identical in all experiments), a secondary peripheral task (in which the role of attention is investigated), and a dual-task condition in which both the central and peripheral tasks are performed concurrently. In each experiment, all trials are organized in the same way irrespective of the experimental condition (i.e., single-task condition or dual-task condition). Only the number of required responses varies between conditions.

Central letter discrimination task. In all experiments, each trial starts with a fixation cross 300 $\pm$ 100 ms before the onset of the central stimulus. At 0 ms, the central stimulus (a combination of five letters) is presented. The five letters (Ts and Ls, either all identical, or one differing from the other four), appear at nine possible locations within 1.2° eccentricity. Each letter is randomly rotated. After the central stimulus onset asynchrony (SOA; the time between the appearance of the central stimulus and the onset of the central mask), each stimulus letter is masked by the letter "F" rotated according to the random orientation of the stimulus letter. For a given subject, the central SOA is the same for both single-task and dual-task conditions. All trial types are presented with equal probability. Subjects are instructed to respond by pressing "S" on the keyboard if the five letters are the same, or "D" if one of the letters differs from the other four. Figure 1 illustrates schematically the set-up of a sample trial of the dual-task paradigm. In earlier studies it was found that the central task performance is quite a sensitive measure to indicate the allocation of attentional resources (Braun, 1994). Subjects' performances on this central task decreased drastically if the SOA was slightly decreased (Li et al., 2002).

Peripheral task. In each peripheral task, the stimulus is always presented 53 ms after the central stimulus onset. Subjects respond to these tasks in a speeded fashion. They are instructed to continuously hold down the mouse button and release it as fast as possible (within 1000 ms) when they have detected the target. For a given trial, the location of the peripheral stimulus is randomly determined, keeping a distance of 6° eccentricity (Figure 1).

Training procedure. Each novel subject to the dual-task paradigm underwent a training process. It usually took more than 10 hours for a new subject to coordinate his/her motor responses well enough to answer both a speeded peripheral task and the central task. The central SOA, starting at 500 ms, was decreased after each block where the performance of this task exceeded 85%

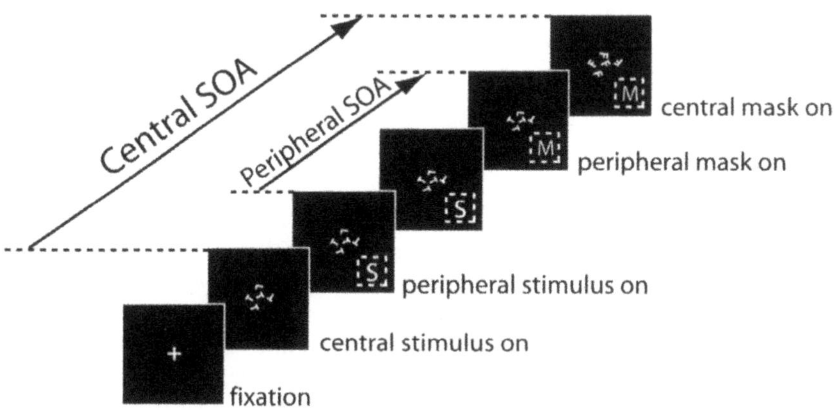

Figure 1. Dual-task experimental set-up for a single trial. A fixation cross of 1° visual angle is flashed for 300 ms at the onset of each trial, started by the subject. After that, the central letter discrimination task stimuli are presented for a central SOA amount of time. The central stimuli are then masked by an appropriate perceptual mask. Central SOA is determined individually for each subject so that the performances of the central letter discrimination task centre around 80% correct. Then, 53 ms after the onset of the central stimuli, a second stimulus is presented randomly at a peripheral location at 6° eccentricity. For different experiments reported in this paper, peripheral stimuli vary. Each peripheral stimulus is then masked by its corresponding perceptual mask, after peripheral SOA amount of time. It is important to note that to ensure that spatial attention is properly withdrawn by the central task under the dual-task condition, the peripheral mask always onsets earlier than or at the same time as the central stimulus mask.

correct. The training procedure was terminated after the subject's performance had stabilized and the central SOA was below 250 ms. This value is chosen to limit the possibility of switching attention or eye movement during stimulus presentation. Central task and peripheral task always received the same amount of training.

Data analysis. For each subject in a given experiment, we obtained two baseline performances: Central letter discrimination with attention (single-task condition) and peripheral recognition task with attention (single-task condition). Each of these two performances consists of performances of 9–15 blocks (depending on the experiment) of 96-trial experiments (Experiment 3 has 48 trials per block). We also obtain the corresponding performances for the central letter discrimination task with attention (primary task) and peripheral recognition task without attention under the dual-task condition. Each of these two performances are based on 9–15 blocks (depending on the experiment) of 96 trials each (Experiment 3 has 48 trials per block). *T*-tests are computed for each experiment to compare single- and dual-task performances. A statistical significance *p*-value of .05 is used for all statistical tests.

To visualize results, we summarize each of the experiment using a ''normalized performance'' figure. The ''normalized performance'' for each task is obtained in the following way. The averages of the two baseline performances are linearly scaled to 100% such that chance level performance remains at 50%. Then the same scaling factor for each subject is used to obtain normalized performance levels of the two tasks under the dual-task condition. In other words,

$$\text{Normalized performance} = 0.5[(P_d - 0.5) / (P_s - 0.5)] + 0.5$$

where P_d and P_s refer to performance in the dual-task and single-task conditions, respectively. It is important to point that since the 100% baseline is the average performance of a given task under the single-task condition (with attention), it is possible that the normalized performance, same task's performance under the dual-task condition (without attention) might sometimes be larger than 100%. This simply means that the actual performance has a higher average under the dual-task condition than under the single-task condition. Statistical tests will determine whether this difference is significant or not.

EXPERIMENT 1: NATURAL SCENE CATEGORIZATION IN THE NEAR ABSENCE OF ATTENTION DOES NOT REQUIRE TRAINING

In our study reported in Li et al. (2002), we had demonstrated the human visual system's amazing efficiency in natural scene categorization with little or no spatial attention. In these experiments, an average training period of 10–15 hours on the dual task was necessary for each subject. It is likely that this training helps sharpen the executive control necessary for performing different tasks, particularly when they are carried out simultaneously (Pashler, 1998; Shapiro, 2001). However, training sometimes also decreases the attentional demand on perceptual processing (Joseph, Chun, & Nakayama, 1998). So, could it be that this superb efficiency in natural scene processing is mainly due to the training process that each subject had received in these experiments? We had argued that this is unlikely since the same amount of training was applied both to the natural scene categorization and the seemingly simpler synthetic stimuli tasks (rotated single T versus L, bisected disc versus its mirror image). Our data showed, however, a large discrepancy between the attentional requirements of these two types of tasks. This is difficult to explain by the same training process.

Here we further investigated the effects of training (or lack of it), particularly its influence on the natural scene stimuli. If training indeed helped performing natural scene categorization with little attention, this might be achieved through learning specific visual features critical for performing this task. It is important to note that all the data collected in the testing phase was from a set of novel images that the subjects were never trained on. Therefore simple image-based learning or memorization cannot account for the observed results. The testing stimuli, however, were drawn from the same set of images as the training images. Could subjects have, therefore, learned to categorize animal versus nonanimal (or vehicle versus nonvehicle) images because the same image types were presented repetitively? In a computational framework, it is conceivable that a specific set of "animal filters" (or "vehicle filters") were sharply tuned and enhanced during this training period. But if this were the case, training on a specific task would only help to tune the specific "filters" for that particular categorization task. If we tested on a different natural scene categorization task, we should be able to observe a difference in performance.

Method

We tested this hypothesis in two separate experiments. In the first experiment, we divided a group of six subjects into two groups. Both groups received the same amount of training in all tasks. Specifically, both groups were trained on the central letter discrimination task (see Procedure section) and a natural scene categorization task under both single- and dual-task conditions. Recall that the "single-task" condition refers to the situation where attention is available to perform the current peripheral task, while the "dual-task" condition refers to the situation where attention is drawn to the centre, leaving the peripheral task in the near absence of spatial attention. There are 96 trials in any given block of task under all conditions. In Group I, the three subjects were trained on the "animal vs. nonanimal" categorization task. They were instructed to categorize natural scenes with or without animals in a go/no-go fashion by releasing a mouse button. The task was speeded so that any response after 1000 ms was automatically registered as a "no target" answer. The test data was then acquired by having them perform "vehicle vs. nonvehicle" categorization without any additional training. Similarly to the animal categorization task, subjects responded by releasing the mouse button when a target, i.e., vehicle(s), was detected. In Group II, the three subjects were trained on vehicle categorization and tested on animal categorization (Figure 2). Previous experiments (Li et al., 2002) have already established that a trained natural scene categorization task requires little attention. We are, therefore, interested in seeing whether such performance can be transferred from one type of categorization (e.g., animal) to another (e.g., vehicle). Namely, will the performance of vehicle categorization without attention be comparable to the performance of animal categorization without attention for Group I subjects, and vice versa for Group II?

One might argue that in the above manipulation, even though the tested natural scene category was not trained, it was nevertheless learned during training because natural scene photographs shared much commonalities (Olshausen & Field, 1996). When one is trained on one type of natural scene categorization, say "animal scenes", it is possible that similar image statistics help to tune the "filters" on other types of natural scene categories (e.g., "vehicle scenes"). If this were the case, however, such performance should not hold for a recognition task that does not share similar stimulus statistics. For the second experiment in this section, we tested this hypothesis with another four subjects who were previously trained on the dual-task paradigm in a task that did not involve natural scene photographs. Specifically, these subjects performed a face gender discrimination task in a dual-task paradigm (Reddy, Wilken, & Koch, 2004). The gender discrimination task utilizes very different stimuli that bore little commonalities with the natural scene images (Troje & Bulthoff, 1996). After the subjects completed their training on this task, we tested them on both animal and vehicle categorization tasks (Figure 3).

Results

The bottom panels of Figure 2 illustrate the normalized performance from the cross-training experiment between two groups of subjects. In Group I, three subjects were trained on animal categorization and then tested on vehicle categorization. Their central performances during the training phase show that under the dual-task condition, they had successfully maintained their attention at the central task (single central task, average over nine blocks: 75.9 $\pm$ 3.5%, 85.4 $\pm$ 3.9%, and 83.2 $\pm$ 2.5% for each subject respectively; dual central task, average over nine blocks: 70.3 $\pm$ 5.5%, 86.1 $\pm$ 2.7%, and 80.8 $\pm$ 4.1% respectively); *t*-test results: $t(16) < 1.75$, $p > .05$ for each subject. During the testing condition, only one subject has a slight drop in central task performance under the dual-task condition (single central task, average over nine blocks: 75.9 $\pm$ 3.6%, 85.4 $\pm$ 3.9%, and 83.2 $\pm$ 2.5% for each subject respectively; dual central task, average over nine blocks: 71.2 $\pm$ 2.9%, 87.6 $\pm$ 3.7%, and 81.3 $\pm$

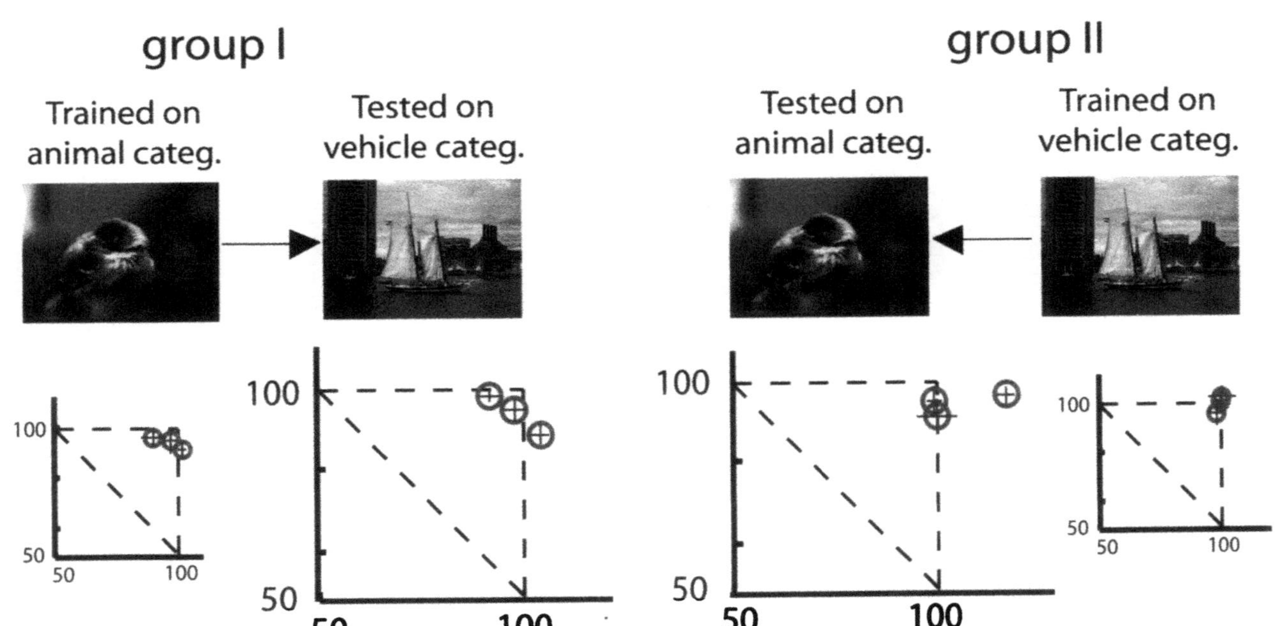

Figure 2. Experiment 1: Cross-training experiment. Six subjects are divided into two groups of three each. In Group I, illustrated on the left, the subjects are trained with central letter discrimination and animal categorization task, as well as the dual-task condition using these two tasks. After training, they are immediately tested on the vehicle categorization task in both the single-task condition (with attention) and the dual-task condition (without attention). Results are shown in normalized performance plots. In Group II, illustrated on the right, the subjects are trained with central letter discrimination and vehicle categorization task, as well as the dual-task condition using these two tasks. After training, they are immediately tested on the animal categorization task in both the single-task condition (with attention) and the dual-task condition (without attention). Results are shown in normalized performance plots. Our results illustrate that subjects need not be trained for the specific natural scene categorization task in order to perform it without spatial attention, suggesting category-specific training is not necessary to carry out this high-level task without spatial attention.

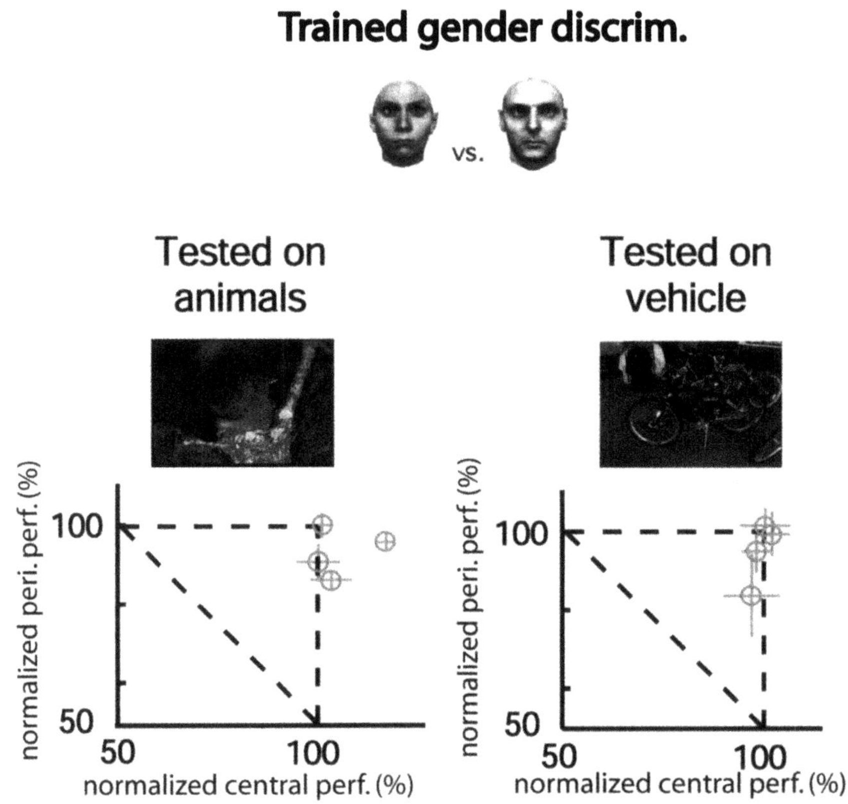

Figure 3. Experiment 1: Gender-training dual-task experiment. In this experiment, four subjects are trained on central letter discrimination task, gender discrimination as well as the dual-task condition using these two tasks. Two examples of the stimuli of the gender discrimination task are shown here (Troje & Bulthoff, 1996). Subjects are instructed to respond whether a hairless face, briefly presented in the periphery and then masked, is female or male (Reddy et al., 2004). After the training process is completed, subjects are tested on two natural scene categorizations without attention: Animal and vehicle. Normalized performances of the single- and dual-task conditions are presented for each natural scene categorization. Our results show that little training on natural scenes is needed to perform natural scene categorization.

4.1%); t-test results: $t(16) = 2.40$, $p = .01$ for the first subject; $t(16) < 1.75$, $p > .05$ for the rest. During the training phase, the subjects performed the animal categorization task without any interference when comparing the performances under the dual-task condition with the single-task condition (single peripheral task, average over nine blocks: 79.5 $\pm$ 0.6%, 84.3 $\pm$ 0.6%, and 77.1 $\pm$ 1.8%; dual peripheral task, average over nine blocks: 77.3 $\pm$ 5.2%, 78.1 $\pm$ 5.6%, and 74.3 $\pm$ 6.7%); t-test results: $t(16) < 1.75$, $p > .05$ for each subject. A similar performance pattern is observed for these three subjects during the testing phase, in which they were directly tested on the vehicle categorization task without any prior training (single peripheral task, average over nine blocks: 83.2 ± 4.5%, 87.3 ± 4.4%, and 80.2 ± 2.6%; dual peripheral task, average over nine blocks: 82.2 ± 2.1%, 79.3 ± 3.9%, and 77.3 ± 3.7%); *t*-test results: $t(16) = 3.22$, $p = .003$ for the second subject; $t(16) < 1.75$, $p > .05$ for the rest. Note that there is a slight drop in the vehicle categorization task under dual-task condition for the second subject. This small decrease, although significant, should be viewed in the light of the results of Lee, Koch, and Braun (1999) and Li et al. (2002): When attention is taken away, performances for a simple rotated T versus L task (or red–green disc versus green–red disc) dropped much more dramatically, often to chance level (50%).

Compared to Group I, Group II subjects went through reversed training and testing categorization tasks. During the training stage, three subjects were trained on vehicle categorization only. All of them have successfully allocated attention at the centre under both the single- and dual-task condition (single central task, average over nine blocks: 68.6 ± 4.6%, 78.5 ± 6.1%, and 88.9 ± 3.2%; dual central task, average over nine blocks: 67.9 ± 2.8%, 78.6 ± 7.2%, and 88.8 ± 3.5%); *t*-test results: $t(16) < 1.75$, $p > .05$ for each subject. Their vehicle categorization task results also show that they were able to perform this task without attention (single peripheral task, average over nine blocks: 72.2 ± 7.9%, 77.4 ± 1.2%, and 81.6 ± 1.6%; dual peripheral task, average over nine blocks: 70.0 ± 4.8%, 78.3 ± 5.0%, and 81.4 ± 4.7%); *t*-test results: $t(16) < 1.75$, $p > .05$ for each subject. During the testing stage, where subjects were tested directly on animal categorization without any prior training (training was done using vehicle categorization), they maintained good performances on central task under both conditions just as they did during the training sessions (single central task, average over nine blocks: 68.6 ± 4.6%, 78.5 ± 6.1%, and 88.9 ± 3.2%; dual central task, average over nine blocks: 68.6 ± 4.2%, 88.3 ± 2.7%, and 88.4 ± 3.4%); *t*-test results: $t(16) = 3.79$, $p = .001$ for the second subject; $t(16) < 1.75$, $p > .05$ for the rest. Similarly, all three subjects performed the animal categorization task under the dual-task condition as well as under the single-task condition (single peripheral task, average over nine blocks: 74.8 ± 5.8%, 84.2 ± 1.8%, and 81.1 ± 4.3%; dual peripheral task, average over nine blocks: 70.4 ± 4.5%, 81.5 ± 3.0%, and 77.8 ± 5.1%); *t*-test results: $t(16) < 1.75$, $p > .05$ for each subject.

Figure 3 illustrates the results from the second experiment in this section. Four new subjects were trained on the dual-task paradigm with the same central letter discrimination task but a peripheral face gender discrimination task (Reddy et al., 2004). Testing of whether animal categorization and vehicle categorization require attention followed after the training phase. Three of the four subjects showed natural scene categorization performances (both animal and vehicle) in the near absence of spatial attention statistically indistinguishable from the same tasks performed with attention available (single peripheral animal categorization task performance, averaged over nine blocks for each: 83.0 ± 3.2%, 81.9 ± 0.6%, 83.3 ± 3.4%, and 76.9 ± 4.3%; dual peripheral animal categorization task performance, averaged over nine blocks: 79.9 ± 4.3%, 77.4 ± 3.7%, 74.3 ± 3.3%, and 77.1 ± 3.4%); *t*-test results: $t(16) = 4.46$, $p = .0002$ for the third subject; $t(16) < 1.75$, $p > .05$ for the rest. Single peripheral vehicle categorization task performance, averaged over nine blocks for each: 81.8 ± 3.9%, 83.0 ± 4.5%, 85.1 ± 2.6%, and 78.1 ± 7.3%; dual peripheral vehicle categorization task performance, averaged over nine blocks: 79.6 ± 2.5%, 80.0 ± 2.7%, 73.6 ± 14.8%, and 77.8 ± 6.4%); *t*-test results: $t(16) < 1.75$, $p > .05$ for each subject. Note that one subject showed a decrease in animal categorization performance when attention was drawn away (single task performance, average over nine blocks: 83.3 ± 3.4%; dual-task performance, average over nine blocks: 74.3 ± 3.3%); *t*-test results: $t(16) = 4.46$, $p = .0002$. Here again, this decrease is much smaller than those observed on known "attentionally demanding" tasks (Lee et al., 1999; Li et al., 2002). Note that the same subject's performance on vehicle categorization was not significantly different in the single- and dual-task conditions.

Both experiments in this section demonstrate that little training is needed for natural scene categorization without attention. Subjects are able to perform natural scene categorization tasks in the near absence of spatial attention without previous training on the specific task or type of stimuli.

Discussion

The first experiment in this section has shown that when trained on animal categorization (or vehicle) in the near absence of attention, subjects can perform vehicle categorization (or animal) with no further training. This effect is even more dramatically demonstrated by the second experiment, in which a totally different type of stimulus is used during training. But as soon as subjects have learned to perform this recognition task under the dual-task paradigm, they can apply this ability to a natural scene categorization task. On the contrary, our previous experiments have shown that single letter discrimination or colour disc discrimination cannot be performed without attention given the same amount of training as the natural scene categorization task (Li et al., 2002). Hochstein and colleagues argue that "higher level" tasks can be more easily transferred than "lower level" tasks (Hochstein & Ahissar, 2002). Could it be that less attentional resource is needed because natural scene categorization is carried out in "higher" areas of the visual system than the other synthetic tasks? We will revisit this point in both Experiments 3 and 5.

EXPERIMENT 2: NATURAL SCENE CATEGORIZATION WITHOUT ATTENTION CAN BE PERFORMED WITHOUT COLOUR

We set out to explore different factors that might contribute to the fast recognition of natural scene categories with little or no attention. A simple question to ask is whether some low-level features might have been useful cues to the categorization task. For example, it has been shown that colour histograms are very informative for natural scene recognition in both human and computer vision (Oliva & Schyns, 2000; Ostergaard & Davidoff, 1985; Price & Humphreys, 1988). In today's computer vision field, some image retrieval algorithms have utilized colour information to categorize different images (Rubner, Tomasi, & Guibas, 1998). Delorme, Richard, and Fabre-Thorpe (2000) have shown that colour is not a critical component in fast categorization, under conditions where attention was not explicitly controlled. In the present experiment, we ask the role of colour information in natural scene categorization with little or no attention, by changing the stimuli to greyscale.

Method

We use the same dual-task paradigm for this experiment as the previous experiment. The central task is an attentionally demanding letter discrimination. The peripheral task is natural scene categorization, using novel greyscale images (examples of the greyscale images are shown in Figure 4). Five subjects participated in this experiment. They were instructed to respond as fast as possible when they detected the presence of an animal in the image shown at a random peripheral position. Subjects performed 15 blocks of dual task and 12 blocks of single task. Each block consisted of 96 trials.

Results

It is first interesting to observe that individual SOAs for the natural scene categorization task are not much different from the ones observed in Li et al. (2002) where colour information was included in the images (average natural scene categorization SOA of five subjects in Li et al., 2002: 61 ms; average SOA of five subjects in current experiment: 85 ms); *t*-test result: $t(8) < 1.86$, $p > .05$. Figure 4 illustrates the normalized performances of the subjects' dual-task performances. Note that each subject has achieved a central-task performance at his/her baseline level in dual-task (single central task, average over 12 blocks for each subject: 76.9 $\pm$ 6.1%, 76.0 $\pm$ 3.2%, 74.2 $\pm$ 4.9%, 75.0 $\pm$ 4.4%, and 74.3 $\pm$ 6.6%; dual central task, average over 15 blocks for each subject: 75.4 $\pm$ 5.7%, 78.4 $\pm$ 4.7%, 71.3 $\pm$ 5.5%, 70.8 $\pm$ 5.5%, and 72.5 $\pm$ 6.3%); *t*-test results: $t(25) < 1.71$, $p > .05$ for each subject. This result assures us that much of attentional

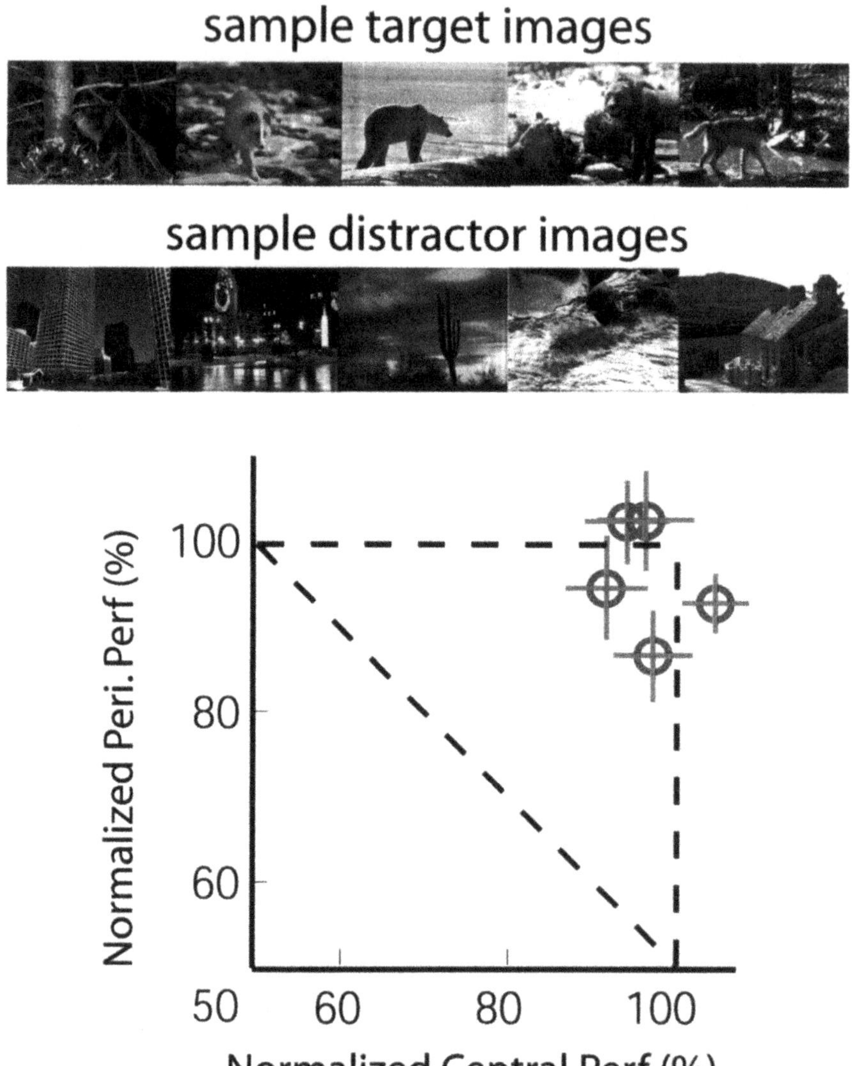

Figure 4. Experiment 2: Natural scene categorization without colour. In this experiment, the peripheral task is animal categorization with greyscale natural scene images. All experimental conditions remain identical to the ones introduced in Li et al. (2002). The only difference is that all peripheral task stimuli as well as the masks are presented in greyscale. The top two rows show some samples of the target stimuli and distractor stimuli respectively. The bottom panel indicates subjects' normalized performances of this task, showing that there is little cost in greyscale natural scene categorization when spatial attention is withdrawn.

resources are allocated to the demanding letter discrimination task under the dual-task condition. Four of the five subjects' peripheral natural scene categorization task performances remain comparable to their respective baseline performances (single peripheral task, average over 12 blocks for each subject: 79.3 $\pm$ 2.7%, 73.2 $\pm$ 7.1%, 76.9 $\pm$ 6.5%, and 74.2 $\pm$ 7.4%; dual peripheral task, average over 15 blocks for each subject: 75.2 $\pm$ 4.7%, 74.4 $\pm$ 5.1%, 71.7 $\pm$ 5.9%, and 72.9 $\pm$ 5.8%); t-test results: $t(25) < 1.71$, $p > .05$ for each subject. Only one subject's natural scene categorization task performance decreases slightly while attention is allocated elsewhere (single peripheral task, average over 12 blocks: 71.9 $\pm$ 5.6%; dual peripheral task, average over 15 blocks: 66.1 $\pm$ 5.3%); t-test results: $t(25) = 3.82$, $p = .0004$. Overall, greyscale natural scene images can be categorized rather efficiently in the near absence of spatial attention.

Discussion

Our results indicate that, when much of attention is engaged elsewhere, subjects can perform a rapid natural scene categorization task without colour. This result suggests that colour information is not critical in performing such a task in the near absence of attention. Dunai, Castiello, and Rossetti (2001) have also found that colour cues are only more informative at a longer timescale in an attentionally demanding detection task. In addition, Delorme et al. (2000) have already shown that colour information is not critical in the initial recognition of the same set of natural scenes that we are using. These findings, together with ours, suggest that natural scene categorization might be carried out by a rapid and efficient process that does not require much of the slower colour information. Torralba and Oliva (2003) suggested that some natural scene images can be categorized based on second-order statistics derived from power spectral analysis. It would be fruitful to test their hypothesis on object recognition such as animal or vehicle categorizations. To conclude, removal of colour information failed to make natural scene categorization an "attentionally demanding" task. Thus, if this type of natural scene processing only relies on low-level features, colour cannot be counted as one such feature.

EXPERIMENT 3: EVIDENCE FOR PARALLEL PROCESSING FOR NATURAL SCENE CATEGORIZATION WITHOUT ATTENTION

We have established so far the amazing robustness of the human visual system when categorizing natural scenes with little attention. In an attempt to search for a "breaking point" of this ability, we investigate the effects of natural scene categorization when the number of peripheral images is increased to two. In other words, instead of searching for a possible target in just one image, the subjects have to now search for a possible single occurrence of the target in two images. We ask whether by effectively halving the "signal to noise ratio" the efficiency of this task decreases? Our rationale is that by identifying the condition in which such natural scene processing is no longer doable, we can start comparing and contrasting different conditions in order to understand the underlying neuronal mechanisms.

Method

The attentionally demanding central letter discrimination task remained the same as in the previous experiments. Five subjects performed a letter discrimination task at the central SOAs individually adjusted for each of them. The peripheral task was a coloured animal scene categorization. Each block consisted of 48 trials. In half of the trials, there were two peripheral natural scene images (''double-image'' condition), with two equally likely configurations—either one of the two images contained a scene with animal(s), or neither image contained an animal. In the other half of the trials, there was only one image, just like in the previous experiments on scene categorization (''single-image'' condition). Subjects are told to respond by lifting the mouse button when they detect the presence of an animal (or animals) in both conditions. These two types of trials were intermixed randomly throughout the experiment. Figure 5 shows the schematic set-up of this experiment. For the double-image condition, the separation between the two images varied randomly between 4° and 12° (each maintaining an eccentricity of 6°, just as in the previous experiments). Subjects were informed before the experiments of the two different possible conditions. No one reported any confusion or difficulty with the instructions. There were a total of 15 blocks for the dual-task condition and 15 blocks for each of the single-task condition.

Results

All five subjects were able to categorize novel natural scenes without spatial attention. The top two panels of Figure 6 illustrate the normalized dual-task performances for the single-image condition and double-image condition respectively. First we observe that all subjects maintained their central letter discrimination task performances under single task and dual-task conditions (single central task performance, average over 15 blocks for each subject: 75.4 ± 3.3%, 71.5 ± 1.3%, 67.7 ± 5.0%, 77.6 ± 4.9%, and 77.1 ± 3.8%; dual central task performance, average over 15 blocks for each subject: 73.8 ± 5.4%, 71.5 ± 5.5%, 67.3 ± 4.3%, 75.0 ± 6.0%, and 77.3 ± 5.9%); *t*-test results: $t(28) < 1.70$, $p > .05$ for each subject. These results indicate that spatial attention was successfully allocated to the central task for all these subjects. Now we are interested in comparing subjects' natural scene categorization

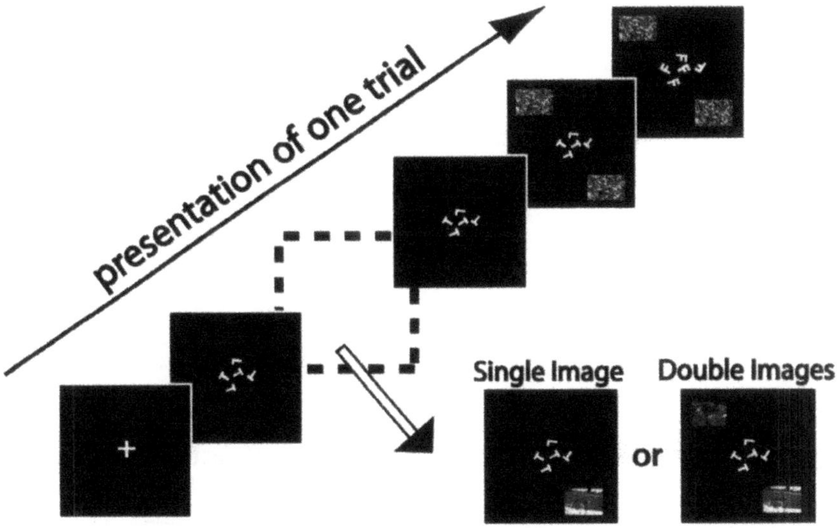

Figure 5. Experimental set-up for single-image versus double-image experiment. We illustrate here the set-up of a single trial for this experiment. The basic procedure is the same as in Figure 1. For a given trial, there are two possible peripheral stimulus presentation set-ups. For 50% of the trials, there are two unrelated natural scene images presented randomly in the periphery. The centre of each image is located at 6° eccentricity with respect to the centre of the screen. Their mutual separation varies randomly. Two similar perceptual masks follow the images after the peripheral SOA. For the other 50% of the trials, there is only one natural scene image presented randomly in the periphery, which is exactly the same case as in Li et al. (2002). At the end of the presentation, a random perceptual mask follows the image stimulus.

performances with or without attention under single-image and double-image conditions. Figure 6 shows the performance pattern of double-image categorization (single task performance, average over 15 blocks for each subject: 64.6 $\pm$ 3.1%, 71.7 $\pm$ 5.8%, 74.7 $\pm$ 7.6%, 75.4 $\pm$ 6.1%, and 73.6 $\pm$ 8.4%; dual-task performance, average over 15 blocks for each subject: 64.0 $\pm$ 6.4%, 68.1 $\pm$ 6.5%, 70.2 $\pm$ 5.8%, 72.3 $\pm$ 6.3%, and 70.7$\pm$6.0%); *t*-test results: $t(28) < 1.70$, $p > .05$ for each subject, as well as the single-image case (single task performance, average over 15 blocks for each subject: 77.1 $\pm$ 5.4%, 80.1 $\pm$ 7.1%, 87.5 $\pm$ 6.9%, 81.9 $\pm$ 4.1%, and 79.2 $\pm$ 11.0%; dual-task performance, average over 15 blocks for each subject: 76.3 $\pm$ 4.9%, 74.0 $\pm$ 4.3%, 79.0 $\pm$ 8.3%, 83.8 $\pm$ 5.9%, and 70.4 $\pm$ 9.0%); *t*-test results: $t(28) = 2.48$, $p = .01$ for the second subject, $t(28) = 1.98$, $p = .03$ for the third subject, $t(28) < 1.70$, $p > .05$ for the rest. All subjects' results show that when there are two images to process, the categorization performances with attention (single-task condition) are statistically no different from the performances without attention (dual-task

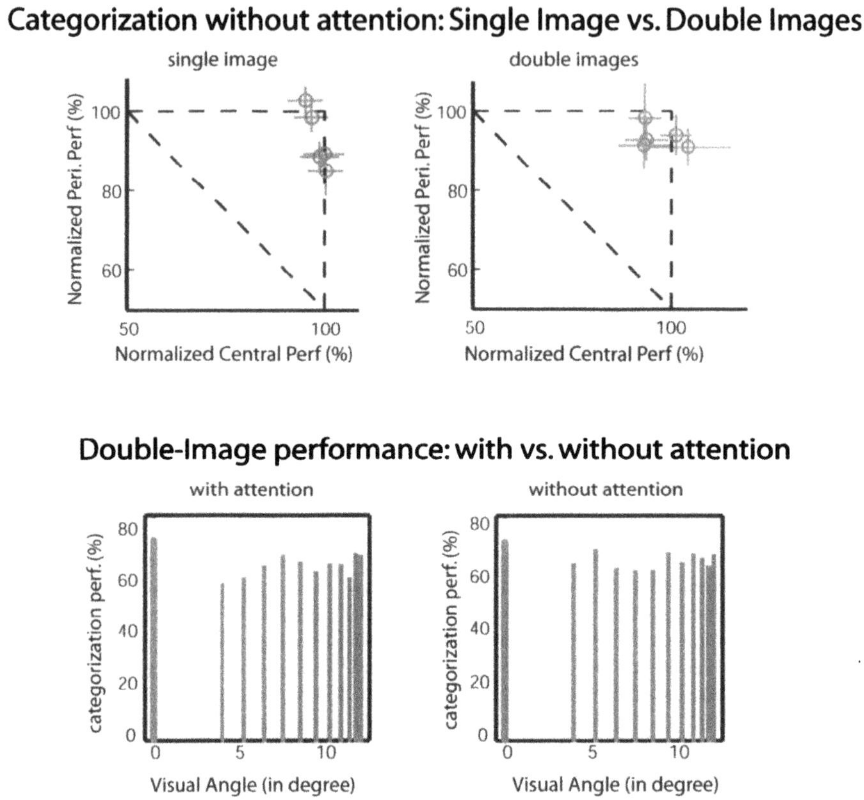

Figure 6. Experiment 3: Results of the single-image versus double-image experiment. The top two panels illustrate normalized performances of the single-image dual-task and the double-image dual-task respectively. Five subjects participated in this experiment. The results show that there is little difference between the one-image case and the two-image case, suggesting that natural scene categorization without attention is a highly parallel process. The bottom two panels break down the performances of the double-image case by the distance separating the two image centres. The two leftmost bars in each panel, placed at the 0° angle, indicate the average single-image performances of the subjects with or without attention respectively. The left panel shows the result when attention is available; whereas the right panel shows when attention is withdrawn. For each attentional condition, there is no apparent pattern of performance difference as a function of visual angle separation.

condition). There is, however, a small but significant drop for one subject when categorizing the single-image without attention compared to with attention. Similarly to the previous arguments, we think this is a rather small effect in the light of the comparative results obtained from synthetic stimuli (Lee et al., 1999; Li et al., 2002). The average baseline performances (i.e., single-task condition when attention is available) show an overall decrease in the double-image categorization condition (single-task condition for single-image case:

81.2 $\pm$ 3.9%; single-task condition for double-image case: 72.0 $\pm$ 4.4%); *t*-test result: $t(8) = 3.04$, $p = .008$. This set-size effect, different from previous results by Rousselet, Fabre-Thorpe, and Thorpe (2002), might simply be attributed to our keeping SOAs constant between the single- and double-image conditions. In addition, stimulus location was totally unpredictable in our study, whereas it was fixed in the experiments of Rousselet et al. But the main result is that, when attention is taken away, subjects were able to perform double-image categorization just as well as they did when attention was available. This result suggests that by halving the "signal-to-noise ratio" of the stimuli, natural scene categorization can still be efficiently carried out in the near absence of attention.

Since we randomly varied the visual angle distance between the two images under the double-image condition, we can ask whether the subjects' performances differ for different visual angle separations. The bottom two panels of Figure 6 show double-image condition performances sorted by the visual angle separation. The left panel corresponds to the condition where attention is available (single-task condition), whereas the right panel corresponds to a reduced attention condition (dual-task condition). The two leftmost bars in each panel, placed at the 0° angles, indicate the average single-image performances of the subjects with or without attention respectively. We investigated the effects of visual angle separation between the two stimulus images with a two-way ANOVA (attentional condition vs. interstimulus separation). In accordance with our previous results, there was no main effect of attention on performances: Single-task condition vs. dual-task condition, $F(1, 96) = 1.2112$, $p > .05$. Furthermore, there was no main effect of the visual angle separations between stimuli either, $F(11, 96) = 0.5817$, $p > .05$. Additionally, there was no significant interaction between these two factors, $F(11, 96) = 1.1654$, $p > .05$. Note that due to the size of the image itself, the minimal separation distance between two images is 4°.

Discussion

Our finding suggests that a natural scene categorization task not only demands little attentional resource, but is also highly parallel. When two images are presented simultaneously, subjects are able to process both of the images in search of a target object in a comparable fashion as when there is only one image. Physiological data from ERP recording also supports this finding. Rousselet et al. (2002) found that subjects are as fast for animal categorization with two images as with one image. Together our results suggest that high-level information can be accessed by the visual system in a parallel fashion with little attentional assistance. It suggests that some categorical information might be able to reach higher areas of the visual hierarchy rather efficiently, without much serial focal-attention selection.

EXPERIMENT 4: MULTIPLE COPIES OF SYNTHETIC STIMULI DO NOT HELP RECOGNITION WITHOUT ATTENTION

We have so far probed in a number of ways to what extent natural scene categorization can be carried out by the human visual system without attention. Our results tell us that such categorization is highly efficient and robust to the lack of attentional resource. In contrast, seemingly much simpler tasks involving synthetic stimuli do not enjoy this freedom of attention (Lee et al., 1999; Li et al., 2002). In the following two experiments, we turn to the question of these synthetic stimuli: What type of manipulation would decrease attentional requirements for these stimuli? In other words, through which dimension of manipulation can we make the synthetic stimuli task ''easier'' to process in the near absence of attention?

One possible hypothesis of this contrast between natural scene images and synthetic stimuli is that an object target (e.g., an elephant) in a natural image might carry multiple ''diagnostic features'' for its detection or recognition. The exact nature of these ''diagnostic features'' is unknown. But it is conceivable that many of the body parts of an animal in the animal categorization task, for example, are potential cues for the detection of the object. The synthetic stimuli, on the other hand, do not enjoy the luxury of multiple potential ''diagnostic features'' (Li et al., 2002). In that study, there were two types of synthetic stimuli, a rotated T versus a rotated L, and a vertical red–green bisected disc versus another vertical green–red disc. In the case of T versus L, the only obvious ''diagnostic feature'' is the T-junction versus the L-junction in the letters respectively. A failure to detect such junction would result in an ambiguous decision. Similarly, for the bisected discs, since the stimuli position is random from trial to trial, it is not possible to determine whether it is a red–green disc or a green–red disc by detecting the colour on half of the disc. The only ''diagnostic feature'' is the junction between the two coloured semicircles. Hence we predict that if the advantage of scene categorization without attention lies in the higher probability of detecting one or more of the possible ''diagnostic features'', then increasing the number of stimuli in the synthetic stimulus task could result in an increase of performance.

Method

We test this hypothesis using the bisected colour discs. Four subjects participated in this experiment. The basic set-up remained the same. The attentionally demanding central task was letter discrimination. In the periphery, subjects were instructed to respond when the bisected disc(s) was(were) arranged in a red–green fashion, as opposed to an equally likely green–red pattern. In half of the trials, there were four identical discs. Subjects were assured of the fact that all stimuli were redundant. In the other half of the trials, there was only one such bisected disc (which is the same condition as in Li et al., 2002). These two types of trials were intermixed randomly within a block of trials. All discs were the same size. Figure 7 illustrates the arrangement of the stimuli. In the single-disc condition, the disc was centred at 6° eccentricity. In the four-disc condition, the centre of the four-disc array was located at 6° eccentricity. Each block consisted of 96 trials. Subjects performed 18 blocks of experiments.

Results

The hypothesis described above predicts that an increased number of peripheral stimuli might result in an increase of dual-task performance for recognition of the colour discs. The intuition is that there are more potential ''diagnostic features'' to be sampled by the visual system when the number of redundant stimuli is greater. Contrary to our prediction, we observe no improvement in dual-task performances for the trials where there are four discs rather than one. For both the single-disc and the four-disc conditions, subjects' dual-task performances of the peripheral colour disc recognition are not significantly better than chance (one-disc performance under dual-task condition, averaged over 18 blocks for each subject: 50.4 $\pm$ 7.63%, 49.3 $\pm$ 6.6%, 48.9 $\pm$ 8.0%, and 52.4 $\pm$ 6.2%; four-disc performance under dual-task condition, averaged over 18 blocks for each subject: 47.9 $\pm$ 8.1%, 49.7 $\pm$ 6.1%, 50.5 $\pm$ 3.4%, and 46.9 $\pm$ 7.4%); t-test results: $t(17) < 1.74$, $p > .05$ for each subject and each task. Note that when the peripheral colour disc recognition task is carried out with attention available, subjects' performances centre around 85% at their individual SOAs (one-disc performance under single-task condition, average over 18 blocks for each subject: 88.8 $\pm$ 8.5%, 81.3 $\pm$ 5.0%, 89.6 $\pm$ 3.3%, and 77.0 $\pm$ 7.5%; four-disc performance, averaged over 18 blocks: 91.7 $\pm$ 6.3%, 87.1 $\pm$ 5.1%, 90.5 $\pm$ 5.4%, and 74.5 $\pm$ 7.5%). It seems that the subjects cannot take advantage of the increased number of possible ''diagnostic features'', even when attention is fully available.

Discussion

We test the hypothesis that independent ''diagnostic features'' might contribute to recognition without attention. The assumption was natural scene categorization might be potentially ''easier'' than the synthetic stimuli recognition due to the multitude of ''diagnostic features''. In other words, different body parts of an animal (or vehicle) might increase the chance of detection while the synthetic stimuli tend to have a very localized, nearly singular point of ''diagnostic feature'' (e.g., bisecting line of the double colour discs). We therefore increased the probable number of features by replicating the number of stimuli from one to four. It is important to point out that this manipulation is not comparable to the one that we did in Experiment 3. In Experiment 3, we left the amount of ''target

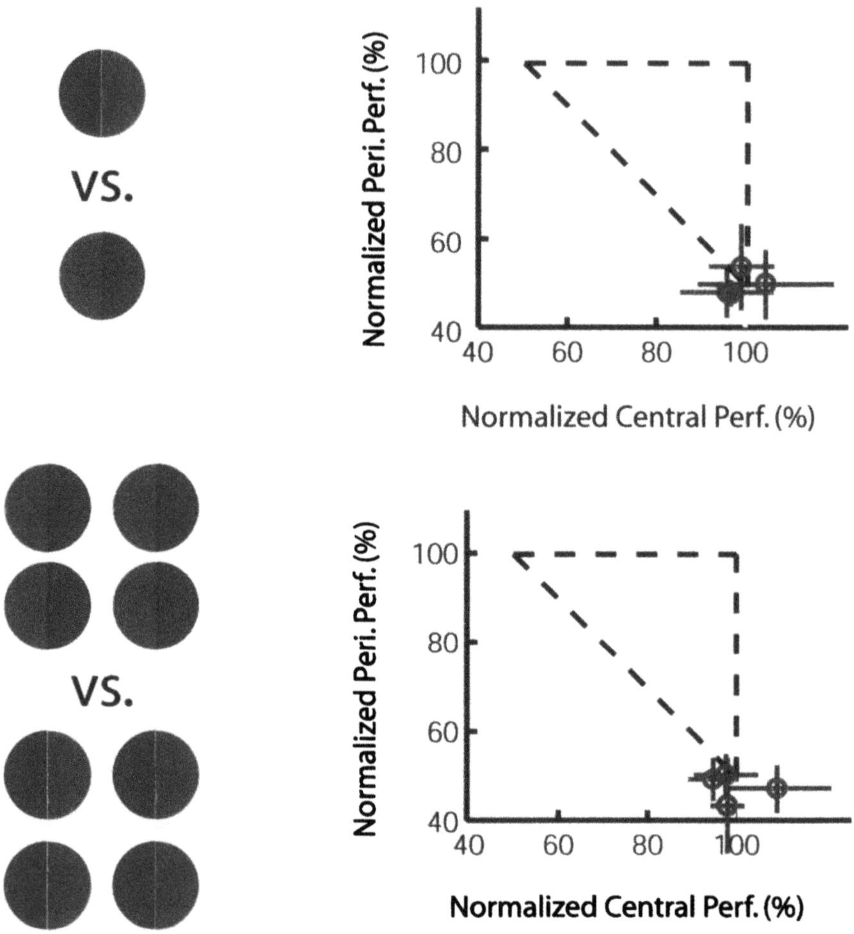

Figure 7. Experiment 4: Multiple-stimuli experiment. The basic set-up of this experiment is also a dual-task paradigm. A colour disc discrimination task is used peripherally. Two types of peripheral stimuli are mixed randomly during the experiment. The top row shows the first peripheral stimulus and the normalized dual-task performances. Peripheral recognition task is a red–green colour disc versus its mirror image, green–red colour disc. Subjects' performances of this task without attention is at chance level compared to their baseline performances, centred around 80% before normalization. The bottom row shows the second type of peripheral stimulus and the corresponding normalized dual-task performances. In this case, the recognition task remains the same as the top row, with the exception that there are four copies of the same stimuli arranged in the indicated pattern. We show here that subjects' performances of this task without attention is also at chance level, no better than the case with one copy of the stimulus.

signal'' (or probability of the presence of an ''animal'' scene) constant while doubling the amount of ''distracting noise'' (or probability of the presence of a ''nonanimal'' scene). Here the absolute number of potential ''diagnostic features'' is increased through having multiple, redundant copies of stimuli. Our results show clearly that such increase of potential diagnostic features did not help at all in recognition of synthetic stimuli with little or no spatial attention. This observation implies that it is unlikely that the bottleneck of such synthetic stimuli recognition without attention is the number of available ''diagnostic features''. Natural scenes might have an overall advantage over the synthetic stimuli used here due to the intrinsic image statistics or different processing mechanisms.

An alternative explanation also deserves further investigation. Assume that features related to the targets and distractors for the stimuli live in a high dimensional ''feature space''. Then it is possible that ''diagnostic features'' in the synthetic stimuli case might lie too closely to the ''distractor features'' of the synthetic stimuli in the ''feature space''. On the other hand, in the rich natural scene stimuli case, the ''diagnostic features'' of the targets might be much more easily isolated from the distractors than the synthetic stimuli case. If this hypothesis were true, simply repeating the number of targets in the synthetic stimuli task would not increase the discriminability of the target from the distractor, just as observed here.

EXPERIMENT 5: EVIDENCE FOR WELL-LEARNED CATEGORIES OF OBJECTS ENTAILING LESS ATTENTIONAL LOAD DURING RECOGNITION

So far our attempt to ''increase'' the difficulty of natural scene recognition without attention by reducing the amount of signal or decreasing the amount of training have not broken down the system dramatically. Similarly, adding copies of stimuli to the synthetic recognition task does not ''ease'' the task difficulty either. We hence want to test whether task ''predictability'' can be an influential factor in the recognition task without attention.

Our observations, however, also point to the direction that it could be the different level of processing that result in such different performances between natural scenes and synthetic stimuli. It has been long known that object categories are encoded in higher level visual areas such as the inferior temporal lobe (IT; Logothetis & Sheinberg, 1997; Tanaka, 1996). The most prominent object category is face for the human visual system (Epstein & Kanwisher, 1998; Rossion & Gauthier, 2002). Haxby, Gobbini, Furey, Ishai, Schouten, & Pietrini (2001) have shown differentiable fMRI patterns in IT and related areas when responding to different type of stimuli of a wide range of visual categories. So could it be that existing neuronal representations of natural scene categories are responsible for such efficient and fast recognition of natural images with little attention? If this is the case, can we find meaningful categories of objects in synthetic stimuli to test this hypothesis?

We test here in this experiment two independent hypotheses by using the letter discrimination task as the peripheral task. The first hypothesis is that stimulus predictability might affect the attentional requirement in recognition. It is conceivable that less attention is required when subjects know beforehand the exact shapes of the stimuli to be discriminated. The second hypothesis is that well-learned object categories can be recognized with significantly less attentional load. Evidence from visual search paradigm using familiar and unfamiliar letter-like patterns indicates that visual search speed is strongly facilitated by more familiar objects (Shiffrin & Schneider, 1977; Wang, Cavanagh, & Green, 1994). Peripheral letter discrimination task in dual-task paradigm has previously been set up in such a way that the single peripheral letter stimulus is randomly rotated on each trial (Braun, 1998; Li et al., 2002). Though these letters can be considered as well-learned categories of objects, letter recognition is better trained for upright letters for obvious reasons (try reading this text upside-down). Hence we might observe some performance difference under the dual-task condition between upright letter discrimination and the original, rotated letter discrimination.

Method

We use the dual-task paradigm to test these hypotheses. As usual, we use the central letter discrimination task as the attention-demanding central task. Three conditions are tested for the peripheral letter discrimination task: Randomly rotated letter, fixed rotation, and upright positions. For all conditions, the letter discrimination task is between T and L (L is the target in a go/no-go set-up where subjects have to release a mouse button when the target is detected). The letter and its mask are located at a random position at 6° eccentricity. For a given block of 96 trials, one of the three different tasks is run and subjects are informed beforehand about the block task. There are 10 blocks tested for each of the three conditions. Figure 8 depicts the three different conditions. Four subjects participated in this experiment. A short training period of 3–4 hours proceeded the actual testing. During this period only the randomly rotated letter discrimination was trained concurrently with the central discrimination task. All three conditions were presented for an equal amount of time in the subsequent real data collection. For each subject, the peripheral letter task SOA was determined based solely on their single-task performance on the randomly rotated letter discrimination task.

Results

When attention is available, the single-task performances for all three different letter discrimination tasks are highly comparable (randomly rotated, average over four subjects and 10 blocks: 77.4 $\pm$ 6.6%; fixed rotation, average over four

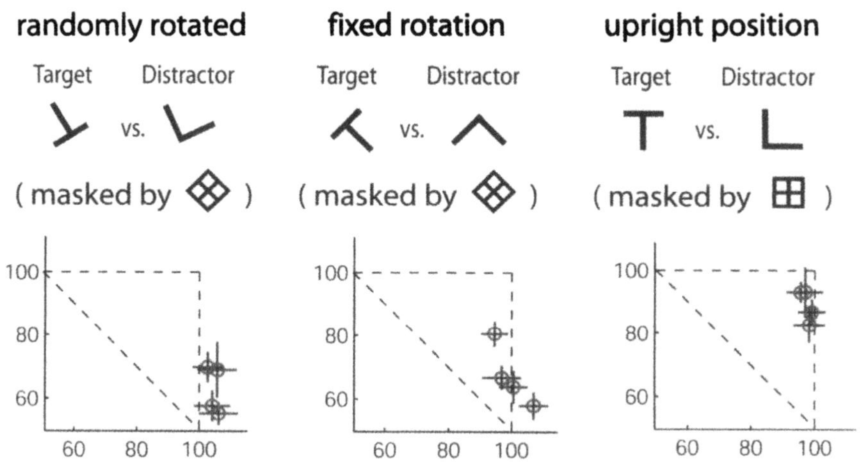

Figure 8. Experiment 5: Rotated versus fixed rotation versus upright letter experiments. The basic set-up of this experiment is also a dual-task paradigm. A letter discrimination task is used peripherally. Three different peripheral conditions and their corresponding performances are shown in three columns. In the first column, the peripheral task is a randomly rotated T versus randomly rotated L, masked by a perceptual mask. Subjects perform barely at or above chance when attention is not available. This result confirms Braun & Julesz (1998) and Li et al. (2002). The second column shows the peripheral task of a fixed rotation of T versus L, indicated in the figure. Subjects' performances of this task without attention are slightly better than the randomly rotated letter condition. In the last column, we show results of subjects' performances for the peripheral task in which the letter T and L are in their upright position, a configuration that is familiar and well learned for all of our subjects. The normalized performances indicate a clear advantage of this condition, showing a near-baseline performance when attention is withdrawn. This result suggests that tasks that are meaningful and well learned can be carried out by the visual system in a much more efficient manner.

subjects and 10 blocks: 78.8 $\pm$ 7.0%; upright, average over four subjects and 10 blocks: 84.9 $\pm$ 4.8%); pair-wise t-test shows no statistically significant difference between each pair of the three different tasks, $t(9) < 1.83$, $p > .05$ for all cases. This suggests that when there is abundant attentional resource, carrying out the letter discrimination task for all rotation conditions is similar. It is important to point out that the fixed rotation and upright conditions have similar performances as the randomly rotated condition. This indicates that the SOA that was determined for each subject for the peripheral task was effective for all condition. We observed, however, different performance results under the dual-task paradigm. For the randomly rotated letter task, subjects' performances were congruent with what is reported in previous studies (Braun & Julesz, 1998; Li et al., 2002). Two subjects' performances were not significantly different from chance (randomly rotated letter task performance under single-task condition, average over 10 blocks for each of the two subjects: 74.2 $\pm$ 3.3% and 85.9 $\pm$ 4.3%; randomly rotated letter task performance under dual-task condition, average over 10 blocks for each of the two subjects: 53.7 $\pm$ 4.8% and 53.7 $\pm$

5.2%); t-test results comparing the dual-task condition with chance: $t(9) < 1.83$, $p > .05$ for each subject. The other two subjects performed slightly better than chance, but significantly lower than their baseline performances obtained when attention was available (randomly rotated letter task performance under single-task condition, average over 10 blocks for each of the two subjects: 70.6 $\pm$ 4.4% and 78.9 $\pm$ 4.1%; randomly rotated letter task performance under dual-task condition, average over 10 blocks for each of the two subjects: 57.8 $\pm$ 7.2% and 61.5 $\pm$ 5.5%); t-test results comparing dual-task condition performances with single-task condition performances: $t(18) > 3.61$, $p < .001$ for each subject. For the fixed but uncommon rotation (nonupright), subjects' performances varied more than the random rotation condition. Three subjects' dual-task performances were comparable to chance level or just slightly higher (fixed-rotation letter task performance under dual-task condition, average over 10 blocks for each of the three subjects: 56.0 $\pm$ 4.5%, 56.0 $\pm$ 6.7%, and 59.1 $\pm$ 4.2%); t-test results: $t(9) < 1.83$, $p > .05$ for the first two subject; $t(9) > 4.30$, $p < .01$ for the third subject. One subject performed at 81% of his baseline level (single-task condition: 78.1 $\pm$ 8.1%; dual-task condition: 67.2 $\pm$ 4.6%). We should then compare this result to the last condition: Upright letter discrimination. Here all subjects performed above 80% of their baseline performance level (single-task condition performance, average over 10 block for each subject: 85.2 $\pm$ 4.4%, 87.2 $\pm$ 5.7%, 78.1 $\pm$ 8.1%, and 89.1 $\pm$ 4.3%; dual-task condition performance, average over 10 blocks for each subject: 72.9 $\pm$ 7.9%, 77.3 $\pm$ 6.1%, 74.2 $\pm$ 9.1%, and 83.6 $\pm$ 5.3%). One subject's performance was not significantly different from her performance when attention was available (single-task condition: 78.1 $\pm$ 8.1%; dual-task condition: 74.2 $\pm$ 9.1%); t-test result: $t(18) < 1.73$, $p > .05$. There is thus a clear pattern that the least amount of attention resource is needed when letters are in their upright positions.

Discussion

Two independent hypotheses were tested in this set of experiments. We found subjects could not recognize rotated letters without attention even when the rotation is fixed to one angle throughout the entire testing period of several hours. This result indicates that stimulus predictability alone cannot reduce much of the attentional load required for such a task.

On the other hand, while keeping everything else exactly the same, recognizing upright letters shows a clear advantage over randomly rotated letters when the attentional resource is scarce. One might argue that a fixed upright rotation is easier for a mental template matching algorithm because the stimuli are much more predictable, whereas such method is less useful for a random rotation. This is why we observe a significantly improved performance for the upright letter condition compared to the random letter condition. But this argument does not support our observation for the fixed but uncommon rotation condition. If the template-matching theory works, it should work for any fixed rotation, not just upright. In fact a modern computer vision algorithm can easily implement such a template-matching method to carry out this task. Our results suggest that there is a clear bias for people to recognize upright letters better than other rotated versions, even highly predictable ones. This is not surprising if we consider how much more learning one receives on upright letter recognition over a lifetime. Similar evidence has also been found in the visual search tasks. Wang et al. (1994) have reported that familiar letters take less time to search than unfamiliar ones. This observation supports our hypothesis that familiar and meaningful categories of objects can be recognized with much less attentional load.

GENERAL DISCUSSION

Natural scene categorization is one of the most evolutionarily relevant tasks of the human visual systems. The superb efficiency of this task has stimulated much research across the fields of neural psychophysics, physiology, and modelling (Biederman, 1972; Li et al., 2002; Potter & Levy, 1969; Thorpe, Fize, & Marlot, 1996; Torralba & Oliva, 2003; VanRullen & Thorpe, 2001). Contrary to the daily experience of the effortlessness of natural scene recognition, it is one of the hardest tasks that modern state-of-art computer vision algorithms have yet to accomplish. This difficulty is mostly due to the vast variability across similar categories of the natural scenes. Unlike low-level tasks such as orientation discrimination or texture recognition, in which much of the tasks can be accomplished through filtering of the primary visual cortex (Malik & Perona, 1990), understanding the categorical information across different examples of natural scenes is usually thought to be a high-level visual task. In an effort to understand the processing of natural scene categorization, we have chosen an approach to study the efficiency of this process in the near absence of attention.

Visual attention is considered to be one of the first and foremost means of controlling the flow of information between different levels of visual processing. Numerous studies have probed the function of attention, demonstrating much attentional control over stimuli with complex and conjugate features (Treisman & Gelade, 1980; Wolfe, 1998; and others). Needless to say, the function of attention is tightly associated with the computational function of recognition in the human visual system. We hope that by manipulating the attentional condition of various natural scene categorization tasks, as well as comparing it with other recognition tasks, much light can be shed in the understanding of the fundamental mechanisms that enable us such a rapid and fast ability of scene categorization.

In this study, we present a series of experiments exploring the mechanisms of attention and recognition for natural scenes, especially in comparison with synthetic stimuli. Based on a previous study (Li et al., 2002), we specifically want to probe the puzzling observation that natural scene categorization seems much more efficient than simple, synthetic stimuli recognition in the near absence of attention. We explored this set of questions by increasing the potential ''difficulty'' of the natural scene categorization task through decreasing signal or lack of training. Meanwhile, we attempted to ''ease'' the synthetic stimuli recognition tasks by increasing the amount of signal, the predictability, or the meaningfulness of the stimuli. Interestingly, most of these manipulations were ineffective, i.e., did not affect the attentional requirements of the tasks.

It is important to point out that we are certainly not claiming that (1) natural scene processing would resist ''all'' attempts to make it attentionally demanding (asking subjects to respond only when there are eight trees in a scene would probably require attention), or that (2) no synthetic stimuli can ever be discriminated without attention (oriented bars can, for example). What we are after is some form of general understanding of the factors that determine attentional requirements for natural scene categorization and its relationship with synthetic stimuli recognition. The experiments that we have chosen are certainly not exhaustive but did allow us to reach interesting conclusions in this respect.

Natural scene categorization is an "easy" task to learn and to perform

Perceptual learning is closely linked to the mechanisms of recognition and attention. Our results show that contrary to common belief, certain seemingly much simpler stimuli are harder to learn to discriminate than complex natural scenes, when attention is not available. In fact, little stimulus specific training is necessary for subjects to perform the natural scene categorization task. Given the current models of visual recognition, this result is highly counterintuitive. Today's start-of-the-art computer algorithms take much more training to recognize natural scenes than simple geometric configurations (Viola & Jones, 2001; Weber, Welling, & Perona, 2000). Hochstein and Ahissar (2002) have coined the term ''easy'' for task conditions where considerable learning transfer occurs. Under this definition, natural scene categorization is a much ''easier'' task given the results of our first set of experiments. They hypothesized that ''easier'' tasks involve higher cortical level processing than lower ones. We will revisit this point in our discussion of ''meaningful categories'', in which our further findings with differently rotated letters also give hint to this possible architecture of the visual processing.

Another piece of indirect evidence of the ''easiness'' for natural scene categorization is its performance pattern without colour information. In Experiment 2 we found that there is virtually no cost in removing colour information of the natural scenes. Again, state-of-the-art computer algorithms for image retrieval often utilize colour features as one of the most informative cues for categorization. Our results show a clear discrepancy between such algorithms and the actual properties of the human visual system. It is suggested that rapid natural scene categorization might take advantage of coarser, achromatic information from the magnocellular pathway earlier than the finer, chromatic parvocellular pathway (Delorme et al., 2000). Several studies on face recognition (Epstein & Kanwisher, 1998) also suggest that there is a response of IT neurons for early phasic component of the stimuli rather than more fine-tuned information. This suggest that categorical recognition might be achieved in higher level visual areas using early waves of neuronal information, where more detailed features such as colours and fine edges have not yet been computed or incorporated.

Parallel processing

The robustness of natural scene categorization is further confirmed through the experiment of double-image recognition (Experiment 3). Our human visual system is surprisingly parallel in processing the complex stimuli of natural scenes (Rousselet et al., 2002). In contrast, Experiment 4 shows that such ability does not apply to stimuli that are defined by their low-level differences, such as the configuration of the red semicircle and a green semicircle. Attention has long been considered to be deployed preferentially to tasks that require much scrutiny and processing. This experiment confirms further that the seemingly much simpler stimuli requires more attention to be categorized.

It has been long suggested that the more a recognition task requires feature conjunctions and binding, the more attention will be needed for this task (Treisman, 1993; Treisman & Gelade, 1980). Therefore only "elementary features" are processed in a parallel fashion (i.e., under visual search, where serial focal attentional scan is not required). Our results suggest the possibility that natural scene categories might belong to the set of "elementary features" while the colour discs or rotated letters do not. But this type of feature is unlikely to be encoded in lower level visual pathways where receptive fields are small and neurons tend to respond to primitive features such as orientations and brightness.

Meaningful categories

In an attempt to understand the efficiency and robustness of natural scene categorization, so far we have gathered much indirect evidence that visual tasks involving higher cortical levels are recognized easier, learned faster, and deploy less attentional resource. In the last set of experiments, we find strong evidence suggesting that meaningfulness and familiarity might participate in determining attentional load and more efficient recognition and learning of the natural scene categorization task. Everything else being equal, an upright letter is discriminated much better than one rotated to a fixed, but uncommon orientation.

There is little low-level difference between these two sets of stimuli, but they do differ in terms of familiarity and meaningfulness. A similar result was recently discovered by Reddy et al. (2004). In their study, they contrast gender discrimination of hairless upright faces versus inverted faces in the near absence of attention. They find that little attention is required to perform the task with upright faces (which are both familiar and meaningful) while the attentional cost is rather high with inverted faces.

This observation is also consistent with the recent development of change blindness studies. Change blindness has shown that attention is critical for our visual awareness (Rensink, O'Regan, & Clark, 1997; Simons & Levin, 1997). Changes of large patches of the visual world can escape our awareness without attending to them. But the amount of attention needed to discern such changes seems to depend also on the meaningfulness of the stimuli. Semantically relevant information is less likely to be neglected in change blindness than less relevant information (Hollingworth & Henderson, 2000).

In short, we hypothesize that natural scene categorization requires little or no attentional cost because of the familiarity and "meaningfulness" of the stimuli and task. Attention acts as a gauge for information processing. When the task or stimuli are unfamiliar, hence are not directly associated with previous neuronal representations, attention helps to select and process features for the recognition task. When there are preexisting neuronal representations for a given task or stimulus, for example natural scene categorization, little attention is needed.

CONCLUSION

We have presented five different experiments in this report. Our results show that natural scene categorization without attention requires little stimulus-specific training. It is robust to lack of colour information or increasing the set size of the stimuli presented. In contrary, multiple redundant copies of synthetic stimuli do not improve the performances of recognition without attention. Some simple tasks, such as single letter discrimination, require much attentional assistance unless the letters are presented in a familiar, upright position. We hypothesize that attention is particularly important for tasks that do not have neuronal representations in the visual pathway. Natural scene categorization, a well-learned and familiarized task for most human observers, does not require much attention.

REFERENCES

Biederman, I. (1972). Perceiving real-world scenes. *Science*, *177*, 77–80.

Brainard, D. H. (1997). The Psychophysics Toolbox. *Spatial Vision*, *10*, 433–436.

Braun, J. (1994). Visual search among items of different salience: Removal of visual attention mimics a lesion in extrastriate area V4. *Journal of Neuroscience*, *14*, 554–567.

Braun, J. (1998). Vision and attention: The role of training. *Nature, 393*, 424–245.

Braun, J., & Julesz, B. (1998). Withdrawing attention at little or no cost: Detection and discrimination tasks. *Perception and Psychophysics, 60*(1), 1–23.

Broadbent, D. E. (1958). *Perception and communication*. London: Pergamon.

Delorme, A., Richard, G., & Fabre-Thorpe, M. (2000). Rapid categorisation of natural scenes is color blind: A study in monkeys and humans. *Vision Research, 40*(16), 2187–2200.

Deutsch, J., & Deutsch, D. (1963). Attention: Some theoretical considerations. *Psychological Review, 70*, 80–90.

Dunai, J., Castiello, U. & Rossetti, Y. (2001). Attentional processing of colour and location cues. *Experiment Brain Research, 138*(4), 520–526.

Epstein, R., & Kanwisher, N. (1998). A cortical representation of the local visual environment. *Nature, 392*(6676), 598–601.

Haxby, J., Gobbini, M. I., Furey, M. L., Ishai, A., Schouten, J. L., & Pietrini, P. (2001). Distributed and overlapping representations of faces and objects in ventral temporal cortex. *Science, 293*, 2425–2430.

Hochstein, S., & Ahissar, M. (2002). View from the top: Hierarchies and reverse heirarchies in the visual system. *Neuron, 36*, 791–804.

Hollingworth, A., & Henderson, J. M. (2000). Semantic informativeness mediates the detection of changes in natural scenes. *Visual Cognition, 7*(1–3), 213–235.

Joseph, J. S., Chun, M. M., & Nakayama, K. (1997). Attentional requirements in a ''preattentive'' feature search task. *Nature, 387*, 805–807.

Joseph, J. S., Chun, M. M., & Nakayama, K. (1998). Reply to Braun. *Nature, 387*, 805–807.

Kahneman, D. (1973). *Attention and effort*. Englewood Cliff, NJ: Prentice-Hall.

Lee, D. K., Koch, C., & Braun, J. (1999). Attentional capacity is undifferentiated: Concurrent discrimination of form, color, and motion. *Perception and Psychophysics, 61*(7), 1241–1255.

Li, F. F., VanRullen, R., Koch, C., & Perona, P. (2002). Rapid natural scene categorization in the near absence of attention. *Proceedings of the National Academy of Sciences, 99*(14), 9596–9601.

Logothetis, N. K., & Sheinberg, D. L. (1997). The role of temporal cortical areas in perceptual organization. *Proceedings of the National Academy of Sciences, 94*, 3408–3413.

Mack, A., & Rock, I. (1998). *Inattentional blindness*. Cambridge, MA: MIT Press.

Malik, J., & Perona, P. (1990). Preattentive texture discrimination with early vision mechanisms. *Journal of the Optical Society of America, A. 7*(5), 923–932.

Neisser, U. (1967). *Cognitive psychology*. New York: Appleton-Century-Croft.

Oliva, A., & Schyns, P. G. (2000). Colored diagnostic blobs mediate scene recognition. *Cognitive Psychology, 41*, 176–210.

Olshausen, B. A., & Field, D. J. (1996). Natural image statistics and efficient coding. *Network: Computation in Neural Systems, 7*(2), 333–339.

Ostergaard, A. L., & Davidoff, J.B. (1985). Some effects of color on naming and recognition of objects. *Journal of Experimental Psychology: Learning, Memory, and Cognition, 11*, 579–587.

Pashler, H. (1998). *The psychology of attention*. Cambridge, MA: MIT Press.

Pelli, D. G. (1997). The VideoToolbox software for visual psychophysics: Transforming numbers into movies. *Spatial Vision, 10*, 437–442.

Posner, M. I., Snyder, C. R. R., & Davidson, B. J. (1980). Attention and the detection and signals. *Journal of Experimental Psychology: General, 109*, 160–174.

Potter, M. C., & Levy, E. I. (1969). Recognition memory for a rapid sequence of pictures. *Journal of Experimental Psychology, 81*(1), 10–15.

Price, C. J., & Humphreys, G. W. (1989). The effects of surface detail on object categorization and naming. *Quarterly Journal of Experimental Psychology, 41A*, 797–828.

Reddy, L., Wilken, P., & Koch, C. (2004). Face–gender discrimination is possible in the near-absence of attention. *Journal of Vision, 4*(2), 106–117.

Rensink, R. A., O'Regan, J. K., & Clark, J. J. (1997). *Psychological Science, 8*(5), 368–373.

Rossion, B., & Gauthier, I . (2002). How does the brain process upright and inverted faces. *Behavioral and Cognitive Neuroscience Reviews*, *1*(1), 63–75.

Rousselet, G., Fabre-Thorpe, M., & Thorpe, S. (2002). Parallel processing in high-level categorization of natural images. *Nature Neuroscience*, *5*, 629–630.

Rubner, Y., Tomasi, C., & Guibas, L. J. (1998). Adaptive color-image embeddings for database navigation. In *Proceedings of the Third Asian Conference on Computer Vision* (Vol. 1, pp. 104–111). London, UK: Springer-Verlag.

Shapiro, K. (Ed.). (2001). *The limits of attention*. Oxford, UK: Oxford University Press.

Shiffrin, R. M., & Schneider, W. (1977). Controlled and automatic human information processing: II. Perceptual learning, automatic attending and a general theory. *Psychological Review*, *84*, 127–188.

Simons, D. J., & Levin, D. T. (1997). Change blindness. *Trends in Cognitive Science*, *1*, 261–267.

Sperling, G., & Dosher, B. (1986). Strategy and optimization in human information processing. In K. R. Boff, L. Kaufman, & J. P. Thomas (Eds.), *Handbook of perception and human performance* (pp. 1–65). New York: Wiley.

Tanaka, K. (1996). Inferotemporal cortex and object vision. *Annual Review of Neuroscience*, *19*, 109–139.

Thorpe, S., Fize, D., & Marlot, C. (1996). Speed of processing in the human visual system. *Nature*, *381*, 520–522.

Torralba, A., & Oliva, A. (2003). Statistics of natural images categories. *Network: Computation in Neural Systems*, *14*, 391–412.

Treisman, A. (1993). The perception of features and objects. In A. Baddeley & L. Weiskrantz (Eds.), *Attention: Selection, awareness and control. A tribute to Donald Broadbent* (pp. 5–35). Oxford, UK: Clarendon Press.

Treisman, A., & Gelade, G. (1980). A feature-integration theory of attention. *Cognitive Psychology*, *12*, 97–136.

Troje, N. F., & Bulthoff, H. H. (1996). Face recognition under varying poses: The role of texture and shape. *Vision Research*, *36*, 1761–1771.

VanRullen, R., & Thorpe, S. J. (2001). The time course of visual processing: From early perception to decision-making. *Journal of Cognitive Neuroscience*, *13*(4), 454–461.

Vinje, W. E., & Gallant, J. L. (2000). Sparse coding and decorrelation in primary visual cortex during natural vision. *Science*, *287*(5456), 1273–1276.

Viola, P., & Jones, M. (2001). Rapid object detection using a boosted cascade of simple features. *Computer Vision and Pattern Recognition*, *1*, 511–518.

Wang, Q., Cavanagh, P., & Green, M. (1994). Familiarity and pop-out in visual search. *Perception and Psychophysics*, *56*, 495–500.

Weber, M., Welling, M., & Perona, P. (2000). Unsupervised learning of models for recognition. In *Proceedings of the sixth European conference of Computer Vision* (Vol. 1842, pp. 18–32). London, UK: Springer-Verlag.

Wolfe, J. M. (1998). Visual search. In H. Pashler (Ed.), *Attention* (pp. 13–74). Hove, UK: Psychology Press.

VISUAL COGNITION, 2005, 12 (6), 925–937

Top-down attentional modulation of spatial frequency processing in scene perception

Emre Özgen
Department of Psychology, University of Surrey, Guildford, UK, and Department of Psychology, Bilkent University, Ankara, Turkey

Paul T. Sowden
Department of Psychology, University of Surrey, Guildford, UK

Philippe G. Schyns
Department of Psychology, University of Glasgow, UK

Christine Daoutis
Department of Psychology, University of Surrey, Guildford, UK

Recent evidence suggests that spatial frequency (SF) processing of simple and complex visual patterns is flexible. The use of spatial scale in scene perception seems to be influenced by people's expectations. However as yet there is no direct evidence for top-down attentional effects on flexible scale use in scene perception. In two experiments we provide such evidence. We presented participants with low- and high-pass SF filtered scenes and cued their attention to the relevant scale. In Experiment 1 we subsequently presented them with hybrid scenes (both low- and high-pass scenes present). We observed that participants reported detecting the cued component of hybrids. To explore if this might be due to decision biases, in Experiment 2, we replaced hybrids with images containing meaningful scenes at uncued SFs and noise at the cued SFs (invalid cueing). We found that participants performed poorly on invalid cueing trials. These findings are consistent with top-down attentional modulation of early spatial frequency processing in scene perception.

The use of spatial scale information in the perception of scenes and other complex images shows flexibility. Evidence suggests that spatial frequency (SF) processing might be influenced by a variety of task-dependent factors such as

Please address all correspondence to: Emre Özgen, Department of Psychology, Bilkent University, Bilkent, Ankara, 06800, Turkey. Email: eozgen@bilkent.edu.tr

This work was supported by BBSRC Grant Nos. 90/S13185 and S13186 awarded to Paul Sowden and Philippe Schyns.

DOI:10.1080/13506280444000580

type of categorization (Schyns, Bonnar, & Gosselin, 2002; Schyns & Oliva, 1999), sensitization through repeated exposure to spatial frequency filtered scenes (Oliva & Schyns, 1997), and attention (Sowden, Özgen, Schyns, & Daoutis, 2003). For example, Schyns and Oliva (1999) used ''hybrid'' face stimuli, which contained both a low-pass (LSF) and a high-pass (HSF) SF filtered face in the same image. They found that participants reported seeing only one of the two components of a hybrid face, which depended on the type of face categorization they were asked to carry out—i.e., they reported seeing the LSF face when categorizing type of expression, and the HSF face when categorizing expressive vs. non-expressive. In a related vein, Oliva and Schyns (1997) repeatedly presented participants with scenes of one type of SF content (either LSF or HSF) combined with noise on the opposite scale, and asked them to categorize them as ''city'' or ''highway''. Subsequently, and unknown to the participants, hybrid scenes were displayed, where both an LSF scene and an HSF scene (one highway, one city) were present. Participants who were sensitized with LSF scenes reported seeing the LSF component, and those sensitized with HSF scenes reported seeing the HSF component of hybrids, but not both.

The precise mechanisms involved in these phenomena are open to study. We propose that sensitization (resulting from cognitive processes such as categorization, explicit top-down cueing, or restriction of information to a region of the SF spectrum) directs attention to those SF channels in early vision whose output is diagnostic for a given task. We found strong psychophysical evidence that a top-down attentional process can sensitize participants to specific SF channels (Sowden et al., 2003). In one experiment, participants made left–right tilt judgements on sinusoidal gratings presented at threshold contrast. We created SF uncertainty (Davis & Graham, 1981; Davis, Kramer, & Graham, 1983; Hübner, 1996a, 1996b) by presenting gratings at one of two possible SFs (e.g., 0.5 and 8 cycles/deg) intermixed randomly, and trained participants to attend to a symbolic sound cue acting top-down (cf. Hübner 1996a, 1996b), signalling the SF of each grating. Subsequently, we interleaved *plaid* stimuli, which consisted of two superimposed gratings (one at each SF) at opposite orientations, to draw an analogy to hybrids. On these trials participants typically reported the orientation of the plaid component corresponding to the cued SF and never perceived both components. This effect of sensitizing participants to SF using sound cues is analogous to the sensitization to spatial scale, resulting from categorization experience, reported by Schyns and Oliva (1999) and Schyns et al. (2002) where task cues the participant to attend to information at specific SFs. In further experiments we found that these effects of expectancy on grating detection were selective for SF in a manner similar to the SF channel tuning observed at early stages of visual analysis.

In sum, there is now considerable evidence that a similar process takes place in the perception of our hybrid scenes and gratings. Further, our work on grating discrimination and detection suggests that attention may modulate the activity of early SF channels resulting in the selective perception of these stimuli. Related to this work, Bonnar, Gosselin, and Schyns (2002) recently found evidence for bottom-up effects involving flexible use of spatial scale. They adapted participants to low-pass or high-pass dynamic noise presented over the entire display area, and subsequently presented them with an ambiguous image (Salvador Dali's painting of *Slave Market with Disappearing Bust of Voltaire*), which had different perceptions depending on whether the fine (HSF) or coarse (LSF) features were attended to. They found that adaptation with LSF noise resulted in the perception of the HSF features of the ambiguous image and vice versa. It appears therefore that a bottom-up adaptation of SF channels forces participants to use the unadapted channels, and determines which scale information is perceived in the ambiguous image.

We have also found evidence that the likely locus of sensitization to spatial scale and flexible scale use might be early in the visual processing stream (Özgen, Sowden, & Schyns, 2002). We repeatedly presented participants with LSF scenes in one part of the visual field, and HSF scenes in another. We observed flexible use of scale when hybrids were presented in line with sensitization specific to each display location. We demonstrated location specificity across distances as small as about 4° visual angle. This evidence is consistent with an early locus as it is well-known that retinotopic mapping is preserved at early stages of visual analysis (De Valois & de Valois, 1988; Tootell, Silverman, Switkes, & de Valois, 1982).

So far sensitization effects have only been demonstrated with scenes using stimulus-driven, or bottom-up designs. It is quite conceivable that when meaningful information is restricted to a band of SFs, the visual system (possibly early vision) sensitizes to this band without the influence of cognitive factors. In a critical band noise-masking paradigm, Solomon and Pelli (1994) found evidence that a single SF channel processes letters irrespective of the spatial scale of the noise. Recent work by Majaj, Pelli, Kurshan, and Palomares (2002) demonstrated that the "sole determinant" of which scale observers attend to was stimulus size: A bottom-up process (i.e., they found that the SF channel used by the observer scaled to stimulus size). Consequently, they argue that observers are not free to choose which channels they use. These findings appear to suggest that although observers *can* attend to different spatial scales, this effect of attention is not top-down but driven by the physical properties of the stimulus. The present study addresses this issue by specifically investigating top-down attentional modulation effects on flexible scale use in scene perception.

In two experiments we explored whether top-down cueing can have the effect of modulating the use of spatial scale much like our previous work with plaids and sine wave gratings (Sowden et al., 2003). Hübner (1996a) makes a distinction between iconic and symbolic cueing (similar to Posner's, 1980, distinction between exogenous and endogenous orienting of attention). Iconic cues are usually same as the stimulus itself and drive attention from the bottom up. Symbolic cues on the other hand are dissimilar to the stimulus, can be cross-modal, and drive attention from the top down. Both types of cueing have been shown to reduce uncertainty effects in SF processing (Davis et al., 1983). We therefore used sound cues to indicate the SF content of each scene, in order to drive a top-down attentional process, by simply manipulating participants' expectations about the stimulus. In both experiments we presented participants with a mixed block of LSF and HSF scenes (highways or cities), combined with noise at the opposite scale, and indicated the scale of each scene by the auditory cues ''coarse'' (LSF) and ''fine'' (HSF). The low- and high-pass scenes were separated by 1.74 octaves along the SF spectrum, ensuring they hit separate SF channels (De Valois & de Valois, 1988). We presented the scenes at threshold contrast, which ensured that the use of cues was necessary. Participants were required to categorize each scene as highway or city. After sufficient cue-dependent sensitization to spatial scale, we started interleaving catch trials. In Experiment 1 these were hybrids, which contained both a LSF and a HSF scene of opposite categories. In Experiment 2, addressing the issue of decision biases, we replaced hybrids with scenes at the opposite scale to that indicated by the cue. In both experiments we observed participants' category reports to indicate which scale they attended to under each cueing condition.

EXPERIMENT 1

In this experiment we first estimated participants' luminance contrast threshold for correctly categorizing scenes as highway or city, for both low- and high-pass SF filtered scenes. We then sensitized participants to our sound cues across 80 trials. On each sensitization trial, we presented either a LSF scene combined with HSF noise or vice versa. A sound cue indicated the scale of each scene. Participants received feedback on their category responses throughout sensitization. Subsequently we tested the effects of sensitization in 64 trials with novel scenes and without feedback. On 25% of the test trials (16 trials) we presented hybrid scenes where both a LSF and a HSF scene of opposite categories were present. We predicted that participants' reports of a scene's category would be biased in accordance with the cue, i.e., the category of the LSF component of a hybrid would be reported more often when the cue was LSF (the sound ''coarse'') and the category of the HSF component would be reported when the cue was HSF (the sound ''fine'').

Method

Participants. Twelve people with normal or corrected vision took part. They were paid a fee (5 GBP) for participation.

Stimuli and apparatus. The stimuli were displayed on an EIZO FlexScan F980 CRT monitor driven by a Cambridge Research Systems 2/5 Visual Stimulus Generator, with a total display area subtending 14.1° (horizontal) $\times$ 10.6° (vertical) visual angle. Viewing distance was secured using a head and chin rest at 161 cm. Stimuli were displayed in the centre of the screen, on a grey (mean luminance) background and measured 4.5 × 4.5°.

Two sets of 80 greyscale images were used: A "highway" set and a "city" set. In addition 64 noise patterns were constructed using Fourier transformation, by randomly shuffling the phase of SFs at all orientations of the scene images, and averaging a city and a highway image for each noise pattern. This meant that each noise pattern contained frequencies found in both a city and a highway. Image resolution was 250 × 250 pixels.

The scenes and noise patterns were SF filtered to form a low-pass and a high-pass example of each. The absolute cut-off SFs for low- and high-pass filters were 1.5 and 5 cycles/deg respectively. As described in the procedure, two types of stimulus were used in the experiment "scene + noise" (S + N) or *sensitization* stimuli (Figure 1a and 1b), and "hybrids" or *test* stimuli (Figure 1c). During the experiment, each stimulus of one of these two types was constructed at runtime by randomly selecting a scene and then combining it with either a noise (S + N) or another scene of the alternative category (hybrid). Runtime combination of images was achieved by alternating between two images at each frame of the monitor's refresh cycle (75 Hz each at 150 Hz refresh rate). The randomization had the restriction that hybrids were constructed using novel scenes (i.e., those that have not been used for the S + N stimuli). For sensitization, 64 scenes in each set were used, while the remaining 16 were used for test trials. For sensitization trials, S + N stimuli were constructed by combining the 64 scenes in each set with the 64 noise patterns. The low-pass (or low spatial frequency—LSF) example of a scene was combined with a high-pass (HSF) example of a noise pattern or vice versa. This resulted in 128 highway and 128 city S + N stimuli. A hybrid was constructed by combining a LSF highway with a HSF city or vice versa, resulting in 16 hybrid stimuli (each hybrid was constructed with entirely new scenes). Noise patterns were additionally used as full-bandwidth backward masks.

As described in the procedure, stimuli were displayed at varying luminance contrast levels. Contrast manipulation was bipolar around the image and screen mean luminance, and was achieved at runtime by using built-in Visual Stimulus Generator functions.

Procedure. The experiment consisted of three phases: Contrast threshold estimation phase, sensitization phase, and test phase. In all three phases the task was the same; participants were presented with a single stimulus, and were required to categorize it as highway or city. Responses were made by pressing appropriate buttons on a gamepad. Participants were shown LSF and HSF examples of highways and cities, and were shown 20 practice trials before they started the experiment.

In the first phase, each participant's luminance contrast threshold for correct categorization of single (no noise added) scenes at a given scale was estimated

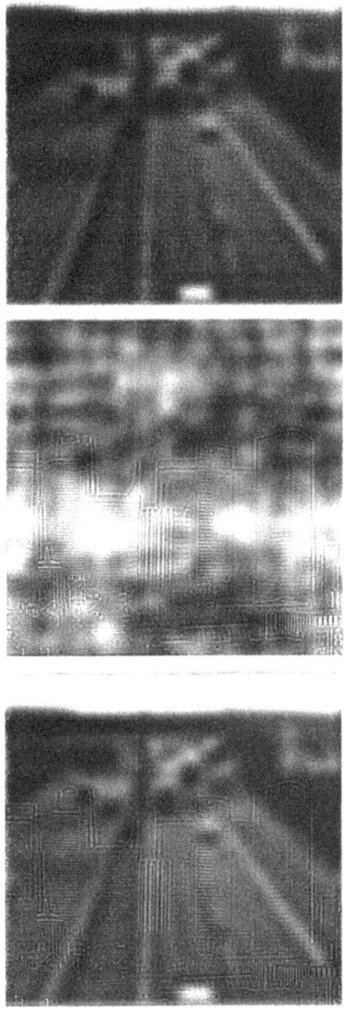

Figure 1. Examples of stimuli used in the present study: (a) low-pass highway combined with high-pass structured noise; (b) high-pass city combined with low-pass structured noise; (c) a *hybrid*: low-pass highway combined with high-pass city.

using the ZEST method (King-Smith, Grigsby, Vingrys, Benes, & Supowit, 1994), which converged on an estimate of threshold at 60% correct, with a beta value of 3.5. Three independent ZEST runs were randomly interleaved, each lasting 48 trials, and the three resulting estimates were averaged to yield the final threshold estimate. One such block of ZEST was completed for each scale (LSF and HSF) separately. The stimuli were randomly selected from among 40

LSF and 40 HSF scenes in each category that were used in the subsequent sensitization phase. The trial sequence for this procedure was as follows. Participants pressed the spacebar to initiate each trial. This was followed by a 1 s pause, then the stimulus was shown for 67 ms, and finally a randomly selected full-bandwidth noise pattern was displayed as a backward mask with contrast fixed at 50%. Participants had unlimited time to respond.

Before the second phase participants were told that for the remainder of the experiment a sound cue would indicate what type of stimulus they would see in the following trial. Using examples, they were instructed that the sound "coarse" would precede LSF scenes, and the sound "fine" would precede HSF scenes. Participants were told that the images might appear quite faint (low contrast) and that these cues would help them in performing well on the task. However, they were also instructed that some of the time the cues may not be accurate. Following the instructions, 80 trials were completed: 40 LSF scenes (combined with HSF noise) and 40 HSF scenes (LSF noise). These were the same 40 scenes used in threshold estimation. The LSF and HSF components (scene or noise) of the images were presented at each participant's threshold contrast level relevant for that scale. The cues were accurate on all trials. In this phase, participants were presented with a feedback tone indicating their accuracy. Again, participants initiated each trial by pressing the space bar which was followed by a 1 s pause, then the sound cue, and another 1 s pause from the onset of the sound cue. This was followed by the stimulus (67 ms), and the backward mask (500 ms). Participants heard the feedback tone immediately after they made their response.

Participants moved on to the subsequent test phase without a break. There were 64 trials in this phase: 48 S + N trials and 16 randomly interleaved hybrids. For the S + N trials 24 novel (not used in previous phases) scenes in each category were used. As described in stimuli, hybrids were constructed with another new set of 16 scenes in each category. The luminance contrast of the scenes and noise patterns at a given scale was set to each participant's threshold appropriate for that scale as in the previous phase. The trial sequence was same as the previous phase except no feedback was given in the test phase. At the end of the experiment participants were briefed on the nature of the experiment. Here we showed them a hybrid scene and asked them whether they had been explicitly aware of such stimuli.

Results

In the sensitization phase, participants correctly categorized scenes at an average of 57.50% (2.6). This was significantly different from chance level of 50%, $t(11) = 2.9$, $p < .05$. There was a slight difference between performance on the LSF (55.2%; 3.25) and HSF (59.8%; 2.45) scenes but this was not statistically significant, $t(11) = 1.84$, $p = .094$.

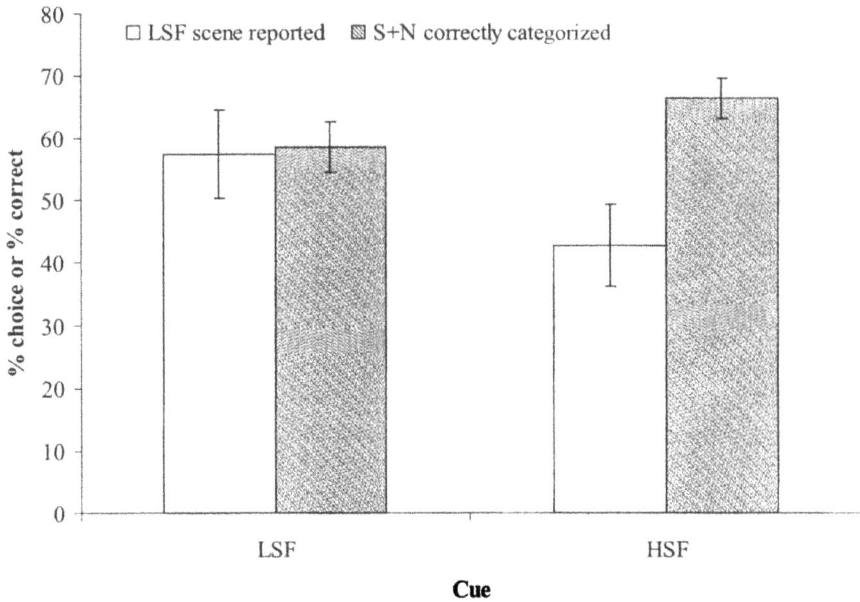

Figure 2. Summary of scene categorization performance in the test phase of Experiment 1. The type of sound cue is shown on the x-axis. Filled bars show percentage correct performance on the S + N trials. Unfilled bars the percentage of times when the LSF component of a hybrid was reported. Error bars show ±1 standard error.

Figure 2 summarizes performance in the test phase. The x-axis shows the cue type. The filled bars show the percentage of correct category responses made to the single scene (S + N) stimuli; unfilled bars show percentage of times when the category of the LSF component of a hybrid was reported. Note that the latter is complementary to percentage HSF reports (LSF reports = 100 – HSF reports). It can be seen that correct categorization on the LSF and HSF scenes appear to be somewhat different. However, this difference was not statistically significant, $t(11) = 1.48, p = .17$. Crucially LSF reports were made more often when the cue was LSF than when the cue was HSF (57.3–42.7%). This difference was statistically significant, $t(11) = 1.9, p < .05$. The average of with-prediction responses across the two cue conditions (LSF reports on LSF cue, 57.3%; HSF reports on HSF cue, 100 – 42.7 = 57.3%) was significantly different from the no-bias level of 50%, $t(11) = 1.9, p < .05$.

Discussion

Just as we predicted, participants' perception of hybrid scenes was influenced by sound cues. When they were cued to attend to the lower end of the SF spectrum, they tended to report seeing the low-pass component scene, and when the cue directed their attention to high SFs, they reported the high-pass scene. This finding is consistent with our hypothesis that SF processing in scene perception can be influenced by top-down attentional modulation.

None of our participants reported being explicitly aware of the existence of hybrid stimuli in the experiment, although some suggested that at times they suspected that there might be two patterns of opposite scale in some images. It should be noted however that there were two patterns of opposite scale in all images throughout the experiment; the second pattern in sensitization trials was noise and in test trials it was another scene. It is therefore impossible to know, based simply on self-reports, whether participants were in fact able to perceive both components of a hybrid scene and chose to report that which was signalled by the cue. This decision bias account is a strong counterargument for our attentional modulation account. The next experiment deals with this issue by using invalid cueing on single-scene stimuli rather than hybrids.

EXPERIMENT 2

Just as in Experiment 1, we sensitized participants using sound cues and then tested them across 64 trials without feedback. The only difference was that in the test phase (again, 25% of the time) we randomly interleaved trials where the sound cue indicated the spatial scale that was opposite to that of the scene component of a scene + noise stimulus. We predicted that when such invalid cues were provided participants would perform worse than when the cues were valid.

Method

Participants. Seventeen people with normal or corrected vision took part in the experiment, and were paid a participation fee (5 GBP).

Stimuli and apparatus. The apparatus was the same as that used in Experiment 1. The same sets of highway, city, and noise stimuli were used. However, in this experiment no hybrids were constructed, and only scene + noise (S + N) stimuli were used throughout. The composition of these stimuli followed the same rules. Separate sets of scenes were used in the sensitization and test phases.

Procedure. The procedure was identical to that of Experiment 1 except for three points. First, through all three phases of the experiment, the noise components of the S + N stimuli were displayed at full contrast. To that end, the initial contrast threshold estimates were made using S + N stimuli (rather than single scenes as it was in Experiment 1) and by manipulating the contrast of only the scene component while holding the noise component at full contrast.

Second, the sensitization phase consisted of 96 trials: 48 randomly selected filtered (24 LSF, 24 HSF) scenes in each category combined with noise at the opposite scale (these were also used in threshold estimation).

Third, although the test phase consisted of 64 trials as in Experiment 1, S + N stimuli were used throughout. This meant that 32 randomly selected novel scenes (16 on each scale) in each category were combined with noise at the opposite scale. However, on 16 randomly selected trials in this phase, the sound cue indicated the opposite scale to what was actually shown (8 LSF, 8 HSF).

Results

Correct categorization performance averaged across the sensitization phase was at 66.9% (2.1). This was significantly different from 50% chance level, $t(16) = 7.96, p < .001$. The LSF and HSF performances in this phase were 70.96% (2.54) and 62.75% (3.5) respectively but this difference did not reach statistical significance, $t(16) = 1.8, p = .08$.

Figure 3 summarizes performance in the test phase. On the x-axis, stimulus scale is indicated. The filled bars show trials with LSF cues and the unfilled bars

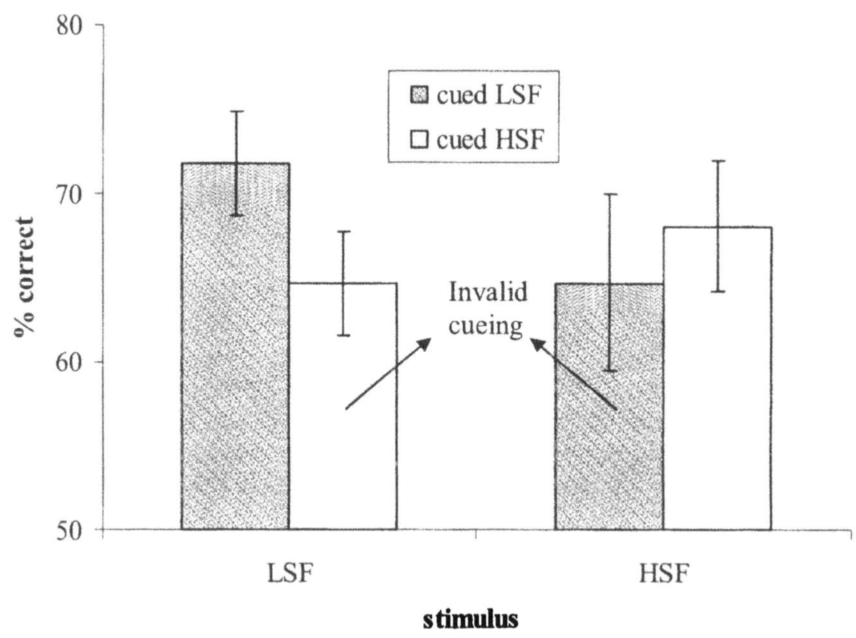

Figure 3. Average percentage correct category responses across the four conditions in the test phase of Experiment 2. Stimulus scale is shown on the x-axis and percentage correct performance is shown on the y-axis. Filled and unfilled bars show performance when the cue was LSF and HSF respectively. Error bars show ±1 standard error.

show those with HSF cues. It can be seen that when the cue was congruent with stimulus scale participants' categorization performance was higher than when it was incongruent. An ANOVA with factors ''scale'' (LSF and HSF) and ''cue validity'' (valid and invalid) revealed a significant main effect of cue validity, $F(1, 16) = 8.96, p < .01$. No other effects were significant in this analysis (largest $F = 0.25$). Performance was equivalent for LSF and HSF scenes on both valid and invalid cueing trials (smallest $p = .5$).

Discussion

Our prediction that invalid sound cues would result in poor performance on scene categorization was supported. When the sound cue indicated the opposite spatial scale to the SF content of a scene, participants performed worse than when the cue was valid. Unlike in Experiment 1 participants were presented with only single scenes and therefore decision biases are less likely to account for the findings of this experiment. The evidence supports our claim that the perception of a SF filtered scene suffers if participants expect to see (and attend to) SFs that are not present in the scene.

GENERAL DISCUSSION

In two experiments we explored whether attention channelled by expectations would influence SF processing in scene perception. We presented participants with low- and high-pass (1.74 octaves apart) SF filtered scenes and asked them to categorize them as highways or cities. Participants expected and attended to low or high SFs through the use of symbolic (top-down) cues; a sound cue indicated the spatial scale of each scene. In Experiment 1, we observed that faced with a hybrid stimulus (one that contains a low- and a high-pass SF filtered scene) participants reported the component of the hybrid that corresponded to the cued and attended scale. However, it could be argued that this might reflect a decision bias and not attentional modulation of perception; participants might perceive both components but report the one indicated by the sound cue. To address this issue, we replaced hybrids with invalid cueing trials in Experiment 2. We presented participants, 25% of the time, with trials where the sound cue signalled the scale opposite to that of the scene. We observed that categorization performance dropped on invalid cueing trials.

These findings are consistent with our attentional modulation account of the effects of expectancy on spatial frequency processing in both complex natural stimuli such as scenes and faces, and in simple psychophysical stimuli such as sine wave gratings and plaids. We argue that diagnosticity can act as an attentional filter in the processing of SF information. Particular types of spatial scale information can be diagnostic for a given type of face categorization (Schyns & Oliva, 1999) even at a specific location within the face stimuli (Schyns et al., 2002), or according to what scale meaningful information is restricted to (Oliva & Schyns, 1997), even at a given retinal location (Özgen et al., 2002). In the present study the location of diagnostic information in the SF spectrum was signalled top-down with sound cues. A similar process of diagnosticity can also be found in psychophysical studies on uncertainty and expectancy. For each trial during a block of gratings varying in SF, the diagnostic information is located at the specific SF of the grating presented on that trial. Uncertainty about the SF of a sinusoidal grating elevates contrast threshold for detection or discrimination (Davis et al., 1983; Sowden et al., 2003). However, crucially, a sound cue that indicates the diagnostic scale makes detection/discrimination more accurate. In all these cases, it is essential to be attending to a particular scale, sometimes at a particular spatial location, in order to accomplish the task. Thus diagnosticity necessitates attention to these parameters. Evidence suggests that attentional modulation of SF channel processing at early stages of visual analysis may indeed be possible; (top-down) attention to the SF of a grating (through the use of symbolic cues) can be restricted to a 1.7 octave range of SFs, suggesting SF tuning such as that typical of early cortical SF channels (Sowden et al., 2003). Task- or categorization-based diagnosticity in the perception of complex patterns might thus drive attention to a particular band of SFs, modulating the activity of SF channels in a similar fashion.

These results on the top-down determination of spatial frequency processing can have far-reaching consequences for human processes of scene recognition. As already discussed, there is evidence that different sorts of categorization information are available from, and used at, different spatial frequency bandwidths and/or orientation and/or image locations (e.g., Oliva & Torralba, 2001; Schyns et al., 2002). This, together with the findings reported here of flexible, retinotopic (Özgen et al., 2002) tuning of spatial frequency channels to different spatial frequencies, suggests a flexible account for the extraction of visual information in scene recognition.

It is now well established that the visual system samples little information in a given visual fixation (at least compared with what is available in the real world surrounding the organism, Archambault, O'Donnell, & Schyns, 1999; Henderson & Hollingworth, 1999; O'Regan, 1992; Rensink, O'Regan, & Clark, 1997; Simons & Levin, 1997). This raises the generic question of how the visual system gets optimally tuned to extract the better cues to categorize the world from sparse information. We believe that the data reported here provide some elements of an answer: The visual system is capable of tuning its information extraction systems (spatial frequency channels) to the spatial frequency bandwidth and image locations that are more informative for the task at hand.

REFERENCES

Archambault, A., O'Donnell, C., & Schyns, P. G. (1999). Blind to object changes: When learning the same object at different levels of categorization modifies its perception. *Psychological Science*, *10*, 249–255.

Bonnar L., Gosselin, F., & Schyns, P. G. (2002). Understanding Dali's *Slave Market with the Disappearing Bust of Voltaire*: A case study in the scale information driving perception. *Perception, 31*, 683–691.

Davis, E. T., & Graham, N. (1981). Spatial frequency uncertainty effects in the detection of sinusoidal gratings. *Vision Research, 21*, 705–712.

Davis, E. T., Kramer, P., & Graham, N. (1983). Uncertainty about spatial frequency, spatial position, or contrast of visual patterns. *Perception and Psychophysics, 33*, 20–28.

De Valois, R. L., & de Valois, K. K. (1988). *Spatial vision*. Oxford, UK: Oxford University Press.

Henderson, J. M., & Hollingworth, A. (1999). High level scene perception. *Annual Review of Psychology, 50*, 243–271.

Hübner, R. (1996a). The efficiency of different cue types for reducing spatial-frequency uncertainty. *Vision Research, 36*, 401–408.

Hübner, R. (1996b). Specific effects of spatial-frequency uncertainty and different cue types on contrast detection: Data and models. *Vision Research, 36*, 3429–3439.

King-Smith, P. E., Grigsby, S. S., Vingrys, A. J., Benes, S. C., & Supowit, A. (1994). Efficient and unbiased modifications of the QUEST threshold method: Theory, simulations, experimental evaluation and practical implementation. *Vision Research, 34*, 885–912.

Majaj, N. J., Pelli, D. G., Kurshan, P., & Palomares, M. (2002). The role of spatial frequency channels in letter identification. *Vision Research, 42*, 1165–1184.

Oliva, A., & Schyns, P. G. (1997). Coarse blobs or fine edges? Evidence that information diagnosticity changes the perception of complex visual stimuli. *Cognitive Psychology, 34*, 72–107.

Oliva, A., & Torralba, A. (2001). Modeling the shape of the scene: A holistic representation of the spatial envelope. *International Journal of Computer Vision, 42*, 145–175.

O'Regan, J. K. (1992). Solving the ''real'' mysteries of visual perception: The world as an outside memory. *Canadian Journal of Psychology, 46*, 461–488.

Özgen, E., Sowden, P. T., & Schyns, P. G. (2002). ''It's a highway—no, wait, it's a city'': Retinal location specificity of flexible scale use. *Perception, 31*(Suppl.), 111.

Posner, M. I. (1980). Orienting of attention. *The Quarterly Journal of Experimental Psychology, 32*, 3–25.

Rensink, R. A., O'Regan, J. K., & Clark, J. J. (1997). To see or not to see: The need for attention to perceive changes on scenes. *Psychological Science, 8*, 368–373.

Schyns, P. G., Bonnar, L., & Gosselin, F. (2002). Show me the features! Understanding recognition from the use of visual information. *Psychological Science, 13*, 402–409.

Schyns, P. G., & Oliva, A. (1999). Dr. Angry and Mr. Smile: When categorization flexibly modifies the perception of faces in rapid visual presentations. *Cognition, 69*, 243–265.

Simons, D. J., & Levin, D. T. (1997). Change blindness. *Trends in Cognitive Sciences, 1*, 261–267.

Solomon, J. A., & Pelli, D. G. (1994). The visual filter mediating letter identification. *Nature, 369*, 395–397.

Sowden, P. T., Özgen, E., Schyns, P. G., & Daoutis, C. (2003). Expectancy effects on spatial frequency processing. *Vision Research, 43*(26), 2759–2772.

Tootell, R. B. H., Silverman, M. S., Switkes, E., & de Valois, R. L. (1982). Deoxyglucose analysis of retinotopic organization in primate striate cortex. *Science, 218*, 902–904.

VISUAL COGNITION, 2005, 12 (6), 938–953

The use of visual information in natural scenes

Maxine McCotter
Department of Psychology, University of Glasgow, UK
Frederic Gosselin
Dépt de Psychologié, Université de Montréal, Canada
Paul Sowden
Department of Psychology, University of Surrey, Guildford, UK
Philippe Schyns
Department of Psychology, University of Glasgow, UK

Despite the complexity and diversity of natural scenes, humans are very fast and accurate at identifying basic-level scene categories. In this paper we develop a new technique (based on *Bubbles*, Gosselin & Schyns, 2001a; Schyns, Bonnar, & Gosselin, 2002) to determine some of the information requirements of basic-level scene categorizations. Using 2400 scenes from an established scene database (Oliva & Torralba, 2001), the algorithm randomly samples the Fourier coefficients of the phase spectrum. Sampled Fourier coefficients retain their original phase while the phase of nonsampled coefficients is replaced with that of white noise. Observers categorized the stimuli into 8 basic-level categories. The location of the sampled Fourier coefficients leading to correct categorizations was recorded per trial. Statistical analyses revealed the major scales and orientations of the phase spectrum that observers used to distinguish scene categories.

Humans are remarkably fast at recognizing and classifying environmental scenes despite a large and varied number of component objects within a scene (Potter, 1975). Recent findings suggest that prior recognition of component objects is not essential for scene recognition, and that the overall gist of the scene may be more important (Henderson & Hollingworth, 1999; Oliva & Schyns, 1997, 2000; Oliva & Torralba, 2001; Sanocki & Epstein, 1997; Schyns

Please address all correspondence to: M. V. McCotter, Department of Psychology, University of Glasgow, 58 Hillhead St., Glasgow, G12 8QB, UK. Email: maxine@psy.gla.ac.uk

This research was supported by BBSRC Grants (nos. SP3185 and S1386) awarded to Philippe Schyns and Paul Sowden. The authors wish to thank Aude Oliva, for lending us the scene stimuli that were used in our experiment, and Bosco Tjan, for useful discussions concerning this research.

DOI:10.1080/13506280444000599

& Oliva, 1994), even though detection of component objects can be achieved in as little as 150 ms (Fabre-Thorpe, Delorme, Marlot, & Thorpe, 2001; Thorpe, Fize, & Marlot, 1996) and in the near absence of attention (Li, VanRullen, Koch, & Perona, 2002).

Studies of the structure of scene categories in memory have identified three particularly useful levels of scene categorization (Rosch, Mervis, Gray, Johnson, & Boyes-Braem, 1976; Tversky & Hemenway, 1983): Superordinate (e.g., artificial/natural), basic (e.g., city/highway), and subordinate level (e.g., a particular example of a city). Gosselin and Schyns (2001b) proposed that basic-level categories are those that minimize the overlap of properties between categories (i.e., strategy length) and maximize the number of properties that are unique to this category (i.e., internal practicability), and therefore give rise to faster recognition (see Gosselin & Schyns, 2001b, for a discussion of the other properties associated with the basic-levelness of a category). Accordingly, the gist of one basic-level scene category should be the scene information that minimizes the overlap of properties with other categories, and maximizes the number of properties specific to this category.

There has so far been no systematic study of the structure of information responsible for basic-level scene categorizations. Part of the problem arises from the complexity and diversity of these stimuli, making it difficult to assess common information use. Here, we develop a new technique (based on *Bubbles*, Gosselin & Schyns, 2001a; Schyns, Bonnar, & Gosselin, 2002) to determine some of the information requirements of basic-level scene categorizations.

INFORMATION FOR SCENE CATEGORIZATION

Before addressing the issue of the information human use for basic-level categorizations, we must address the issue of the information that is available in the image statistics to perform the task. Analysis of a scene in the Fourier domain results in Fourier coefficients, representing the energy and phase relationships of each frequency in the image (Campbell & Robson, 1968; de Valois & de Valois, 1988; see Figures 1b and 1c for example of a Fourier transform). The energy in a Fourier coefficient is the contrast energy of this frequency in the image. Numerous psychophysical studies have shown that the human visual system is selectively sensitive to limited bands of spatial frequencies and to the orientation bandwidth of image components (Campbell & Kulikowski, 1966; Campbell & Robson, 1968; de Valois & de Valois, 1988). In addition to these physiological restrictions on information content, the spectral distribution of natural scenes also imposes constraints on the available information. Natural scenes are known to have a characteristic energy spectrum, with a linear decrease of energy with increasing spatial frequency (Field, 1987; Parraga, Troscianko, & Tolhurst, 1999). The distribution of energy in natural scenes has been shown to characterize their structure, albeit as a first approximation (e.g., Guérin-Dugué &

Oliva, 2000; Oliva & Torralba, 2001; Schwartz & Simoncelli, 2001; Schyns & Oliva, 1994; Simoncelli, 2003; Simoncelli & Olshausen, 2001; Switkes, Mayer, & Sloan, 1978; Tadmor & Tolhurst, 1993; Torralba & Oliva, 2003). For example, the 2-D image of a city has dense vertical and horizontal organization and the occurrence of the horizon line in coastal scenes produces a bias towards horizontal organization. The reasoning is that if these represent distinctive properties of scene categories, then they should be represented with higher energies in their amplitude spectra. Oliva and Torralba (2001) examined the amplitude spectra of artificial and natural scenes to formulate the ''spatial envelopes'' of scene categories. The average slope and dominant orientations of the amplitude spectra corresponded to degrees of scene ''openness'', ''expansion'', ''roughness'', and ''ruggedness''. These characteristic amplitude spectra were compared with basic-level scene categories. For example, mountain scenes scored highly on the ''ruggedness'' parameter, whereas coastal scenes and landscapes scored highly on the ''openness'' parameter. However, this approximation is valid for all visual stimuli compatible with the amplitude spectra of these scenes, and visually meaningful scenes are only a small subset of this set. Most other stimuli are simply noise (see Figure 1e).

This arises because phase relationships describe how spatial frequencies of varying energy linearly contribute to represent the structures of the image (the blobs, contours, and edges; Morrone & Burr, 1988; Oppenheim & Lim, 1981; Piotrowski & Campbell, 1983; see Figure 1d and 1e). The importance of phase is illustrated by the effect of disrupting the phase of spatial frequencies, which renders a scene unrecognizable (compare original image in Figure 1a to image in Figure 1e; see also Sadr & Sinha, 2004; Schwartz, Tjan, & Chung, 2003; Sekuler & Bennett, 2001; Thomson & Foster, 1997; Thomson, Foster, & Summers, 2000). While amplitude spectra vary from scene to scene (Tadmor & Tolhurst, 1993, 1994), the statistics captured by phase information contain the majority of the visual information used to discriminate scenes. Natural image statistics differ primarily from each other in terms of the higher order correlations that structure their phase spectra (Thomson, 1999; Thomson & Foster, 1997; Thomson et al., 2000), which allow sparse linear coding of higher order image statistics across spatial frequencies (Field, 1994; Morrone & Burr, 1988; Olshausen & Field, 1996; Sekuler & Bennett, 2001; Simoncelli & Olshausen, 2001). Consequently, everyday scene categorizations must use the information represented in the phase spectra.

To determine the phase information required for basic-level scene categorizations, we used a technique of selective alteration of phase in the Fourier coefficients while normalizing (whitening) the amplitude of each component frequency (see Figure 2c; Simoncelli & Olshausen, 2001; Tadmor & Tolhurst, 1994; Thomson, 1999). The algorithm randomly samples the Fourier coefficients of the phase spectrum (see Figure 2b). Sampled Fourier coefficients retain their original phase while the phase of nonsampled coefficients is

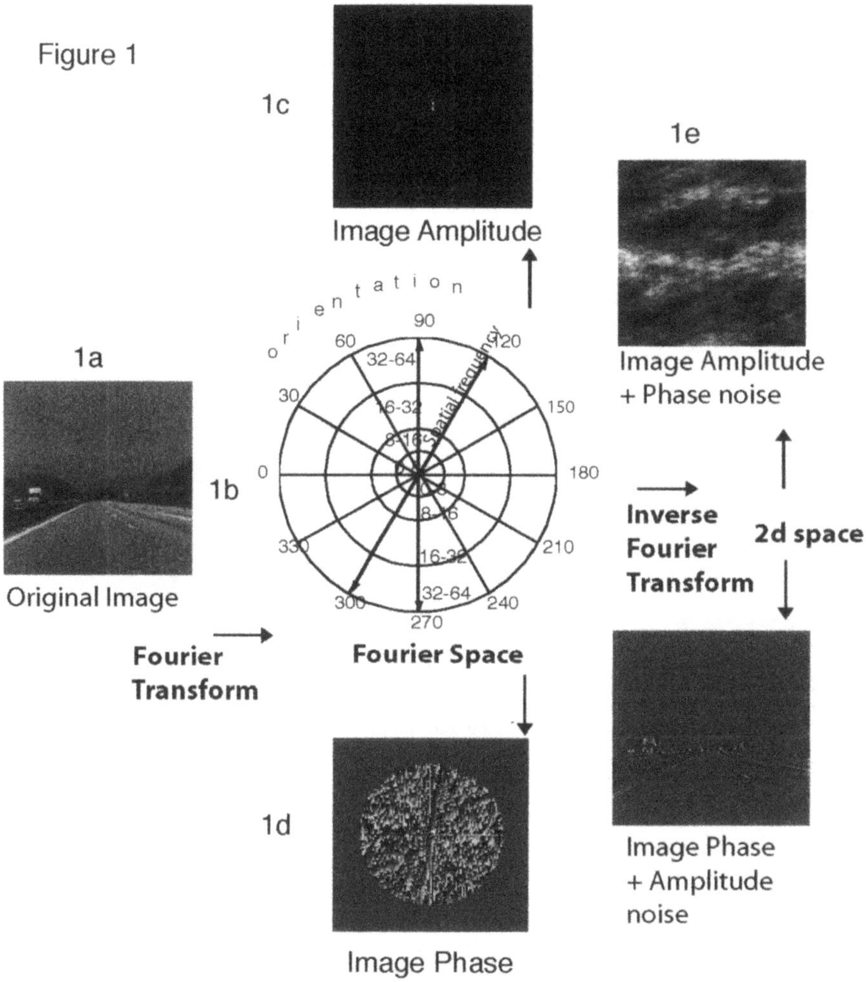

Figure 1. (a) The original image in 2-D space. (b) Representation of Fourier space with spatial frequency bandwidths of 0–8, 8–16, 16–32, and 32–64 cycles per image, and orientations from 0 to 359°. Orientations from 180 to 359° are a mirror symmetric sample of the phase and amplitude components at 0–179°. (c) Representation of image amplitude component in Fourier space. (d) Representation of image phase component in Fourier space. (e) Reconstruction of image amplitude with the phase component scrambled (above), and image phase with the amplitude component scrambled (below) in 2-D space.

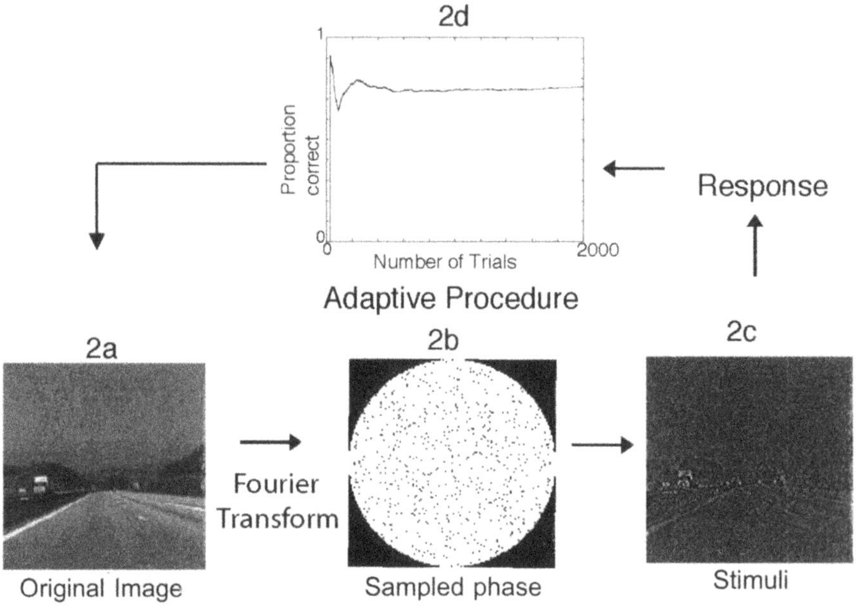

Figure 2. (a) Original image. (b) Random sampling of phase component. (c) Stimuli consisting of sampled phase with remaining phase scrambled and image amplitude replaced by white noise. (d) Adaptive procedure, which determines density of phase sampling per trial.

replaced with that of white noise. An inverse Fourier transform reconstructs the experimental stimulus of that trial (see Figures 1e, 2c, & Figure 3 for examples). To maintain categorization performance below ceiling (at 75% correct), the ratio of image phase to noise phase is adjusted online, on a trial per trial basis, independently for each category. For each category, the Fourier coefficients leading to correct categorizations are registered independently of those leading to incorrect categorizations. To ensure that the experimental task does not trivialize the complexity of real-world scene recognition, we chose a wide a variety of scenes and scene categories from an established database (Oliva & Torralba, 2001).

METHOD

Subjects

Twenty-four male and female observers aged between 18 and 35 took part in the experiment. All observers had normal or corrected to normal vision.

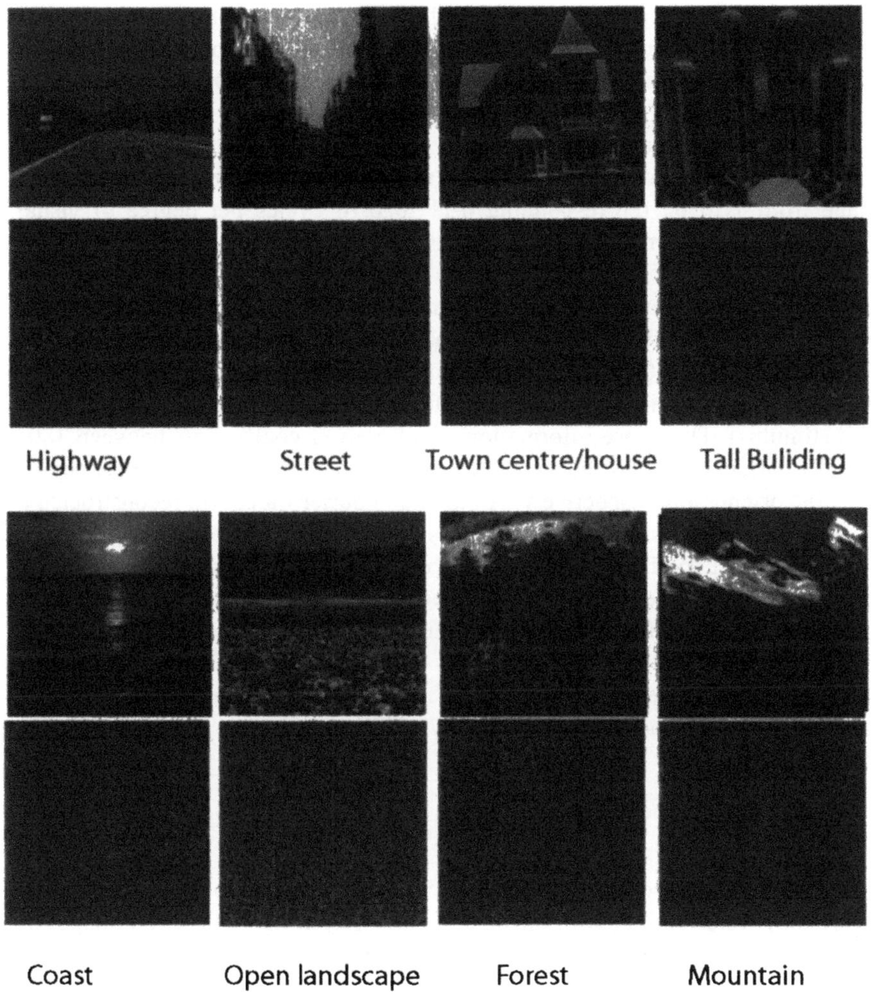

Figure 3. Examples of stimuli from each of the eight scene categories used in the experiment in their original format and in amplitude noise with phase density of .95.

Stimuli

Images from a scene database (Oliva & Torralba, 2001), 2400 in total, were used as stimuli. There were 300 examples of scenes for each category, and eight categories in total (highway, street, town centre/house, tall building, coast/beach, open landscape, forest, and mountain; see Figure, 3). The taxonomy of the scene categories had been validated previously (see Oliva & Torralba, 2001). The 127×127 pixel 256 greyscale images subtended $9.3 \times 9.3°$ of visual angle on the screen. Using a fast Fourier transform, we extracted the Fourier coefficients of each scene and "whitened" its amplitude spectrum by replacing it with the amplitude spectrum of white noise—resulting in an average energy slope equal to zero ($\alpha = 0$) across scenes.

In the whitened images, we introduced phase noise by randomly sampling the Fourier coefficients (see Figure 2b). The sampling range spanned all cycles per image from 1 to 63 (corresponding 0.11 to 6.76 cycles per degree of visual angle; 64 cycles per image = DC, which is not sampled), and for all orientations between 0° and 179°. For each sample, a mirror symmetric sample was constructed to extract spatial frequencies at orientations between 180° and 359°. We randomly transformed the phase ($-\pi$ radians to π radians) of nonsampled Fourier coefficients between 0° and 179° (orientation) at each frequency by replacing it with the phase of white noise (with a different white noise image computed for each stimulus). The phase information of all Fourier coefficients between 180° and 359° orientation was the same as the phase of the coefficients between 0° and 179° orientation, respectively. An inverse Fourier transform reconstructed a sparse experimental stimulus (see Figures 1e, 2c, and 3 for examples).

During the experiment, to maintain categorization accuracy at 75%, we adjusted online the density of the sampled phase, independently for each category (see Figure 2). Phase density was fixed at 95% for the first 50 trials per category to obtain a stable estimate of performance accuracy. Stimuli were constructed and the experiment was run using MATLAB version 5.0 and the Psychophysics Toolbox (Brainard, 1997; Pelli, 1997), on a Macintosh G4 computer.

Procedure

Practice. Observers completed a practice session prior to the experiment to ensure they were familiar with the categorization task. In the task 160 images of scenes from the same scene database (20 examples per category) were used. None of the images used in the practice session were used in the experiment. Observers were presented with a greyscale image of a scene and asked to name it using one of the eight possible categories. Presentation of 160 images constituted one practice block. Each observer repeated practice blocks until they reached a criterion of 95% correct for one block.

Experiment. In a within-subjects design, a total of 2400 experimental stimuli were presented to each observer. Presentation was segmented into four blocks of 600 trials each. Order of presentation of experimental stimuli was randomized across observers. In all, the experiment lasted approximately 2.5–3 hours. Each scene was only presented once. On each experimental trial an observer categorized the sparse stimulus into one of the eight basic-level categories by pressing the appropriate labelled keyboard key. There were no constraints on response time, and the stimulus remained on the screen until their response.

RESULTS AND DISCUSSION

Figure 4a summarizes the average phase density required for observers to reach the 75% categorization correct performance criterion. Figure 4b shows a confusion matrix indicating the errors made with each scene category. Note that performance with coast scenes and landscape scenes fall below the performance criterion, and these scenes were often confused—this was true even when density of phase sampling was at the maximum allowed in the algorithm, 99.5% of phase information in phase, 0.5% of the phase scrambled. Figure 4c shows the average density of phase sampling per trial for each scene category averaged across subjects. Although the highest level of phase sampling was required for coast and landscape scenes, observers' performance remained well above chance level (performance accuracy of 45–50% when performance at chance level equals 12.5% correct for an eight alternative forced choice task).

It is possible that the visual features that typically occur in coast and landscape scenes are particularly sensitive to disruption of amplitude, and thus, cannot tolerate the effects of phase noise to the same extent as the other scene categories. Previous studies have shown that amplitude noise is most disruptive for perception of textured and shaded components occurring in natural scenes (e.g., the border between the coastline and skyline, contours of hills, and surface of a lake) that predominate coast and landscape scene categories (Morgan, Ross, & Hayes, 1991; Tadmor & Tolhurst, 1993). The four artificial scene categories, and the forest and mountain scene categories, consisted mainly of well-defined edges and were less affected by amplitude noise, and, consequently, able to tolerate higher levels of phase noise (see Figure 3 for a comparison of sparse stimuli). It is also likely that the visual characteristics of the basic-level scene categories coast and landscape are not ''redundant'' enough, or have low internal practicability (Gosselin & Schyns, 2001b, 2002). That is, the features typically occurring in coast scenes, (e.g., horizon line between sky and sea, ripples of sea) also occur frequently in landscape scenes (horizon line between sky and landscape, ripples of lake), leading to confusions between the two categories (e.g., 38.5% of responses to coast scenes were in the landscape category; see Figure 4b).

Now, we turn to examine the spectral information (spatial scale and orientation) that was effective for scene categorization. For each trial, we recorded the location of all the sampled Fourier coefficients together with the accuracy of the observer (correct or incorrect). Across the trials of a category, regularities should emerge in these paired locations and accuracies if the corresponding phase information represents a discriminative property of this scene category. The dual information of correct and total Fourier coefficients was kept

Figure 4a

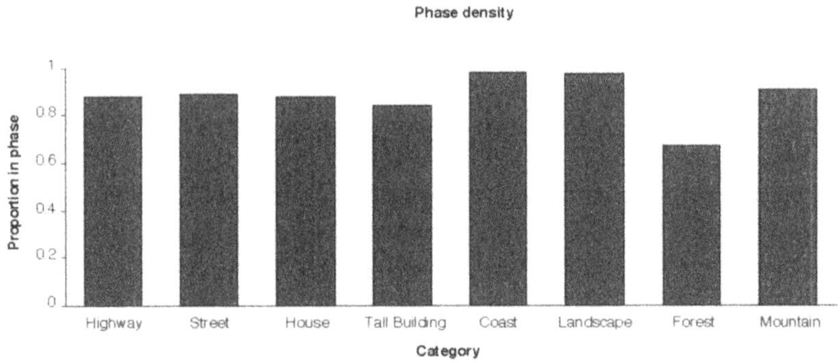

Figure 4b

	Highway	Street	House	Tall Building	Coast	Landscape	Forest	Mountain
Highway	70.97	15	2.78	.89	1.06*	5.92	2.48	.9
Street	5.16	70.76	14.02	3.32	.25	1.29	4.19	1.02*
House	.67	12.4	71.4	10.37	.33*	.85	3.4	.59
Tall Building	.52	3.81	10.9	75.01	.35	1.02*	7.1	1.29*
Coast	1.59*	1.14	.89*	.56	43.35	38.05	5.32	9.11
Landscape	4.77	2.60	1.57	.59*	9.79	49.72	15.24*	15.71
Forest	.33	1.95	4.37	3.6	.68	4.41*	78.13	6.52
Mountain	.73	1.3*	1.86	1.46*	2.9	9.92	10.7	71.13

Figure 4c

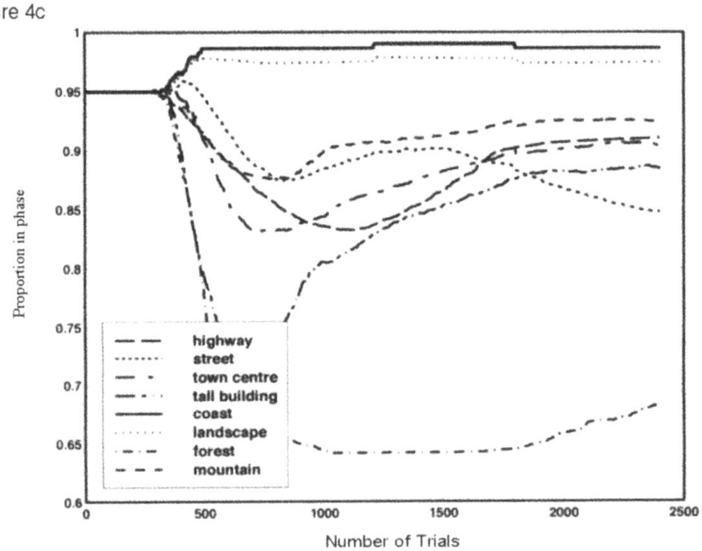

Figure 4. (a) Proportion of phase spectrum required for 75% accuracy per scene category. (b) Distribution of responses per scene category in percentages, including error responses. (c) Mean phase density (specified by a gradient descent algorithm) across trials for each scene category.

separately for each for the eight basic-level categories, and averaged across all observers. For each category, we then computed the proportion of correct over total (correct/total), for each Fourier coefficient. This proportion is the observer probability that a given coefficient leads to a correct categorization. To the extent that the amplitude information of this Fourier coefficient was whitened, the probability isolates the contribution of phase information.

For each scene category, we have a total of 12,644 Fourier coefficients to examine, hardly a small dimensional space! To simplify the data, we averaged the proportions associated with each Fourier coefficient according to 12 orientations of spatial frequency (from 0° to 179° by increment of 15° intervals) and four spatial frequency bandwidths (0–8, 8–16, 16–32, and 32–64 cycles per image; see Figure 1b for illustration). This segmented the data into 48 dimensions, each representing a different bandwidth and orientation given, respectively, by the radius length and angle in the semidiscs of Figure 5. A vertical orientation in the Fourier spectrum corresponds to horizontally orientated components in a scene, for example, the horizon line in a coast or highway scene.

We then transformed the segmented data for each category into 48 *Z*-scores (by computing an average and standard deviation from the 48 averaged proportions, independently for each category). A *Z*-score > 1.65 ($p < .05$) was considered "diagnostic". Figure 5 represents in red these diagnostic regions. All bandwidths and orientations were transformed into *Z*-scores, but all significant *Z*-scores (> 1.65) occurred in the 0–8 and 8–16 cycles per image bandwidths (0–1.74 cycles per degree), thus Figure 5 shows only these bandwidths.

How do these diagnostic regions in Fourier space correspond to 2-D image features in natural scenes? A striking aspect of our results is that the diagnostic bandwidths of the phase spectra for all scene categories occurred at relatively low spatial frequencies. Low spatial frequency information can provide a quick and rough estimate of a scene sufficient for fast recognition (Schyns & Oliva, 1994). For example, the localized structure (phase) of the components of a highway scene should provide sufficient information to discriminate a highway from the localized structure of components in mountain scene, even if image energy is obscured by white noise. To better relate the diagnostic orientations in the phase spectra to image features, we compared our *Z*-score data (Figure 5) with the averaged energy spectrums of scene categories taken from the same database of scenes (Oliva & Torralba, 2001). Remember that a vertical orientation in the Fourier spectrum corresponds to horizontally orientated components in a scene, for example, the horizon line in a coast or highway scene. The diagnostic phase spectra (present study) and the energy spectra (Oliva & Torralba, 2001) for coast and highway scenes were biased to vertical orientations. The visibility of the horizon line was described as the degree of "openness" in energy spectra. Horizontal phase and amplitude components correspond to vertically structured components in a scene, for example, the outline of a house

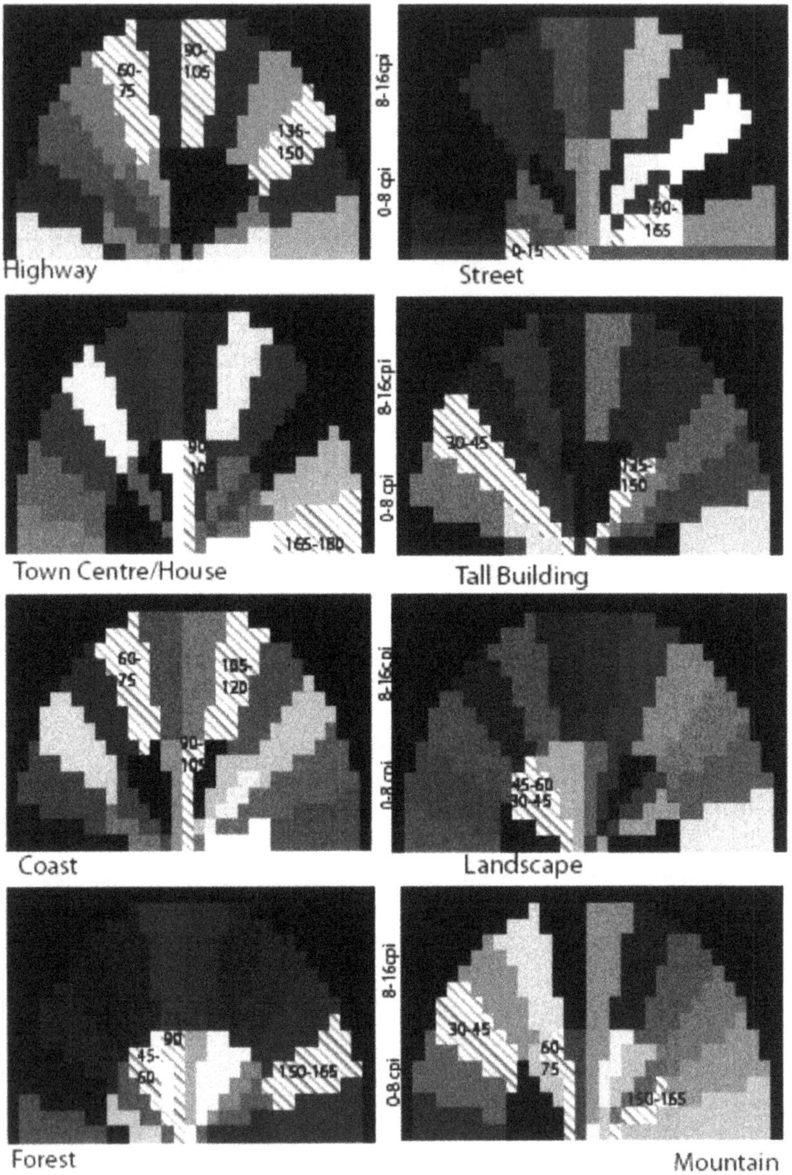

Figure 5. Plot of diagnostic phase in Fourier space (0–16 cycles per image, with orientation in degrees) for each scene category.

or building. In the present study, horizontal orientations in the phase spectra were diagnostic for town centre/house and street scene categories. This coincides with properties found in the averaged energy spectra of artificial (e.g., "urban close up" and "city centre") scene categories and is also described by Oliva and Torralba (2001) as degree of "roughness" in the energy spectra. Previous studies of image statistics have also shown that vertically structured features are common to artificial scene categories (Baddeley, 1997; Switkes et al., 1978; van der Schaaf & van Hateren, 1996). Diagonal orientations featured in the diagnostic phase spectra of mountain scenes, highways, streets, and tall buildings. Diagonal orientations in the Fourier spectrum correspond to scenes containing sloping edges, for example, the outline of a mountain, or perspective view of a street. Diagonal orientations occurred in the averaged energy spectra of mountain scenes (described as degree of "ruggedness") and described the degree of "expansion" in artificial scenes (e.g., vanishing lines in the perspective view of a scene; Oliva & Torralba, 2001). These comparisons suggest that the diagnostic phase spectra for each scene category coincide with characteristic amplitude spectra of the same scene categories reported by Oliva and Torralba.

However, a direct correlation of the results of Oliva and Torralba (2001) with our findings is not practical for the following reasons: First, we used the phase spectra in our study, not the energy spectra. While established methods exist for averaging the energy spectra of a set of images (e.g., van der Schaaf & van Hateren, 1996), averaging the phase information of a collection of images in a scene category does not provide a meaningful description. Previous studies of image phase in natural scenes have used higher order statistics to describe image phase (e.g., measures of skewness or kurtosis; Thomson, 1999). Thus, we cannot compare the "average" image phase of scene categories with our diagnostic phase spectra. Second, our data is not correlated directly with the energy spectrums of Oliva and Torralba because their study reveals the amplitude information *available* in the scene categories, whereas our diagnostic scene spectrums reveal the *potent* phase information—the subspace of available information used effectively—in these scene categories (Gosselin & Schyns, 2002).

How effective is this diagnostic phase information for discrimination of one scene category from another, and to what extent do diagnostic regions overlap between scene categories? The third phase of the analysis tested the effectiveness of the diagnostic regions of the phase spectrum to distinguish the images used in the experiment. To this end, we reconstructed the images used in the experiment using only the "diagnostic" regions of the phase spectrum, replacing nondiagnostic regions with the phase of white noise, and cutting off frequencies above 16 cycles per image (e.g., a coast scene with the diagnostic spectrum of *coast*). For each scene picture (e.g., one coast) we constructed seven distractors with the diagnostic spectra of the other scene categories (e.g., one coast with the phase of landscape, forest, highway, etc.). We then correlated the reconstructed scenes (both diagnostic and distractors) with the original images for each of the eight categories. A t-test (paired samples, $df = 7$) applied to the correlation coefficients of the diagnostic reconstructed scenes, and the correlation coefficients of the distractors, revealed higher correlations with the original image for images reconstructed with diagnostic phase spectra than for distractors, $t = 2.617, p < .05$. The correlational data demonstrates that even in high levels of phase noise, the diagnostic phase spectra for each scene category distinguished each scene from the nondiagnostic distractors. This implies that different regions of the phase spectra are diagnostic for different scene categories.

To examine the extent to which scene categories shared the same diagnostic orientations and bandwidths, the diagnostic regions for each scene category (see Figure 5) were added together and each region expressed as a proportion of the maximum possible overlap (from 0 to 8 categories). Figure 6 shows the frequency of diagnostic regions common to more than one scene category. Diagnostic regions shared by more than one category have a value above 0.125, those diagnostic for one scene category only have a value of 0.125 and regions used by none of the scene categories have a value of zero. The asterisked boxes in the table in Figure 4b indicate which scene categories overlapped. If observers were using one common area of the phase spectrum nonspecific to scene category, the number of regions shared by scene categories should be relatively high. Figure 6 shows that no one region is shared by more than two scene categories. This low level of overlap suggests that the local structures and edges described by the diagnostic phase spectra of a scene category are not common to many other

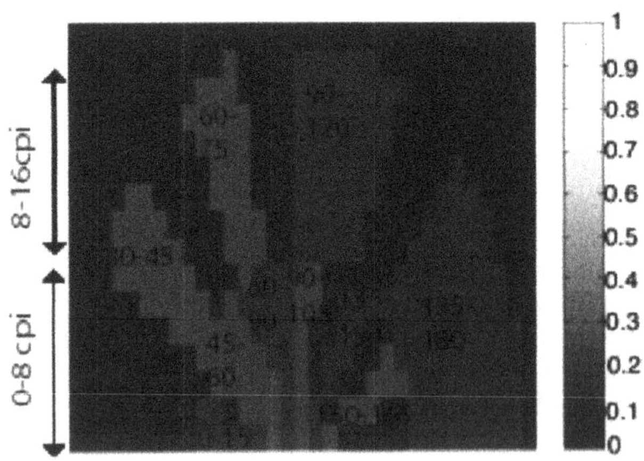

Figure 6. The diagnostic phase of all scene categories weighted by the frequency of occurrence for each bandwidth and orientation (0–16 cycles per image).

scene categories. For example, diagonal orientations at 30–45° and at 8–16 cycles per image are diagnostic for a mountain scene. This phase information should outline the sloping edge of the mountain, and differentiate it from images containing sloping edges that describe component features not specific to the "mountain" category (e.g., a highway).

CONCLUSION

In sum, we applied the bubbles technique (Gosselin & Schyns, 2001a) to the phase spectra of scenes to determine the spectral information that is effective for scene categorization. Analyses of the spectral information that led to correct categorizations produced diagnostic regions of the phase spectra that were category specific. According to the properties associated with basic-level categories (Gosselin & Schyns, 2001b), it is likely that these diagnostic orientations and bandwidths contain the scene information that minimizes the overlap of properties with other basic-level categories, and maximizes the number of properties specific to this category.

REFERENCES

Baddeley, R. (1997). The correlational structure of natural images and the calibrations of spatial representations. *Cognitive Science*, *21*, 351–372.

Brainard, D. H. (1997). The psychophysics toolbox. *Spatial Vision*, *10*, 433–436.

Campbell, F. W., & Kulikowski, J. J. (1966). Orientation selectivity of the human visual system. *Journal of Physiology*, *187*, 437–445.

Campbell, F. W., & Robson, J. G. (1968). Application of Fourier analysis to the visibility of gratings. *Journal of Physiology*, *197*, 551–556.

De Valois, R. L., & de Valois, K. K. (1988). *Spatial vision*. Oxford, UK: Oxford University Press.

Fabre-Thorpe, M., Delorme, A., Marlot, C., & Thorpe, S. (2001). A limit to the speed of processing in ultra rapid visual categorisation of novel natural scenes. *Journal of Cognitive Neuroscience*, *13*, 171–180.

Field, D. J. (1987). Relations between the statistics of natural images and the response properties of cortical cells. *Journal of the Optical Society of America*, *4*, 2379–2394.

Field, D. J. (1994). What is the goal of sensory coding? *Neural Computation*, *6*, 559–601.

Gosselin, F., & Schyns, P. G. (2001a). Bubbles: A technique to reveal the use of information in recognition. *Vision Research*, *41*, 2261–2271.

Gosselin, F., & Schyns, P. G. (2001b). Why do we SLIP to the basic level? Computational constraints and their implementation. *Psychological Review*, *108*, 735–758.

Gosselin, F., & Schyns, P. G. (2002). RAP: A new framework for visual categorization. *Trends in Cognitive Sciences*, *6*, 70–77.

Guérin-Dugué, A., & Oliva, A. (2000). Classification of scene photographs from local orientations features. *Pattern Recognition Letters*, *21*, 1135–1140.

Henderson, J. M., & Hollingworth, A. (1999). High level scene perception. *Annual Review of Psychology*, *50*, 243–271.

Li, F., VanRullen, R., Koch, C., & Perona, P. (2002). Rapid natural scene categorization in the near absence of attention. *Proceedings of the National Academy of Sciences*, *99*(14), 9596–9601.

Morgan, M. J., Ross, J., & Hayes, A. (1991). The relative importance of local phase and local amplitude in patchwise image reconstruction. *Biological Cybernetics*, *65*, 113–119.

Morrone, M. C., & Burr, D. C. (1988). Feature detection in human vision: A phase dependent energy model. *Proceedings of the Royal Society of London, Series B*, *235*, 221–245.

Oliva, A., & Schyns, P. G. (1997). Coarse blobs or fine edges? Evidence that information diagnosticity changes the perception of complex visual stimuli. *Cognitive Psychology*, *34*, 72–107.

Oliva, A., & Schyns, P. G. (2000). Diagnostic colors mediate scene recognition. *Cognitive Psychology*, *41*, 176–210.

Oliva, A., & Torralba, A. (2001). Modeling the shape of the scene: A holistic representation of the spatial envelope. *International Journal of Computer Vision*, *42*, 145–175.

Olshausen, B. A., & Field, D. J. (1996). Emergence of simple cell receptive field properties by learning a sparse code for natural images. *Nature*, *381*, 607–609.

Oppenheim, A. V., & Lim, J. S. (1981). The importance of phase in signals. *Proceedings of the IEEE*, *69*, 529–541.

Parraga, C. A., Troscianko, T., & Tolhurst, D. J. (1999). The human visual system is optimized for processing the spatial information in natural visual images. *Current Biology*, *10*, 35–38.

Pelli, D. (1997). The video toolbox software for visual psychophysics: Transforming numbers into movies. *Spatial Vision*, *10*, 437–442.

Piotrowski, L., & Campbell, F. W. (1982). A demonstration of the visual importance and flexibility of spatial frequency amplitude and phase. *Perception*, *11*, 337–346.

Potter, M. C. (1975). Meaning in visual search. *Science*, *187*, 965–966.

Rosch, E., Mervis, C. B., Gray, W. D., Johnson, D. M., & Boyes-Braem, P. (1976). Basic objects in natural categories. *Cognitive Psychology*, *8*, 382–439.

Sadr, J., & Sinha, P. (2004). Object recognition and random image structure evolution. *Cognitive Science*, *28*, 259–287.

Sanocki, T., & Epstein, W. (1997). Priming spatial layout of scenes. *Psychological Science*, *8*, 374–378.

Schwartz, N. Z., Tjan, B. S., & Chung, S. T. L. (2003). *Spatial frequency phase noise in fovea and periphery*. Paper presented at the annual meeting of the Vision Sciences Society.

Schwartz, O., & Simoncelli, E. P. (2001). Natural signal statistics and sensory gain control. *Nature Neuroscience*, *4*, 819–825.

Schyns, P. G., Bonnar, L., & Gosselin, F. (2002). Show me the features! Understanding recognition from the use of visual information. *Psychological Science*, *13*, 402–409.

Schyns, P. G., & Oliva, A. (1994). From blobs to boundary edges: Evidence for time and spatial scale dependant scene recognition. *Psychological Science*, *5*, 195–200.

Sekuler, A. B., & Bennett, P. J. (2001). Visual neuroscience: Resonating to natural images. *Current Biology*, *11*, R733–R736.

Simoncelli, E. P. (2003). Vision and the statistics of the environment. *Current Opinion in Neurobiology*, *13*, 144–149.

Simoncelli, E. P., & Olshausen, B. A. (2001). Natural image statistics and neural representation. *Annual Review of Neuroscience*, *24*, 1193–1216.

Switkes, E., Mayer, M. J., & Sloan, J. A. (1978). Spatial frequency analysis of the visual environment: Anisotropy and the carpentered environment hypothesis. *Vision Research*, *18*, 1393–1399.

Tadmor, Y., & Tolhurst, D. J. (1993). Both the phase and the amplitude spectrum may determine the appearance of natural images. *Vision Research*, *33*, 141–145.

Tadmor, Y., & Tolhurst, D. J. (1994). Discrimination of changes in the second order statistics of natural and synthetic images. *Vision Research*, *34*, 541–554.

Thomson, M. G. A. (1999). Higher order structure in natural scenes. *Journal of the Optical Society of America, A*, *16*, 1549–1553.

Thomson, M. G. A., & Foster, D. H. (1997). Role of second and third order statistics in the discriminability of natural images. *Journal of the Optical Society of America*, *14*, 2081–2090.

Thomson, M. G. A., Foster, D. H., & Summers, R. J. (2000). Human sensitivity to phase perturbations in natural images: A statistical framework. *Perception*, *29*, 1057–1069.

Thorpe, S. J., Fize, D., & Marlot, C. (1996). Speed of processing in the human visual system. *Nature*, *381*, 520–522.

Torralba, A., & Oliva, A. (2003). Statistics of natural image categories. *Network: Computation in Neural Systems*, *14*, 391–412.

Tversky, B., & Hemenway, K. (1983). Categories of environmental scenes. *Cognitive Psychology*, *15*, 121–149.

Van der Schaaf, A., & van Hateren, J. H. (1996). Modelling the power spectra of natural images: Statistics and information. *Vision Research*, *36*, 2759–2770.

VISUAL COGNITION, 2005, 12 (6), 954–978

The cortical basis of visual scene processing

Russell A. Epstein
Department of Psychology, University of Pennsylvania, Philadelphia, USA

Several lines of evidence suggest that the human brain contains special-purpose machinery for processing information about visual scenes. In particular, a region in medial occipitotemporal cortex—the ''parahippocampal place area'', or PPA—represents the geometric structure of scenes as defined primarily by their background elements. Neuroimaging studies have demonstrated that the PPA responds preferentially to scenes but not to the objects within them, while neuropsychological studies have shown that damage to this region leads to an impaired ability to learn new scenes. More recent evidence suggests that the PPA encodes novel scenes in a viewpoint-specific manner and that these codes are more reliable in good navigators than bad navigators. The PPA may be part of a larger network of regions involved in processing navigationally relevant spatial information. The role of this region in place recognition and gist comprehension is also discussed.

How are visual scenes processed in the human brain? To answer this question, we first need to define the term ''scene'' in a way that distinguishes scenes from other kinds of visual stimuli. Henderson and Hollingworth (1999) provide the following definition: ''[A] semantically coherent (and often nameable) view of a real-world environment comprising background elements and multiple discrete objects arranged in a spatially licensed manner''. Implicit in this definition is a contrast between scenes and nonscene objects such as cars, faces, and coffee cups, which do not have background elements and are not ''environments''. A simple heuristic is that objects are spatially compact entities that one acts upon, while scenes are spatially distributed entities that one acts within. Thus, to answer the question of how scenes are processed in the cortex, we first need to determine whether there is special-purpose cortical machinery for processing scenes that can be distinguished from the machinery involved in processing nonscene objects.

Please address all correspondence to: Russell Epstein, Department of Psychology, University of Pennsylvania, 3720 Walnut St., Philadelphia, PA 19104-6241, USA.
Email: epstein@psych.upenn.edu

I thank Nancy Kanwisher and Sharon Thompson-Schill for comments on this manuscript.

DOI:10.1080/13506280444000607

However, the project does not end there. According to Henderson and Hollingworth's (1999) definition, a scene has at least four primary aspects: (1) it is semantically coherent (i.e., it has a "gist"), (2) it depicts a real-world environment, (3) it contains background elements such as walls, and (4) it contains objects that are spatially organized in a meaningful way relative to each other and relative to the background elements. It is quite possible that there may be separate mechanisms for identifying a scene as a particular kind of environment or situation ("desert", "picnic", "party"), identifying the real-world location of the depicted environment ("North side of campus, next to the library"), processing the overall spatial structure of the scene as defined by the background elements, and processing individual objects and their locations. Thus, once we have identified regions of cortex involved in scene processing, we then need to find out which aspect of the scene the region processes.

Here I review a series of neuroimaging and neuropsychological experiments that suggest that a particular region of the brain—parahippocampal cortex—is critical for at least one aspect of scene processing: Representing the spatial structure of the scene as defined primarily by its background elements. I will argue that this region is part of a larger cortical network involved in processing navigationally relevant spatial information. Indeed, one of the lessons of this work is that scenes are special because they depict environments, and thus convey critical information about where one can move.[1]

THE PARAHIPPOCAMPAL PLACE AREA RESPONDS SELECTIVELY TO VISUAL SCENES

As a first step towards uncovering the cortical basis of scene processing, Kanwisher and I used functional magnetic resonance imaging (fMRI) to identify a region of the brain that responded selectively to visual scenes (Epstein & Kanwisher, 1998). In our first experiment, nine subjects were scanned while viewing greyscale photographs of faces, objects, and scenes presented in 16 s long "blocks". In all nine subjects, significantly greater activation was observed when subjects viewed scenes than when they viewed faces or objects in a bilateral region of posterior parahippocampal cortex (Figure 1). The locus of this activation—straddling the collateral sulcus near the parahippocampal/lingual boundary—was remarkably consistent across subjects. We named this region the "parahippocampal place area" (PPA) because it appeared to respond strongly to

[1] A word about terminology. I will use the term "scene" in two ways: First, to refer to a *view* of a section of the world such as a room or a tabletop (as in Henderson and Hollingworth's, 1999, definition), and second, to refer to the section of the world itself. The meaning should be clear from the context; however, to forestall confusion, I will occasionally use the term "layout" to refer to the real-world entity. The term "place" also refers to the real-world entity, but with a different emphasis: Places are locations in the world, and thus have a spatial relationship to each other that cannot always be ascertained from a single view.

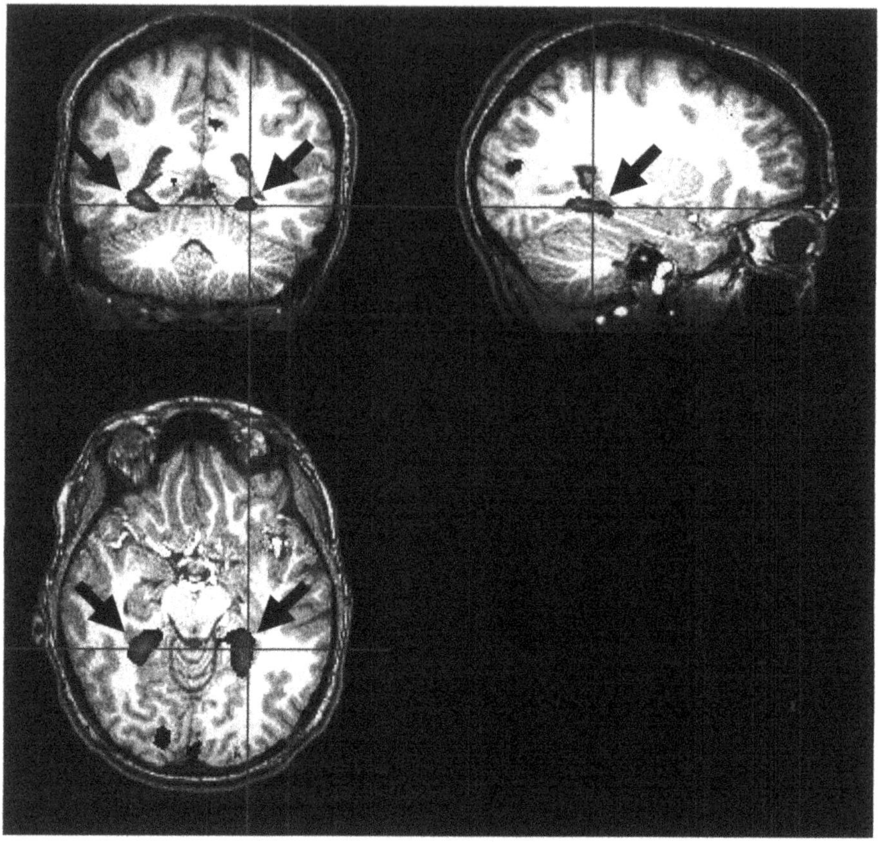

Figure 1. The parahippocampal place area in one subject. Voxels responding significantly more strongly ($t > 3.5$) to scenes than to objects during blocked viewing are highlighted. Right hemisphere is on the right. The PPA is generally found straddling the collateral sulcus near the parahippocampal/lingual boundary in both hemispheres.

photographs of places (i.e., indoor and outdoor scenes). Importantly, the region of activation did not include the hippocampus proper.

We then used a region of interest analysis to examine the response in the PPA more closely. In this analysis, the PPA is defined for each subject using functional data from a set of independent ''localizer'' scans. All contiguous voxels in the parahippocampal region that exceed a predefined statistical criterion for responding significantly more strongly to scenes than to faces and objects in these localizer scans are included. We then measured the timecourse of MR activity in these voxels while subjects viewed scrambled and intact photographs of scenes, faces, houses, and objects in a second set of scans acquired within the same scan session (see Figure 2 for examples of stimuli for this and subsequent experiments). Subjects either viewed the photographs passively (half of the

Figure 2. Examples of stimuli used in the experiments. (Grid lines were only overlaid on stimuli in the first experiment.)

scans) or performed a "1-back" repetition detection task in which they pressed a button whenever they saw two identical photographs in a row.

The average percentage signal change (from a fixation baseline) for each of the eight stimulus conditions in this experiment is shown in Figure 3. (Results for subsequent experiments are also plotted in this figure.) The PPA responded more strongly to scenes than to faces, objects, or houses. This preferential response to scenes held even after the response to the scrambled versions of the photographs was subtracted out (i.e., the scene vs. house × intact vs. scrambled interaction was significant). Thus, the greater response to scenes than to other stimuli is unlikely to be attributable to low-level differences between the stimulus classes (e.g., texture density, average lightness), as many of these differences should be present in the scrambled images as well. The pattern of the PPA response did not differ between passive viewing and 1-back runs (data not shown). Although the mean PPA response to intact buildings (0.95%) was larger than the mean response to intact objects (0.59%), the scene vs. house × intact vs. scrambled interaction was not significant.

This experiment demonstrated two novel and important facts. First, there is a swath of territory in ventromedial occipitotemporal cortex (the PPA) that responds preferentially to visual scenes. A number of other research groups have subsequently confirmed this basic result (e.g., Kohler, Crane, & Milner, 2002;

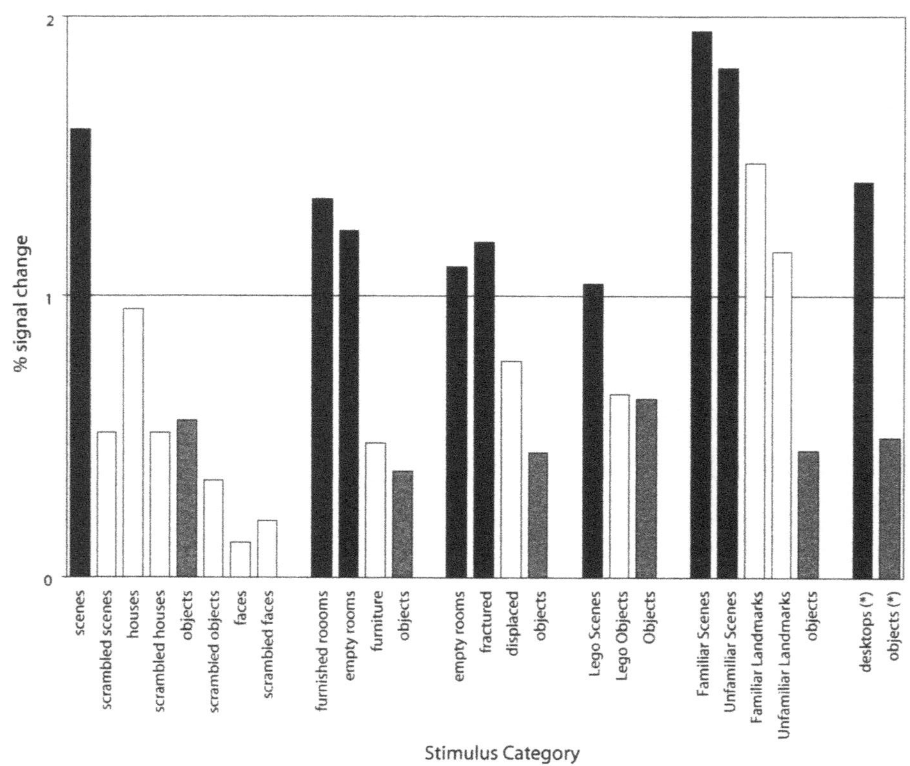

Figure 3. Response of the PPA to different stimulus classes in several experiments. Values are percentage signal change relative to a fixation baseline. Stimulus classes appearing within the same experiment are grouped together. Across experiments, the PPA response to scene-like stimuli (black bars) was significantly higher than the response to object-like stimuli (grey bars). Buildings gave an intermediate response. (*) indicates stimuli presented in colour.

Maguire, Frith, & Cipolotti, 2001). Second, this region responds automatically to scenes, even when subjects are not actively involved in navigation. Earlier experiments had demonstrated increased parahippocampal response when navigation tasks were compared to less-navigationally demanding control tasks (Aguirre, Detre, Alsop, & D'Esposito, 1996; Ghaem et al., 1997; Maguire, Frackowiak, & Frith, 1997). For example, Aguirre et al. (1996) found that parahippocampal activity was greater when subjects actively navigated through a virtual reality maze than when they simply moved through an endlessly looping virtual reality corridor. Neuropsychological evidence (Aguirre & D'Esposito, 1999, see discussion below; Habib & Sirigu, 1987) also implicated parahippocampal cortex in navigation-related processing. The current results allowed us to go beyond these earlier results by assigning a particular role to parahippocampal cortex in navigation: Processing information about the local scene.

THE PPA PROCESSES INFORMATION ABOUT THE SPATIAL STRUCTURE OF THE LOCAL ENVIRONMENT

What aspect of scenes drives the strong PPA response to them? There are many differences between scenes, objects, and faces that could potentially explain this pattern. For example, the PPA might be responding to the greater visual and/or semantic complexity of the scenes, or it might be computing the relative locations of the various objects in the scenes. Alternatively, the PPA might be responding to the background elements of the scene, which define the geometry of the surrounding environment. In this section, I will present evidence that this last explanation is correct: The PPA processes information about the shape of local space. I will refer to this as the *spatial layout hypothesis*.

PPA response is strongly affected by the presence of background elements in a scene, but not by the presence or absence of discrete objects (Epstein & Kanwisher, 1998, Exp. 2). We demonstrated this by scanning subjects while they viewed greyscale photographs of (1) furnished rooms, (2) empty rooms, which were the same rooms from the same viewpoint but empty of all furniture, and (3) furniture arrays, which consisted of the furniture and objects from the rooms on a white background (Figure 2). Except for the use of different stimuli, the procedure for this experiment was identical to the one described above. The task in this experiment (and in all of the experiments described in this section and the next) was passive viewing in half of the scans and 1-back repetition detection in the other half; data from both tasks were averaged together. As before, the PPA was defined for each subject as the set of contiguous voxels in the parahippocampal region that responded significantly more strongly to scenes than to faces and objects in a separate set of localizer scans.

Strikingly, the PPA responded much more strongly to the furnished and empty rooms than to the object arrays (see Figure 3). Furthermore, the response to the empty rooms was not significantly different from the response to the furnished rooms. Thus, the response of the PPA depended critically on the presence of background structures such as walls and floors that define the geometry of the three-dimensional space, but not on the presence or absence of individual objects within that space. This result provides strong support for the hypothesis that the PPA processes information about the spatial structure of the local environment.

In addition to furnished and empty rooms, subjects also viewed photographs of a college campus and natural landscapes in this experiment. The PPA response to these outdoor scenes was comparable to the response to the rooms (Epstein & Kanwisher, 1998). Thus, the PPA responds generally to all kinds of scenes, irrespective of whether they depict an indoor or outdoor location, or a natural or manmade environment. Subsequent experiments have shown that the PPA also responds strongly to desktop scenes (Epstein, Graham, & Downing, 2003; see Figures 2 and 3), scenes made out of Lego blocks (Epstein et al., 1999; see Figures 2 and 3), and even fairly minimal scenes created by rendering software (Downing, Epstein, & Kanwisher, unpublished data). This pattern of results is consistent with the spatial layout hypothesis: Despite their visual dissimilarity, all of these types of scenes convey spatial information. Importantly, this pattern is difficult to explain in terms of Haxby's object form topography model (Haxby, Gobbini, Furey, Ishai, Schouten, & Pietrini, 2001), which posits that regions of ventral temporal cortex respond preferentially to stimuli that share the same visual features. It is unclear what features are shared by the widely varying stimuli that activate the PPA, except for the ''feature'' of conveying information about the geometry of local space (see Gauthier, 2000 for discussion).

A further experiment provided additional support for the spatial layout hypothesis. Five subjects viewed images of empty rooms that were ''fractured'' into their component surfaces (Epstein & Kanwisher, 1998, Exp. 3). In one condition (fractured rooms), the relative positions of the resulting surfaces were preserved. In another condition (fractured + rearranged), the surfaces were rearranged so that they no longer defined a coherent space. Subjects also viewed the original unfractured images of the rooms (intact rooms). Percentage signal change in the PPA was significantly lower in the fractured + rearranged condition (0.8%) than in the fractured condition (1.2%). Insofar as the stimuli in both conditions contained the exact same image components, the difference in response must be attributed to the fact that the surfaces compose a coherent three-dimensional space in the fractured condition but not in the fractured + rearranged condition.

The idea that there might be a special-purpose mechanism for processing the geometric structure of scenes as defined by their background elements is consistent with at least one line of behavioural results. Cheng (1986) and Margules and Gallistel (1988) explored the cues that rats used to reorient themselves after disorientation. They trained rats to find food in one corner of a rectangular box and then removed them from the box and slowly spun them around. When returned to the box, the rats searched for food with equal probability in the correct corner and in the geometrically equivalent diagonally opposite corner, indicating that the rats knew where the food was relative to the shape of the room. From these findings, Cheng and Margules and Gallistel hypothesized that an informationally encapsulated "geometric module" in the rat's brain uses information about the overall shape of the environment to orient the animal. Evidence for the primacy of geometric cues has also been reported for rhesus monkeys (Gouteux, Thinus-Blanc, & Vauclair, 2001) and human infants (Hermer & Spelke, 1994, 1996), although the efficacy of nongeometric cues during reorientation remains controversial in the human case (Learmonth, Newcombe, & Huttenlocher, 2001). Although it would be too simplistic to suggest that the PPA is the anatomical locus of the geometric module, it would be surprising if the representation of spatial layout in the PPA did not play an important role in this geometry-based reorientation process.

THE PPA RESPONSE DOES NOT DEPEND ON THE "SEMANTIC" QUALITIES OF THE SCENE

The above experiments demonstrate that the PPA responds strongly to visual scenes and that the critical factor driving this response is the presence in the stimulus of information about the spatial structure of the surrounding environment. However, the scene stimuli used in these experiments were mostly photographs of real places in the world. Thus, the PPA response to scenes could reflect either (1) the operation of mechanisms directly involved in analysing the spatial structure of the scene (irrespective of its "semantic" qualities, such as whether or not it depicts a familiar place), or (2) the operation of higher level navigational or recognition mechanisms that relate this spatial structure (which might be computed elsewhere in the brain) to stored representations of different places in the world. Note that these two possibilities are by no means mutually exclusive. However, in experiments designed to test whether the PPA was sensitive to the "semantic" (i.e., nonperceptual) qualities of the stimulus, we failed to observe any such sensitivity.

First, we demonstrated that the level of PPA response is not affected by real-world familiarity with the environment depicted in the scene (Epstein, Harris, Stanley, & Kanwisher, 1999). Four MIT students and four Tufts University students were scanned while viewing scenes and buildings from the MIT and Tufts campus. The MIT students had never visited Tufts, and the Tufts students had never visited MIT. All subjects reported that they recognized most or all of the scenes and buildings from their own campus and none from the other campus. We predicted that the PPA should respond more strongly to the familiar than to the unfamiliar places if it were involved in retrieval of information about either (1) the real-world location of the place depicted in the photograph, or (2) the meaning, identity, or significance of the place. Strikingly, no such pattern was observed. The PPA responded equally strongly to photographs of familiar in unfamiliar places, even though semantic and topographical information about the depicted place could only be retrieved in the familiar condition (Epstein et al., 1999, Exp. 1). This result suggests that the PPA supports processes that operate solely on the spatial information that is locally available within the scene. A further experiment demonstrated that the PPA responded strongly to scenes even when they depict layouts made out of Lego blocks, which clearly did not depict places in the world and thus were unlikely to engage place recognition mechanisms (Epstein et al., 1999, Exp. 2). Note that these results do not speak to the issue of whether scenes from familiar and unfamiliar environments are represented in the same manner at the *neural* level; they simply demonstrate that both kinds of scenes activate the PPA to the same extent.

In contrast to this lack of a familiarity effect for scenes, the PPA *did* respond more strongly to photographs of familiar buildings than to photographs of unfamiliar buildings when the images were abstracted from their surroundings and placed on a white background. This pattern may have to do with the ambiguous nature of buildings, which are neither completely "objects" nor completely "scenes". Several experiments have shown that the response to buildings is intermediate between the response to scenes and the response to other objects. Indeed, the PPA is often described (not quite accurately) as a "building" area (Aguirre, Zarahn, & D'Esposito, 1998; Haxby et al., 2001; Ishai, Ungerleider, Martin, Schouten, & Haxby, 1999). Importantly, this preference for buildings over household objects such as tools, shoes, and blenders is not due to the larger real-world size of the buildings: The PPA responds no more strongly to large movable objects such as ships and airplanes than to smaller objects (Epstein & Kanwisher, unpublished data). The key factor driving the strong PPA response to buildings may be the fact that buildings are stable objects that usually play an important role in defining the spatial structure of the environment. In essence, a photograph of a building is simultaneously a photograph of a discrete object and a photograph of a partial scene. Familiarity with a building might bias the brain to process it more as a partial scene (i.e., as a stimulus that tells you something about the geometry of surrounding space) than as a potentially movable object. This hypothesis is consistent with behavioural literature from rats indicating that these animals only use objects as landmarks if they have been shown by experience to be stable over time (Biegler & Morris, 1993).

Not only is the overall level of PPA response unaffected by familiarity with the place depicted in the scene, it is also unaffected by the navigational relevance of the scene. When subjects viewed colour photographs of outdoor scenes and tabletop layouts, the response to the tabletop scenes was just as strong as the response to the outdoor scenes (Epstein et al., 2003; see Figure 3). This was despite the fact that the photographs of tabletop scenes were all taken from quite close up so that they did not provide any information about the surrounding room. This result is consistent with the idea that the PPA supports an analysis of the spatial structure of visual scenes as defined primarily by background elements rather than an analysis of the semantic qualities of the scene. Note that these results do not contradict the general notion that the PPA is involved in navigation: The PPA may support a spatial processing mechanism whose primary purpose is representation of scenes for navigation but which engages automatically to all scenes, including desktop scenes.

Interestingly, while the magnitude of the PPA response was not modulated by extraexperimental experience with the *place* depicted in the scene, it was modulated by within-experimental experience with particular scene photographs. In particular, response to repeated scene photographs was reduced compared to response to novel scene photographs (Epstein et al., 1999, Exp. 4). This result is particularly illuminating when considering the functional relationship between the PPA and the hippocampus. In the rat hippocampus, cells are found that represent the animal's position relative to other locations in the world. Although these cells are partially controlled by information about the local visual scene (see below), they do not directly represent the scene. In contrast, the PPA appears to represent the spatial structure of the currently visible scene, but does not represent where that scene is in the world. Consequently, the magnitude of the PPA response is affected by familiarity with particular scene images, but not by familiarity with the environment in which those scenes are embedded. Clearly, these two different kinds of spatial representations are complementary, and both are necessary for successful navigation.

THE PPA PLAYS A CRITICAL ROLE IN THE LEARNING OF NEW SCENES

Damage to parahippocampal cortex is associated with deficits in spatial navigation and scene processing. In particular, patients with damage to this region can have trouble recognizing scenes, landmarks, and places (Habib & Sirigu, 1987; Landis, Cummings, Benson, & Palmer, 1986; Whiteley & Warrington, 1978), and occasionally show a more general impairment in the learning of new topographical information (Aguirre & D'Esposito, 1999). They are also impaired on an analogue to the Morris water maze task in which they have to remember the location of a hidden platform within an unfamiliar room (Bohbot, Kalina, Stepankova, Spackova, Petrides, & Nadel, 1998) as well as on a spatial memory task in which they have to remember the location of a target on the screen (Ploner et al., 2000). In a study inspired by the fMRI results described above, my colleagues and I tested two such patients and found that they were selectively impaired at encoding new scenes but not new objects (Epstein, DeYoe, Press, Rosen, & Kanwisher, 2001).

The two patients suffered strokes that damaged the parahippocampal territory where the PPA is usually found (one bilaterally, one in the right hemisphere). Both patients reported an almost complete inability to learn new topographical information but only a mild general memory impairment. Interestingly, both patients were able to recall topographical information learned prior to their injury; for example, they were able to draw detailed maps of prior residences that contrasted favourably to their less detailed and less accurate maps of more recently encountered environments. Based on the neuroimaging data, we predicted that both patients should have a scene-processing deficit. A key question was whether such a deficit would manifest itself in perceptual tasks or memory tasks. Comparison of the results from two very similar recognition memory tasks suggested that mnemonic processing was particularly impacted:

In the *different-views matching task*, subjects viewed two photographs of layouts made out of Lego blocks sequentially presented with a 10 s interstimulus interval and reported whether they depicted the same layout from different viewing angles or different layouts. Previous fMRI studies demonstrated that the PPA responds strongly to these stimuli but not to ''objects'' made out of the same materials (Epstein et al., 1999; see Figure 2). In order to perform this task successfully, subjects must form a representation of the Lego layout, hold it in short-term memory for 10 s, and match it to a second Lego layout that is not visually identical. The patients were unimpaired at this task compared to aged-matched normal controls. Furthermore, they did no better on the task when stimuli were photographs of Lego objects than when they were photographs of Lego layouts.

In the *continuous visual memory task*, subjects viewed a continuous sequence of photographs and reported whether or not they had seen each one before in the experiment. When the photographs repeated themselves, they did so either immediately (1-back), with two intervening items (3-back), or with four intervening items (5-back). In different versions of the test, stimuli were either photographs of Lego layouts or Lego objects. The patients were relatively impaired for the Lego layouts but not for the Lego objects, but only at the longest interval (i.e., in the 5-back but not the 1-back case).

Taken together, these experiments demonstrated that: (1) PPA damage results in a specific deficit for processing scenes but not objects, as predicted from the fMRI data; and (2) this deficit is found in recognition tasks only when the scene must be encoded into memory and retrieved after the presentation of a number of intervening items. Thus, these experiments suggest that the PPA may play a critical role in encoding information about the geometry of local space into memory.

A recent report by Mendez and Cherrier (2003) has confirmed and extended these results. They examined a patient who suffered damage to the right lingual/

parahippocampal region after a posterior cerebral artery stroke. This patient could describe routes without difficulty and could recognize landmarks such as buildings. However, he was severely impaired at recognizing scenes that did not contain distinguishing landmarks. This dissociation of recognition performance between landmarks and scenes was observed both in familiar environments (such as his home neighbourhood) and in novel environments. This report provides convincing evidence that one can be impaired specifically at the encoding and recognition of scenes without being impaired at the encoding and recognition of nonscenic sources of topographical information.

THE PPA ENCODES SCENES IN A VIEWPOINT-SPECIFIC MANNER

The above experiments strongly suggest that the PPA mediates a representation of the spatial structure of the current visual scene, but they do not reveal the nature of this representation. For example, they do not indicate whether scene representations in the PPA are viewpoint specific or viewpoint invariant. In the viewpoint-specific case, the PPA might represent the spatial relationship between the observer and the surfaces in the scene—a relationship that is different for different viewpoints. In the viewpoint-invariant case, the PPA might represent the intrinsic geometric arrangement of the surfaces without reference to the location or orientation of the observer. Either account could explain the strong PPA response to scenes in the previous experiments. For example, when subjects view many different photographs of empty rooms within a stimulus block, the PPA might be calculating the shape of each room, or it might be calculating the implied location of the observer relative to the walls of each room.

In order to distinguish between these two possibilities, we must have some way of determining whether the PPA considers two views of the same scene to be representationally identical (i.e., the viewpoint-invariant case) or representationally distinct (i.e., the viewpoint-specific case). In the viewpoint-invariant case, different views of the same place would activate the same subset of neurons in the PPA, while in the viewpoint-specific case they would activate different neural subsets. Although the spatial resolution of fMRI is far too coarse for direct examination of neural firing patterns, information about these patterns can be obtained indirectly by looking at neural adaptation effects (Grill-Spector & Malach, 2001). When two identical stimuli are presented sequentially, neurophysiological experiments have demonstrated that response to the second item is reduced compared to the response to the first item (Baylis & Rolls, 1987; Brown, Wilson, & Riches, 1987; Brown & Xiang, 1998; Li, Miller, & Desimone, 1993; Miller, Li, & Desimone, 1991; Muller, Metha, Krauskopf, & Lennie, 1999; Ringo, 1996). This reduction in neural response can be observed using fMRI (Buckner et al., 1998; Grill-Spector, Kushnir, Edelman, Avidan,

Itzchak, & Malach, 1999; Grill-Spector, Kushnir, Edelman, Itzchak, & Malach, 1998; Henson, Shallice, & Dolan, 2000; Kourtzi & Kanwisher, 2001; Schacter & Buckner, 1998; Stern et al., 1996; van Turennout, Bielamowicz, & Martin, 2003; Vuilleumier, Henson, Driver, & Dolan, 2002). For example, if two scenes are presented sequentially within a 1–2 s long ''event'', the overall fMRI response to the event will be smaller if the scenes are representationally identical than if they are representationally distinct, because the two stimuli will engage the same set of neurons in the ''same'' case (leading to neural adaptation) but a nonoverlapping set of neurons in the ''different'' case. Importantly, this effect should also be observed when the stimuli in the ''same'' case are not physically identical but are ''the same'' in terms of the kind of information represented by a cortical region (Grill-Spector & Malach, 2001; Kourtzi & Kanwisher, 2001).

My colleagues and I (Epstein et al., 2003) used this logic to investigate the viewpoint-specific vs. viewpoint-invariant question in the PPA. We scanned subjects while they viewed events consisting of two successively presented photographs of tabletop layouts presented for 300 ms each with a 100 ms interstimulus interval. The two photographs could have one of four possible relationships to each other: (1) They could be completely identical (no-change condition); (2) they could depict the same layout from the same viewing angle, but differ in the identity of a prominent central object (object-change condition); (3) they could depict the same layout with the same central object, but shown from different viewpoints (viewpoint-change condition); or (4) they could show the same central object placed in different spatial contexts (place-change condition) (Figure 4). We predicted that the PPA would respond more strongly to a change in scene layout (i.e., to the place-change condition) than to a change in viewpoint (i.e., to the viewpoint-change condition) if it represented scenes in a viewpoint-invariant manner, but would respond equally to both kinds of changes if it represented scenes in a viewpoint-specific manner.

The results indicated viewpoint-specific scene representations: PPA response to viewpoint- and place-changes was almost exactly the same. In other words, as far as the PPA is concerned, two photographs of the same scene taken from different viewpoints were as representationally distinct as two photographs of different scenes. In contrast, neural adaptation led to significant reduction of response when viewpoint was repeated in the object-change and no-change conditions—even though the visual change was quite large in the object-change case. The finding that scene representations in the PPA are encoded in a viewpoint-specific manner is consistent with a number of behavioural studies showing viewpoint-specific memory and priming effects for scenes (Christou & Bulthoff, 1999; Chua & Chun, 2003; Diwadkar & McNamara, 1997; Mou & McNamara, 2002; Nakatani, Pollatsek, & Johnson, 2002; Shelton & McNamara, 2001; but see Sanocki & Epstein, 1997). Interestingly, subsequent experiments suggested that the complete viewpoint specificity observed in this experiment might relate to the unfamiliarity of the stimuli and that some viewpoint

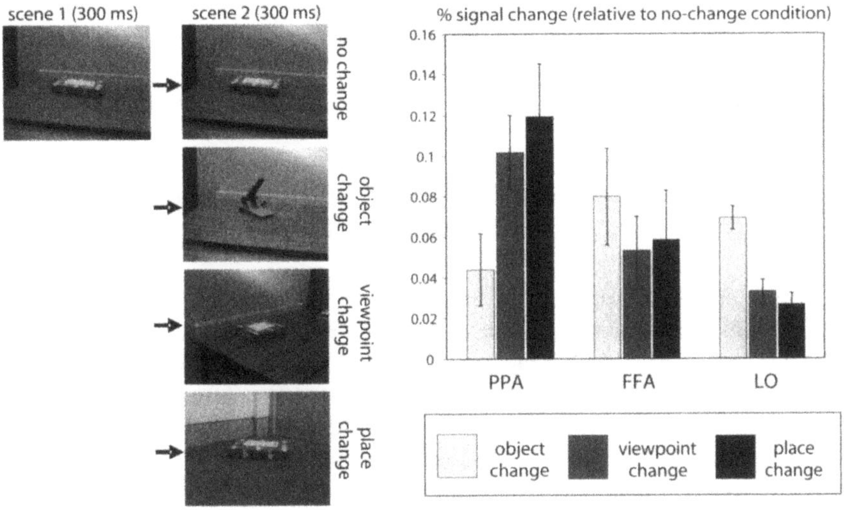

Figure 4. Scene representations in the PPA are encoded in a viewpoint-specific manner. Left: Examples of stimuli. PPA activity was measured in response to "events" in which object, viewpoint, or place information could change. Right: PPA response to viewpoint and place changes was equivalent, and greater than response to object changes. In contrast, the lateral occipital complex (LO) responded more strongly to object changes, and the fusiform face area did not distinguish between the conditions.

invariance can develop as scenes become more familiar (Epstein, Higgins, & Thompson-Schill, 2005—see next section for discussion).

In contrast to the greater PPA response to spatial changes, the "lateral occipital complex" (LO; Malach et al. 1995) responded significantly more strongly to object changes than to viewpoint or place changes, consistent with previous work implicating this region in the perception of object shape (Grill-Spector, Kourtzi, & Kanwisher, 2001; Kanwisher, Woods, Iacoboni, & Mazziotta, 1997b; Kourtzi & Kanwisher, 2001; Malach et al., 1995). Thus, we observed a double dissociation between the PPA and LO: The former responds more strongly to spatial changes, while the latter responds more strongly to object changes. These results suggest a scheme by which ventral occipitotemporal cortex subserves at least two anatomically and functionally distinct processing mechanisms: One that processes information about the spatial relationship between the observer and the background elements of the scene, and one that processes information about the objects within the scene. Consistent with this claim, Levy, Hasson, Avidan, Hendler, and Malach (2001) have noted the PPA is well-positioned to receive inputs from early visual areas that process information about the periphery of the visual field. In contrast, object processing areas such as the fusiform face area (FFA; Kanwisher, McDermott, & Chun,

1997a) are well-positioned to receive inputs from the centre of the visual field. Levy et al. (2001; see also Epstein et al., 2003) also provided evidence that the PPA responds to some extent even to nonscenic information in the periphery, although this effect is secondary to the main ''categorical'' effect.

THE QUALITY OF PPA SCENE REPRESENTATIONS CORRELATES WITH NAVIGATIONAL ABILITY

As noted above, a number of neuroimaging studies have obtained parahippocampal activation when subjects perform navigational tasks (Aguirre et al., 1996; Ghaem et al., 1997; Maguire et al., 1997). The claim that the PPA is involved in processing information about the spatial structure of the local scene is consistent with these earlier results. In a recent study, my colleagues and I (Epstein et al., 2005) used fMRI adaptation to establish a more direct link between PPA scene representations and navigation. Specifically, we showed that the strength of fMRI adaptation effects in the PPA correlates with self-reported navigational ability, as assessed by Hegarty's Santa Barbara Sense of Direction scale (SBSOD; Hegarty, Richardson, Montello, Lovelace, & Ilavanil, 2002). These results suggest that one component of successful navigation is the encoding of accurate representations of the local scene.

The experiment consisted of two parts, conducted within the same scan session. The first part was similar to the previous experiment: Subjects viewed pairs of successively presented photographs of indoor and outdoor scenes, which could either be identical (no-change condition), different views of the same place (viewpoint-change condition), or photographs of two different places (place-change). In the second part, subjects viewed single scene photographs and made indoor/outdoor judgements. These photographs could be identical to those used in the first part of the experiment (''old views''), previously unseen views of previously seen places (''new views''), or completely novel places (''new places''). Thus, pairwise comparisons of the three conditions in the first part of the experiment allowed us to examine short-term fMRI adaptation effects (operating over the course of hundreds of milliseconds), while pairwise comparison of the three conditions in the second part of the experiment allowed us to examine long-term fMRI adaptation effects (operating over the course of minutes).

The first part of the experiment replicated our earlier results implicating the PPA in viewpoint-specific scene processing: The PPA responded equivalently to viewpoint and place changes but response in the no-change condition was reduced. The second part of the experiment revealed both viewpoint-specific (new view > old view) and viewpoint-invariant (new place > new view) long-term adaptation effects in the PPA. Interestingly, these effects were significantly correlated with navigational ability as assessed by the SBSOD score. For example, SBSOD scores accounted for approximately 44% of the variance in the magnitude of the viewpoint-specific effect in the right PPA. Both viewpoint-specific and viewpoint-invariant adaptation effects were larger in good navigators than in bad navigators (despite the fact that the overall level of response to scenes did not vary as a function of navigational ability). These results suggest that navigational competence relates directly to the quality of scene representations in the PPA. Specifically, good navigators encode representations that distinguish between new and old views of a previously seen place, while bad navigators do not.

Beyond the correlation with navigational ability, the finding that long-term adaptation in the PPA was partially viewpoint invariant is intriguing because it suggests that scene representations may become less viewpoint specific as one gains experience with different views of a place. A similar learning process has been postulated for nonscene objects, but to my knowledge no evidence for the formation of viewpoint-invariant object representations over the course of an experimental session has been observed. The above results suggest that the PPA might be the locus of a mechanism for learning places from individual views; however, this hypothesis must be tested through further experiments. For example, it would be interesting to see if a greater degree of viewpoint invariance is found for photographs of familiar environments than was found for the photographs of unfamiliar environments used in the present experiment.

THE PPA MAY BE PART OF A LARGER NETWORK OF SPATIAL-PROCESSING REGIONS

The PPA is not the only region of the brain that processes navigationally relevant spatial information. *Place cells* in the rat hippocampus fire when the animal is in a particular location but are quiescent otherwise (O'Keefe & Dostrovsky, 1971). Furthermore, these cells often exhibit considerably different place fields in different environments, indicating that the rat hippocampus not only represents the location of the animal within the local environment but also distinguishes between different local environments (Best, White, & Minai, 2001). Results such as these led O'Keefe and Nadel (1978) to postulate that the hippocampus supports a cognitive map (Tolman, 1948) that represents the spatial structure of the environment in allocentric (i.e., world-centred) coordinates. Supporting this contention is the fact that rats with hippocampal lesions are severely impaired on tasks that require recall of the relationship between different world locations, such as swimming from a random starting position to a hidden fixed platform in a water-filled pool (Morris, Garrud, Rawlins, & O'Keefe, 1982).

Some recent studies have identified place cells in the primate hippocampus (Matsumura, Nishijo, Tamura, Eifuku, Endo, & Ono, 1999), while others have identified *spatial view cells* in the same region that fire when the animal looks at particular locations in the world (Rolls, Robertson, & Georges-Francois, 1997). Place cells have also been identified in the human hippocampus using intracranial recordings (Ekstrom et al., 2003). Spatial view cells and place cells are similar in two critical respects. First, they both respond to locations in an allocentric manner: Their firing does not depend on the bodily orientation, eye position, or head position of the animal (Georges-Francois, Rolls, & Robertson, 1999). Second, they respond to locations *per se* rather than particular visual cues experienced in those locations. For example, place cells fire in the dark and retain their firing pattern with some degree of accuracy as the animal moves through an unlit environment, while spatial view firing fields are unchanged when the walls of the experimental chamber are obscured by curtains (Robertson, Rolls, & Georges-Francois, 1998). Thus, both the rat and monkey hippocampus appear to be involved in supporting an amodal (i.e., nonvisual) representation of the locations of ''places'' in a world-centred coordinate frame. Interestingly, a separate set of neurons in the postsubiculum, anterior thalamus, mamillary bodies, and retrosplenial cortex appears to be involved in representing the orientation of the animal in the world (Taube, 1998). These *head direction (HD) cells* are similar to place and spatial view cells in that they represent orientation in allocentric coordinates (i.e., they represent orientation relative to a world-centred coordinate frame, rather than bearings to specific landmarks) and can be updated using both visual and nonvisual cues (Stackman & Taube, 1997).

The parietal lobes are also critically involved in processing spatial information. Neurophysiological data indicate that parietal regions represent the locations of behaviourally relevant targets in a number of different coordinate frames tied to effector and receptor surfaces such as the eye, hand, or mouth (Andersen, Snyder, Bradley, & Xing, 1997; Colby & Goldberg, 1999). For example, neurons in the medial intraparietal region (MIP) respond when the animal reaches for a target and show selectivity both for the retinotopic location of the target and for the reaching hand, while neurons in the lateral intraparietal region (LIP) and area 7a encode the locations of eye-movement targets in retinotopic coordinates (Colby, Duhamel, & Goldberg, 1996; Gottlieb, Kusunoki, & Goldberg, 1998). Although parietal involvement in mediating movements of the whole body through space has been comparatively less studied in monkeys, human neuropsychological and neuroimaging data implicate several parietal subregions in active navigation. For example, medial parietal regions are active during scene viewing and both medial and inferior parietal regions are active during navigational tasks (Burgess, Maguire, Spiers, & O'Keefe, 2001; Maguire, Burgess, Donnett, Frackowiak, Frith, & O'Keefe, 1998). Patients with damage to parietal regions can show a severe deficit in navigating through the immediate environment, often bumping into walls and other obstacles (Stark, Coslett, & Saffran, 1996). These data suggest that parietal regions play a critical *online* role in guiding locomotion—for example, by tracking the locations of navigationally relevant surfaces as one moves. Importantly, this kind of information is constantly changing, and thus parietal representations may be too transient to support navigational planning (Milner & Goodale, 1995). However, they may be important as inputs to more durable spatial representations supported by other regions.

How is the PPA connected to these other spatial processing regions? In the monkey, there are strong anatomical projections from parietal regions LIP and 7a to parahippocampal cortex region TF (Suzuki & Amaral, 1994) and indirect projections from parietal cortex to parahippocampal regions TF and TH via retrosplenial cortex (Morris, Paxinos, & Petrides, 2000; Morris, Petrides, & Pandya, 1999; Suzuki, 1996). Parahippocampal cortex also receives input from ventral stream visual areas TE and TEO, putting it in an ideal position to combine parietal information about the locations of different surfaces with ventral stream information about the overall "shape" of the surrounding environment. Parahippocampal cortex projects to entorhinal cortex, which is the primary source of input to the hippocampus. Thus, the anatomical data suggest that a pathway from parietal cortex to parahippocampal cortex to the hippocampus may be critical for processing navigationally relevant spatial information. Supporting this contention, Burgess et al. (2001) observed increased activation in all of these regions when recall of the spatial aspects of an event ("Where did this event take place?") was contrasted with recall of the non-spatial aspects (e.g., "Who was involved?").

Indeed, functional studies suggest that these regions form a processing hierarchy in humans. Maguire et al. (1997) found both hippocampal and parahippocampal activation when subjects retrieved routes through complex environments, but only parahippocampal activation when they recalled individual landmarks or scenes. Furthermore, King, Burgess, Hartley, Vargha-Khadem, and O'Keefe (2002) found that a patient with bilateral hippocampal atrophy was strongly impaired at recalling the locations of objects within a virtual-reality courtyard when he was tested using a viewpoint different from the one used at encoding, but not when he was tested using the same viewpoint used at encoding. Taken together, these results indicate that the PPA supports viewpoint-specific scene representations while the hippocampus may be critical for encoding viewpoint-independent scene representations and understanding the routes between different places. This formulation fits both the cognitive map theory, and also the more general idea that the hippocampus is critical for encoding sequences of locations or events (Eichenbaum, Dudchenko, Wood, Shapiro, & Tanila, 1999; Fortin, Agster, & Eichenbaum, 2002; Levy, 1996).

Neurophysiological data from animals also support this scheme. Neither place cells nor HD cells represent visual cues (as evidenced by their firing in darkness), but both place fields and HD fields are strongly controlled by certain kinds of visual cues when they are present. In particular, the relationship between the animal and the overall configuration of distant visible landmarks (Knierim, Kudrimoti, & McNaughton, 1995; Muller & Kubie, 1987; O'Keefe & Speakman, 1987) or barriers (O'Keefe & Burgess, 1996) can be a critical determinant of place cell response. Similar results have been obtained for HD cells—indeed, place and HD fields usually move in a tightly coupled manner when landmarks locations change (Knierim, Kudrimoti, & McNaughton, 1998). These results suggest that both place and HD cells receive input from a common representation of the geometry of local scene (Golob & Taube, 1997). Although the locus of such a representation has not been identified in animals, the human data suggest that parahippocampal cortex may be a source of such geometric information.

UNRESOLVED ISSUES AND CONCLUSION

The evidence discussed above strongly suggests that parahippocampal cortex plays a critical role in processing information about the spatial structure of scenes as defined primarily by their background elements. Thus, we have identified the neural basis for processing information about one major aspect of visual scenes. What about the other aspects? Can we say anything, for example, about the cortical regions involved in place recognition, or in processing gist? Note that both of these operations involve linking visual to nonvisual information. In place recognition, the aim is to determine the *identity* of the place depicted in the scene, both in conceptual terms ("What is the name of this particular place?") and spatial terms ("Where is this place relative to other places in the world?"). In gist recognition, the aim is to determine the *category* of the scene ("What *kind* of scene is this?").

Place recognition appears to involve multiple regions of the brain. Nakamura et al. (2000) used PET to measure the neural response to photographs of faces and places that were either personally familiar or unfamiliar to the subjects. In occipital-temporal regions, the well-established face/place division was observed: Both familiar and unfamiliar faces activated fusiform gyrus, while both familiar and unfamiliar places activated parahippocampal cortex and the parietal/occipital junction. In contrast, the temporal pole responded to familiar but not unfamiliar stimuli, irrespective of stimulus type. (Similar results were obtained by Gorno-Tempini and Price, 2001.) These results fit well with McCarthy, Evans, and Hodges' (1996) report of a patient with anterior temporal damage who is impaired not only at place recognition but also in his ability to recall semantic information about different places. Thus, place recognition (in the sense of being able to identify different places in the world) may require the linking of parahippocampal representations of scene geometry with long-term semantic memory stores in anterior temporal cortex. Interestingly, it is unclear whether long-term information about the actual *locations* of places is stored in anterior temporal cortex: McCarthy et al.'s patient could draw an accurate map of his home town (even though he couldn't recognize locations in that town from photographs), while Teng and Squire (1999) reported that a patient with extensive anterior and medial temporal damage could retrieve information about the relative locations of landmarks in his former hometown. (In both patients, this location information was acquired before their injury.) Thus, at least for highly familiar environments, retrieving the relative locations of different places might involve different brain regions than those involved in retrieving the names and meanings of the places (see also Aguirre & D'Esposito, 1997).

As for gist processing, comparatively little is known about how this is implemented in the brain. Behavioural studies have demonstrated that we can interpret scenes (''picnic'', ''beach'', ''airplane flying over clouds'') quite quickly (Potter, 1975)—almost as quickly as we can identify individual objects (Thorpe, Fize, & Marlot, 1996). The mechanisms behind this impressive ability are unclear. In many cases, ''gist'' might be identifiable from a few key objects or materials such as a cactus or camel in the desert. The visual system might operate in parallel on several different objects within a scene in order to classify the gist of the scene in approximately the same amount of time that it takes to classify the individual objects. In other cases, gist recognition—at least insofar as it involves classifying a scene into a preexisting category such as ''beach'' or ''desert''—might rely on mechanisms that analyse the overall shape, texture, or colour distribution within a scene (Oliva & Schyns, 1997, 2000; Oliva & Torralba, 2001) and are independent of the mechanisms for object form analysis (Steeves, Humphrey, Culham, Menon, & Goodale, 2002).

Given the strong response of the PPA to scenes, one might expect that it plays an important role in processing scene gist. For example, the PPA might combine information about the overall spatial layout of the scene with a summary of either object or feature information to come up with a unified interpretation of the scene's meaning. To date, this hypothesis has not been tested. However, recent results from Bar and Aminoff (2003) are consistent with the idea that the PPA may be involved in rapidly summarizing object information within a scene. These researchers found that parahippocampal cortex responded more strongly objects that were strongly associated with particular contexts than to objects that had no such association. Interestingly, the ''contexts'' in this experiment could either be particular kinds of places (''farm'', ''kitchen'') or more abstract situations or ideas (''music'', ''birthday''). Although preliminary, these results point to a role for the PPA in processing scene gist that bears further investigation.

Finally, the fourth aspect of visual scenes is that they contain objects. The mechanisms involved in identifying objects and specifying their locations have been extensively studied. Indeed, as it is generally formulated, the standard division of the visual system into what vs. where pathways *assumes* that object recognition and localization are the two main goals of vision (Mishkin, Ungerleider, & Macko, 1983). However, just as actors cannot act without a stage, objects cannot appear except within the context of a scene. Thus, one of salutary aspect of studying scene perception is that it expands our conception of what vision is for. Vision scientists have spent many years studying the actors; now it is time to direct some attention to the stage.

REFERENCES

Aguirre, G. K., & D'Esposito, M. (1997). Environmental knowledge is subserved by separable dorsal/ventral neural areas. *Journal of Neuroscience, 17*, 2512–2518.

Aguirre, G. K., & D'Esposito, M. (1999). Topographical disorientation: A synthesis and taxonomy. *Brain, 122*, 1613–1628.

Aguirre, G. K., Detre, J. A., Alsop, D. C., & D'Esposito, M. (1996). The parahippocampus subserves topographical learning in man. *Cerebral Cortex, 6*, 823–829.

Aguirre, G. K., Zarahn, E., & D'Esposito, M. (1998). An area within human ventral cortex sensitive to "building" stimuli: Evidence and implications. *Neuron, 21*, 373–383.

Andersen, R. A., Snyder, L. H., Bradley, D. C., & Xing, J. (1997). Multimodal representation of space in the posterior parietal cortex and its use in planning movements. *Annual Review of Neuroscience, 20*, 303–330.

Bar, M., & Aminoff, E. (2003). Cortical analysis of visual context. *Neuron, 38*, 347–358.

Baylis, G. C., & Rolls, E. T. (1987). Response of neurons in inferotemporal cortex in short-term and serial recognition memory tasks. *Experimental Brain Research, 65*, 614–622.

Best, P. J., White, A. M., & Minai, A. (2001). Spatial processing in the brain: The activity of hippocampal place cells. *Annual Review of Neuroscience, 24*, 459–486.

Biegler, R., & Morris, R. G. M. (1993). Landmark stability is a prerequisite for spatial but not discrimination-learning. *Nature, 361*, 631–633.

Bohbot, V. D., Kalina, M., Stepankova, K., Spackova, N., Petrides, M., & Nadel, L. (1998). Spatial memory deficits in patients with lesions to the right hippocampus and to the right parahippocampal cortex. *Neuropsychologia, 36*, 1217–1238.

Brown, M. W., Wilson, F. A. W., & Riches, I. P. (1987). Neuronal evidence that inferomedial temporal cortex is more important than hippocampus in certain processes underlying recognition memory. *Brain Research, 65*, 158–162.

Brown, M. W., & Xiang, J. Z. (1998). Recognition memory: Neuronal substrates of the judgement of prior occurrence. *Progress in Neurobiology, 55*, 149–189.

Buckner, R. L., Goodman, J., Burock, M., Rotte, M., Koutstaal, W., Schacter, D., et al. (1998). Functional-anatomic correlates of object priming in humans revealed by rapid presentation event-related fMRI. *Neuron, 20*, 285–296.

Burgess, N., Maguire, E. A., Spiers, H. J., & O'Keefe, J. (2001). A temporoparietal and prefrontal network for retrieving the spatial context of lifelike events. *Neuroimage, 14*, 439–453.

Cheng, K. (1986). A purely geometric module in the rat's spatial representation. *Cognition, 23*, 149–178.

Christou, C. G., & Bulthoff, H. H. (1999). View dependence in scene recognition after active learning. *Memory and Cognition, 27*, 996–1007.

Chua, K. P., & Chun, M. M. (2003). Implicit scene learning is viewpoint dependent. *Perception and Psychophysics, 65*, 72–80.

Colby, C. L., Duhamel, J. R., & Goldberg, M. E. (1996). Visual, presaccadic, and cognitive activation of single neurons in monkey lateral intraparietal area. *Journal of Neurophysiology, 76*, 2841–2852.

Colby, C. L., & Goldberg, M. E. (1999). Space and attention in parietal cortex. *Annual Review of Neuroscience, 22*, 319–349.

Diwadkar, V. A., & McNamara, T. P. (1997). Viewpoint dependence in scene recognition. *Psychological Science, 8*, 302–307.

Eichenbaum, H., Dudchenko, P., Wood, E., Shapiro, M., & Tanila, H. (1999). The hippocampus, memory, and place cells: Is it spatial memory or a memory space? *Neuron, 23*, 209–226.

Ekstrom, A. D., Kahana, M. J., Caplan, J. B., Fields, T. A., Isham, E. A., Newman, E. L., et al. (2003). Cellular networks underlying human spatial navigation. *Nature, 425*, 184–188.

Epstein, R., DeYoe, E. A., Press, D. Z., Rosen, A. C., & Kanwisher, N. (2001). Neuropsychological evidence for a topographical learning mechanism in parahippocampal cortex. *Cognitive Neuropsychology, 18*, 481–508.

Epstein, R., Graham, K. S., & Downing, P. E. (2003). Viewpoint-specific scene representations in human parahippocampal cortex. *Neuron, 37*, 865–876.

Epstein, R., Harris, A., Stanley, D., & Kanwisher, N. (1999). The parahippocampal place area: Recognition, navigation, or encoding? *Neuron, 23*, 115–125.

Epstein, R. A., Higgins, J. S., & Thompson-Schill, S. L. (2005). Learning places from views: Variation in scene processing as a function of experience and navigational ability. *Journal of Cognitive Neuroscience, 17*, 73–83.

Epstein, R., & Kanwisher, N. (1998). A cortical representation of the local visual environment. *Nature, 392*, 598–601.

Fortin, N. J., Agster, K. L., & Eichenbaum, H. B. (2002). Critical role of the hippocampus in memory for sequences of events. *Nature Neuroscience, 5*, 458–462.

Gauthier, I. (2000). What constrains the organization of the ventral temporal cortex? *Trends in Cognitive Sciences, 4*, 1–2.

Georges-Francois, P., Rolls, E. T., & Robertson, R. G. (1999). Spatial view cells in the primate hippocampus: Allocentric view not head direction or eye position or place. *Cerebral Cortex, 9*, 197–212.

Ghaem, O., Mellet, E., Crivello, F., Tzourio, N., Mazoyer, B., Berthoz, A., et al. (1997). Mental navigation along memorized routes activates the hippocampus, precuneus, and insula. *Neuroreport, 8*, 739–744.

Golob, E. J., & Taube, J. S. (1997). Head direction cells and episodic spatial information in rats without a hippocampus. *Proceedings of the National Academy of Sciences of the United States of America, 94*, 7645–7650.

Gorno-Tempini, M. L., & Price, C. J. (2001). Identification of famous faces and buildings: A functional neuroimaging study of semantically unique items. *Brain, 124*, 2087–2097.

Gottlieb, J. P., Kusunoki, M., & Goldberg, M. E. (1998). The representation of visual salience in monkey parietal cortex. *Nature, 391*, 481–484.

Gouteux, S., Thinus-Blanc, C., & Vauclair, J. (2001). Rhesus monkeys use geometric and nongeometric information during a reorientation task. *Journal of Experimental Psychology: General, 130*, 505–519.

Grill-Spector, K., Kourtzi, Z., & Kanwisher, N. (2001). The lateral occipital complex and its role in object recognition. *Vision Research, 41*, 1409–1422.

Grill-Spector, K., Kushnir, T., Edelman, S., Avidan, G., Itzchak, Y., & Malach, R. (1999). Differential processing of objects under various viewing conditions in the human lateral occipital complex. *Neuron, 24*, 187–203.

Grill-Spector, K., Kushnir, T., Edelman, S., Itzchak, Y., & Malach, R. (1998). Cue-invariant activation in object-related areas of the human occipital lobe. *Neuron, 21*, 191–202.

Grill-Spector, K., & Malach, R. (2001). fMR-adaptation: A tool for studying the functional properties of human cortical neurons. *Acta Psychologica, 107*, 293–321.

Habib, M., & Sirigu, A. (1987). Pure topographical disorientation—A definition and anatomical basis. *Cortex, 23*, 73–85.

Haxby, J. V., Gobbini, M. I., Furey, M. L., Ishai, A., Schouten, J. L., & Pietrini, P. (2001). Distributed and overlapping representations of faces and objects in ventral temporal cortex. *Science, 293*, 2425–2430.

Hegarty, M., Richardson, A. E., Montello, D. R., Lovelace, K., & Ilavanil, S. (2002). Development of a self-report measure of environmental spatial ability. *Intelligence*, *30*, 425–447.

Henderson, J. M., & Hollingworth, A. (1999). High-level scene perception. *Annual Review of Psychology*, *50*, 243–271.

Henson, R., Shallice, T., & Dolan, R. (2000). Neuroimaging evidence for dissociable forms of repetition priming. *Science*, *287*, 1269–1272.

Hermer, L., & Spelke, E. (1996). Modularity and development: The case of spatial reorientation. *Cognition*, *61*, 195–232.

Hermer, L., & Spelke, E. S. (1994). A geometric process for spatial reorientation in young children. *Nature*, *370*, 57–59.

Ishai, A., Ungerleider, L. G., Martin, A., Schouten, H. L., & Haxby, J. V. (1999). Distributed representation of objects in the human ventral visual pathway. *Proceedings of the National Academy of Sciences of the United States of America*, *96*, 9379–9384.

Kanwisher, N., McDermott, J., & Chun, M. M. (1997a). The fusiform face area: A module in human extrastriate cortex specialized for face perception. *Journal of Neuroscience*, *17*, 4302–4311.

Kanwisher, N., Woods, R. P., Iacoboni, M., & Mazziotta, J. C. (1997b). A locus in human extrastriate cortex for visual shape analysis. *Journal of Cognitive Neuroscience*, *9*, 133–142.

King, J. A., Burgess, N., Hartley, T., Vargha-Khadem, F., & O'Keefe, J. (2002). The human hippocampus and viewpoint dependence in spatial memory. *Hippocampus*, *12*, 811–820.

Knierim, J. J., Kudrimoti, H. S., & McNaughton, B. L. (1995). Place cells, head direction cells, and the learning of landmark stability. *Journal of Neuroscience*, *15*, 1648–1659.

Knierim, J. J., Kudrimoti, H. S., & McNaughton, B. L. (1998). Interactions between idiothetic cues and external landmarks in the control of place cells and head direction cells. *Journal of Neurophysiology*, *80*, 425–446.

Kohler, S., Crane, J., & Milner, B. (2002). Differential contributions of the parahippocampal place area and the anterior hippocampus to human memory for scenes. *Hippocampus*, *12*, 718–723.

Kourtzi, Z., & Kanwisher, N. (2001). Representation of perceived object shape by the human lateral occipital complex. *Science*, *293*, 1506–1509.

Landis, T., Cummings, J. L., Benson, D. F., & Palmer, E. P. (1986). Loss of topographic familiarity: An environmental agnosia. *Archives of Neurology*, *43*, 132–136.

Learmonth, A. E., Newcombe, N. S., & Huttenlocher, J. (2001). Toddlers' use of metric information and landmarks to reorient. *Journal of Experimental Child Psychology*, *80*, 225–244.

Levy, I., Hasson, U., Avidan, G., Hendler, T., & Malach, R. (2001). Center–periphery organization of human object areas. *Nature Neuroscience*, *4*, 533–539.

Levy, W. B. (1996). A sequence predicting CA3 is a flexible associator that learns and uses context to solve hippocampal-like tasks. *Hippocampus*, *6*, 579–590.

Li, L., Miller, E. K., & Desimone, R. (1993). The representation of stimulus-familiarity in anterior inferior temporal cortex. *Journal of Neurophysiology*, *69*, 1918–1929.

Maguire, E. A., Burgess, N., Donnett, J. G., Frackowiak, R. S. J., Frith, C. D., & O'Keefe, J. (1998). Knowing where and getting there: A human navigation network. *Science*, *280*, 921–924.

Maguire, E. A., Frackowiak, R. S. J., & Frith, C. D. (1997). Recalling routes around London: Activation of the right hippocampus in taxi drivers. *Journal of Neuroscience*, *17*, 7103–7110.

Maguire, E. A., Frith, C. D., & Cipolotti, L. (2001). Distinct neural systems for the encoding and recognition of topography and faces. *Neuroimage*, *13*, 743–750.

Malach, R., Reppas, J. B., Benson, R. R., Kwong, K. K., Jiang, H., Kennedy, W. A., et al. (1995). Object-related activity revealed by functional magnetic-resonance-imaging in human occipital cortex. *Proceedings of the National Academy of Sciences of the United States of America*, *92*, 8135–8139.

Margules, J., & Gallistel, C. R. (1988). Heading in the rat—Determination by environmental shape. *Animal Learning and Behavior*, *16*, 404–410.

Matsumura, N., Nishijo, H., Tamura, R., Eifuku, S., Endo, S., & Ono, T. (1999). Spatial- and task-dependent neuronal responses during real and virtual translocation in the monkey hippocampal formation. *Journal of Neuroscience*, *19*, 2382–2393.

McCarthy, R. A., Evans, J. J., & Hodges, J. R. (1996). Topographic amnesia: Spatial memory disorder, perceptual dysfunction, or category specific semantic memory impairment? *Journal of Neurology, Neurosurgery, and Psychiatry*, *60*, 318–325.

Mendez, M. F., & Cherrier, M. M. (2003). Agnosia for scenes in topographagnosia. *Neuropsychologia*, *41*, 1387–1395.

Miller, E. K., Li, L., & Desimone, R. (1991). A neural mechanism for working and recognition memory in inferior temporal cortex. *Science*, *254*, 1377–1379.

Milner, A. D., & Goodale, M. A. (1995). *The visual brain in action*. Oxford, UK/New York: Oxford University Press.

Mishkin, M., Ungerleider, L. G., & Macko, K. A. (1983). Object vision and spatial vision—2 cortical pathways. *Trends in Neurosciences*, *6*, 414–417.

Morris, R., Paxinos, G., & Petrides, M. (2000). Architectonic analysis of the human retrosplenial cortex. *Journal of Comparative Neurology*, *421*, 14–28.

Morris, R., Petrides, M., & Pandya, D. N. (1999). Architecture and connections of retrosplenial area 30 in the rhesus monkey (Macaca mulatta). *European Journal of Neuroscience*, *11*, 2506–2518.

Morris, R. G., Garrud, P., Rawlins, J. N., & O'Keefe, J. (1982). Place navigation impaired in rats with hippocampal lesions. *Nature*, *297*, 681–683.

Mou, W., & McNamara, T. P. (2002). Intrinsic frames of reference in spatial memory. *Journal of Experimental Psychology: Learning, Memory, and Cognition*, *28*, 162–170.

Muller, J. R., Metha, A. B., Krauskopf, J., & Lennie, P. (1999). Rapid adaptation in visual cortex to the structure of images. *Science*, *285*, 1405–1408.

Muller, R. U., & Kubie, J. L. (1987). The effects of changes in the environment on the spatial firing of hippocampal complex-spike cells. *Journal of Neuroscience*, *7*, 1951–1968.

Nakamura, K., Kawashima, R., Sato, N., Nakamura, A., Sugiura, M., Kato, T., et al. (2000). Functional delineation of the human occipito-temporal areas related to face and scene processing—A PET study. *Brain*, *123*, 1903–1912.

Nakatani, C., Pollatsek, A., & Johnson, S. H. (2002). Viewpoint-dependent recognition of scenes. *Quarterly Journal of Experimental Psychology A*, *55*, 115–139.

O'Keefe, J., & Burgess, N. (1996). Geometric determinants of the place fields of hippocampal neurons. *Nature*, *381*, 425–428.

O'Keefe, J., & Dostrovsky, J. (1971). The hippocampus as a spatial map: Preliminary evidence from unit activity in the freely-moving rat. *Brain Research*, *34*, 171–175.

O'Keefe, J., & Nadel, L. (1978). *The hippocampus as a cognitive map*. Oxford, UK: Clarendon Press/Oxford University Press.

O'Keefe, J., & Speakman, A. (1987). Single unit activity in the rat hippocampus during a spatial memory task. *Experimental Brain Research*, *68*, 1–27.

Oliva, A., & Schyns, P. G. (1997). Coarse blobs or fine edges? Evidence that information diagnosticity changes the perception of complex visual stimuli. *Cognitive Psychology*, *34*, 72–107.

Oliva, A., & Schyns, P. G. (2000). Diagnostic colours mediate scene recognition. *Cognitive Psychology*, *41*, 176–210.

Oliva, A., & Torralba, A. (2001). Modeling the shape of the scene: A holistic representation of the spatial envelope. *International Journal of Computer Vision*, *42*, 145–175.

Ploner, C. J., Gaymard, B. M., Rivaud-Pechoux, S., Baulac, M., Clemenceau, S., Samson, S., et al. (2000). Lesions affecting the parahippocampal cortex yield spatial memory deficits in humans. *Cerebral Cortex*, *10*, 1211–1216.

Potter, M. C. (1975). Meaning in visual search. *Science*, *187*, 965–966.

Ringo, J. L. (1996). Stimulus specific adaptation in inferior temporal and medial temporal cortex of the monkey. *Behavioural Brain Research*, *76*, 191–197.

Robertson, R. G., Rolls, E. T., & Georges-Francois, P. (1998). Spatial view cells in the primate hippocampus: Effects of removal of view details. *Journal of Neurophysiology*, *79*, 1145–1156.

Rolls, E. T., Robertson, R. G., & Georges-Francois, P. (1997). Spatial view cells in the primate hippocampus. *European Journal of Neuroscience*, *9*, 1789–1794.

Sanocki, T., & Epstein, W. (1997). Priming spatial layout of scenes. *Psychological Science, 8*, 374–378.

Schacter, D. L., & Buckner, R. L. (1998). Priming and the brain. *Neuron, 20*, 185–195.

Shelton, A. L., & McNamara, T. P. (2001). Systems of spatial reference in human memory. *Cognitive Psychology, 43*, 274–310.

Stackman, R. W., & Taube, J. S. (1997). Firing properties of head direction cells in the rat anterior thalamic nucleus: Dependence on vestibular input. *Journal of Neuroscience, 17*, 4349–4358.

Stark, M., Coslett, H. B., & Saffran, E. M. (1996). Impairment of an egocentric map of locations: Implications for perception and action. *Cognitive Neuropsychology, 13*, 481–523.

Steeves, J. K. E., Humphrey, G. K., Culham, J. C., Menon, R. S., & Goodale, M. A. (2002). *Scene classification and parahippocampal place area activation in an individual with visual form agnosia.* Paper presented at the Vision Sciences Society, Sarasota, FL.

Stern, C. E., Corkin, S., Gonzalez, R. G., Guimaraes, A. R., Baker, J. R., Jennings, P. J., et al. (1996). The hippocampal formation participates in novel picture encoding: Evidence from functional magnetic resonance imaging. *Proceedings of the National Academy of Sciences of the United States of America, 93*, 8660–8665.

Suzuki, W. A. (1996). Neuroanatomy of the monkey entorhinal, perirhinal and parahippocampal cortices: Organization of cortical inputs and interconnections with amygdala and striatum. *Seminars in the Neurosciences, 8*, 3–12.

Suzuki, W. A., & Amaral, D. G. (1994). Perirhinal and parahippocampal cortices of the macaque monkey—Cortical afferents. *Journal of Comparative Neurology, 350*, 497–533.

Taube, J. S. (1998). Head direction cells and the neurophysiological basis for a sense of direction. *Progress in Neurobiology, 55*, 225–256.

Teng, E., & Squire, L. R. (1999). Memory for places learned long ago is intact after hippocampal damage. *Nature, 400*, 675–677.

Thorpe, S., Fize, D., & Marlot, C. (1996). Speed of processing in the human visual system. *Nature, 381*, 520–522.

Tolman, E. C. (1948). Cognitive maps in rats and men. *Psychological Review, 55*, 189–208.

Van Turennout, M., Bielamowicz, L., & Martin, A. (2003). Modulation of neural activity during object naming: Effects of time and practice. *Cerebral Cortex, 13*, 381–391.

Vuilleumier, P., Henson, R. N., Driver, J., & Dolan, R. J. (2002). Multiple levels of visual object constancy revealed by event-related fMRI of repetition priming. *Nature Neuroscience, 5*, 491–499.

Whiteley, A. M., & Warrington, E. K. (1978). Selective impairment of topographical memory: A single case study. *Journal of Neurology, Neurosurgery, and Psychiatry, 41*, 575–578.

VISUAL COGNITION, 2005, 12 (6), 979–1002

Minimal use of working memory in a scene comparison task

Daniel A. Gajewski and John M. Henderson
Michigan State University, East Lansing, MI, USA

Eye movement behaviour in hand-eye tasks suggests a preference for a "just in time" processing strategy that minimizes the use of working memory. In the present study, a scene comparison task was introduced to determine whether the preference holds when the task is primarily visual and when more complex naturalistic scenes are used as stimuli. In two experiments, participants made same or different judgements in response to simultaneously presented pairs of scenes that were identical or differed by one object. The number of fixations per scene glance and the number of fixations intervening between glances to corresponding objects suggest that frequently one object at a time is encoded and maintained in visual working memory. The same pattern of results was observed in a third experiment using word and object arrays. Overall, the results suggest a strong general bias toward minimal use of visual working memory in complex visual tasks.

Because the highest level of acuity is limited to a rather small portion of the visual field, a region roughly the size of one's thumb nail when viewed at arm's length, saccadic eye movements are needed to direct the eyes toward points of interest in a scene. The fact that our visual experience seems to include more detail than can be provided by a single fixation suggests a role for memory in scene perception; however, the exact role played by memory in scene perception is a topic of debate. Some theorists have proposed that the online visual representation of a scene includes a highly detailed representation of the object

---

Please address all correspondence to: Daniel A. Gajewski, Department of Psychology, Michigan State University, East Lansing, MI 48824-1117, USA. Email: dan@eyelab.msu.edu

This research was supported by a fellowship from the MSU graduate school on behalf of the NSF IGERT program in Cognitive Science to Daniel A. Gajewski. This work was also supported by the National Science Foundation (BCS-0094433 and ECS-9873531) and the Army Research Office (DAAD19-00-1-0519). Aspects of this data were presented at the 43rd annual meeting of the Psychonomic Society, Kansas City, MO, and the third annual meeting of the Vision Sciences Society, Sarasota, FL. The opinions expressed in this paper are those of the authors and do not necessarily represent the views of the Department of the Army or any other governmental organization. Reference to or citations of trade or corporate names does not constitute explicit or implied endorsement of those entities or their products by the author or the Department of the Army.

DOI:10.1080/13506280444000616

of the current fixation, abstract visual memory representations of the three–four most recently attended items, as well as a representation of the structure or layout of the scene (e.g., Hollingworth & Henderson, 2002; Irwin & Andrews, 1996; Irwin & Zelinsky, 2002; but see Rensink, 2000). Others have suggested that the online visual representation is limited to information that is relevant to the ongoing task (Ballard, Hayhoe, Pook, & Rao, 1997; Hayhoe, Bensinger, & Ballard, 1998). According to a framework called *active vision* (Aloimonos, Weiss, & Bandopadhay, 1987), the purpose of vision is not to build a general representation of the environment. Rather, visual information is believed to be extracted from the environment in the service of specific behavioural goals. Importantly, this framework favours "just in time" processing strategies where the acquisition of visual information is delayed until the point in time when it is needed.

A body of work done by proponents of this framework supports a preference to minimize the use of visual working memory. For example, Ballard, Hayhoe, and colleagues (Ballard, Hayhoe, & Pelz, 1995; Hayhoe et al., 1998) have explored this issue using a block-copying task. Participants were given a display that was divided into three regions—a model region, a source region, and a workspace region. The model region contained a pattern of coloured blocks to be copied and the source region provided a set of coloured blocks to be moved into the third area, the workspace region. Inferences about the information retained at any moment during the task were made on the basis of the eye movement patterns. The most frequently observed eye movement pattern began with a saccade to the model, then to the source area, and then to the model again before directing gaze to the workspace in order to guide the placement of the selected block into the appropriate space in the emerging replica. The fact that the same block was fixated twice at different points during the operation suggested that different information was acquired during each glance. That is, the first glance, prior to fixating the source area, appears to have been used to determine the colour of the block that was needed, and the second glance, prior to positioning the selected block into the workspace, appears to have been used to determine the spatial position of the selected block. The authors considered this pattern memoryless because it suggested a preference to delay the acquisition of information until the point in time when it was needed, thereby minimizing the use of visual working memory. Importantly, this preference could not be attributed to capacity limitations because when the distance between the model and the workspace was increased, a manipulation designed to increase the cost associated with acquiring information, the frequency of the memoryless pattern decreased. The preferred strategy, therefore, appears to reflect issues of cognitive economy.

A question that follows from this work is the extent to which these findings will generalize to a task that is primarily visual. The work of Ballard, Hayhoe, and colleagues (Ballard et al., 1995; Hayhoe et al., 1998) arose out of a computational framework that is at least partially concerned with the functional use of visual information to guide the programming of motoric interactions with the world. As a result, the focus has been on tasks that involve the coordinated use of the hand and the eyes. The suggestion is that the algorithmic complexity of hand–eye tasks can be reduced by allowing the point of fixation to serve as the centre of an external reference frame and by representing in memory only the information needed at each step in the operation (Ballard et al., 1997). The crux of this theoretical perspective is that eye movements play a role in variable binding because the fixation can be used as a deictic device that points to objects in the world that participate in the execution of behavioural programs. Thus, any task that requires the organization of sequential operations, even tasks that do not require the coordination of their visual and motor components, should similarly reveal preferences to minimize the use of memory. However, the possibility remains that the preferred strategy will differ when the visual task does not demand this kind of interaction with the environment. The bias towards minimal short-term representation in the block-copying task might be driven by its motor component.

In addition to testing the generality of the minimal memory preference in a task that is primarily visual, our interest is in the use of visual working memory in the perception of scenes. Therefore, we tested the minimal memory hypotheses using stimuli that approach the complexity of natural viewing situations. Participants were asked to make same-or-different judgements about side-by-side copies of rendered scenes. The pairs of scenes were either identical or differed by one object. Two primary measures were used to make inferences about the use of memory in this task. The first measure was intended to provide an indication of how many objects are looked at in one scene prior to directing the eyes to the other scene. If the preferred strategy is to minimize the use of visual working memory, gaze shifts from one scene to the other should be frequent, with the number of objects fixated during each glance of a scene close to one. The second measure was intended to provide an indication of the number of objects fixated between the first glance to an object in one scene and the first glance to the corresponding object in the other scene. If the minimal memory strategy is preferred, participants should most often direct the eyes immediately from one object in one scene to the corresponding object in the other scene, holding just the one object in memory in order to make the comparison.

The appropriate use of the number of objects fixated as an index of memory use, however, depends upon two assumptions: (1) Objects are the level of analysis in scene perception; and (2) objects are the unit of capacity of visual working memory. The first assumption could arguably be satisfied by the demands of the task given that it is at the level of the object that two scenes may differ; however, an object level of analysis in scene viewing is also supported by empirical work. Nelson and Loftus (1980), for example, showed that the likelihood that an object will be encoded sufficiently to support discrimination in a recognition task declines sharply with increased distance from the centre of the nearest fixation. In addition, changes made to objects in change detection tasks are reliably detected only if the object is fixated before the change (Hollingworth & Henderson, 2002). The second assumption is satisfied by several studies suggesting that integrated objects are the unit of capacity and that between three and four objects can be held in visual working memory (Irwin, 1992; Irwin & Andrews, 1996; Luck & Vogel, 1997; Vogel, Woodman, & Luck, 2001; but see Wheeler & Treisman, 2002). For example, Luck and Vogel (1997; Vogel et al., 2001) tested memory for briefly presented arrays of bars that varied in colour and orientation. Performance declined with increasing array size but was unaffected by the need to remember only the colour, only the orientation, or the conjunction of both features. This result was also found when the items in the array were conjunctions of four features. Thus, it was the number of objects that limited performance, not the number of features. These studies suggest that if memory is used maximally in the scene comparison task, at least three–four objects will be fixated in one scene prior to directing gaze to the other scene.

The interest in the number of objects fixated, however, does not come without complications when naturalistic scenes are used as stimuli. Ideally, one would like to take the fixation data for each participant and count the number of objects fixated during a given scene glance. In order to count the number of objects fixated, one would have to assign each fixation to an object. We were attracted to using scenes because of the complexity offered by them, but the things that make scenes complex, such as partial occlusions and the clustering of objects, also make it difficult to assign fixations to objects. In addition, a fair amount of subjectivity would be introduced by this approach. The issue would be more easily resolved if object gazes were comprised of only one fixation. One could then simply count the number of fixations per scene glance. However, scene-viewing studies suggest that this is not the case. Experiments run in this laboratory typically find between 1.5 and 2 fixations per object on average. Henderson, Weeks, and Hollingworth (1999), for example, report an average of 1.7 fixations per gaze on target objects. Our solution was to take the number of fixations during the first glance to strategically placed critical objects as an index of the number of fixations that generally contribute to an object glance in this task. Using this estimate, predictions based on the number of objects fixated per scene glance can be expressed in terms of numbers of fixations per scene glance.

In this study there were two measures of primary interest. The first dependent measure was the mean number of fixations made in each scene before gaze is directed to the other scene. In order to test competing hypotheses about the use of memory in this task, models were generated for both minimal and maximal use of memory. Figure 1A illustrates the type of viewing pattern one would expect if the use of memory were minimized. The two squares on the left represent a pair of scenes comprised of objects, and the grid on the right represents the number of objects fixated during successive glances to the scenes. If memory is minimized, participants should most frequently begin by fixating one object in one of the scenes and then immediately fixate the corresponding object in the other scene to make the comparison. Then, if the number of shifts between scenes is also minimized, participants should select the next item from the same scene. If this pattern continues, the expected number of objects per scene glance is simply the number of objects fixated divided by the number of scene glances. In this case its 1.75. Figure 1B illustrates the expected viewing pattern given a preference to make full use of memory. Given the evidence for a capacity of three–four items, predictions for maximal use of memory were made conservatively on the basis of three objects. In this scenario, participants should fixate three objects in the first scene before directing their eyes to the corresponding objects in the other scene, and three new objects should be fixated before returning to the original scene. The expected number of objects per scene glance here is 5.25.

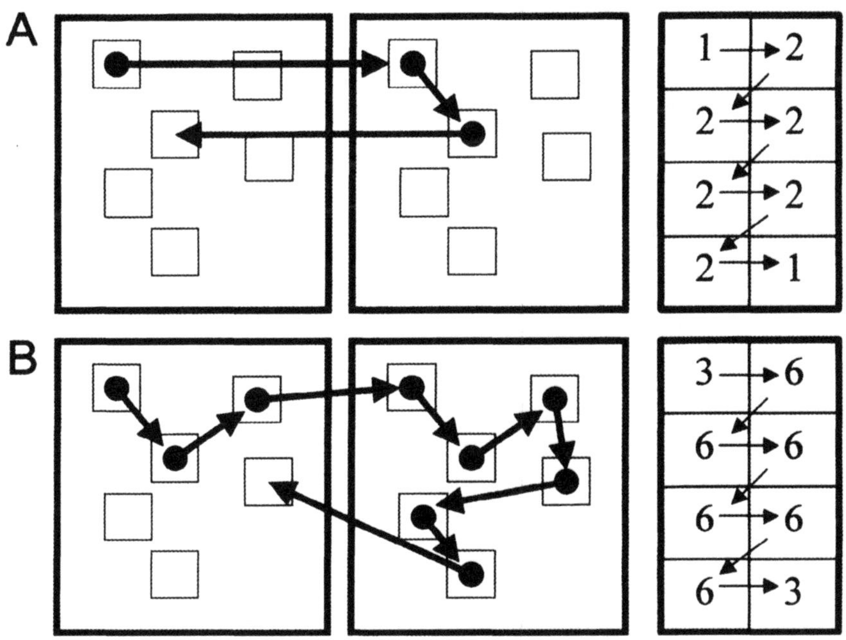

Figure 1. (A) The two large squares on the left represent a pair of scenes comprised of objects. The expected viewing pattern given minimal use of memory is illustrated by the flow of the arrows. The right side of the figure represents a model of the number of objects fixated during each successive scene glance. In this eight-glance model, each scene is entered four times and the expected number of objects fixated per scene is 1.75. (B) An eight-glance model was similarly generated based on the full use of memory. Here, the expected number of objects fixated is 5.25.

Note that the expected values generated by these models depend on the number of times each scene is viewed. In these eight-glance models, each scene is entered four times. As the number of glances to each scene increases, the expected number of objects fixated per scene approaches two and six for minimal and maximal memory, respectively. Therefore, the average number of scene glances was determined from the data in order to appropriately formulate predictions. Finally, in order to compare the predicted values to those experimentally derived, the expected number of objects fixated per scene glance was multiplied by the mean number of fixations per object.

The second measure of interest is the number of fixations that intervened between the first fixation on the first critical object fixated and the first fixation on the corresponding object in the other scene. This analysis provides an indication of how frequently participants went from one object directly to the corresponding object in the other scene. If the use of working memory is minimized, subjects should most frequently look directly from one object to the corresponding object in the other scene. If the number of fixations per object is between one and two, the most frequent number of intervening fixations should be between zero and one. If the use of memory is maximized, however, subjects should most frequently fixate at least two objects prior to fixating the corresponding object in the other scene. The most frequent number of intervening fixations, therefore, should be between two and three.

EXPERIMENT 1

In Experiment 1, participants made same-or-different judgements in response to 32 pairs of scenes. In three conditions the pairs of scenes differed in terms of one object in one of the scenes: An object was removed (deletion), replaced with an item from a different basic-level conceptual category (different-type), or replaced with a different exemplar from the same category (different-token). In the fourth condition, both scenes were identical (same). All four conditions occurred with equal frequency. Judgements of same versus different were indicated with a button press. Each image remained visible for as long as was needed to make the decision; however, participants were instructed to press the appropriate button as soon as the decision was made.

Method

Participants. Twelve Michigan State University undergraduate students participated in the experiment for course credit. All participants had normal vision and were naive with respect to the hypotheses under investigation.

Stimuli. Thirty-two scene images were computer-rendered from three-dimensional (3-D) wire-frame models using a commercial rendering program. Wire-frame models were acquired commercially, donated by 3-D graphic artists, or developed inhouse. Each model depicted a typical, human-scaled environment. Base scenes were rendered from these models. To create the different-type, different-token, and deletion conditions, the target objects were replaced or removed in the models, and the scenes were rerendered. All scene images subtended 7.8° × 5.9° visual angle at a viewing distance of 1.13 m. Each image was comprised of two versions of scenes side-by-side along the middle of the display with a small gap in between. The remainder of the display was a neutral grey. Figure 2 shows a sample stimulus image. Target objects subtended 1.14° on average along the longest axis and 0.71° along the shortest axis. The objects used for the different-type and different-token conditions were chosen to be similar in size to the no-change target object in the corresponding scene. Target objects were placed so that they were offset from the centres of the scenes. They were placed in uncluttered regions of the scenes so that fixations on them could be easily identified. Nothing about the target objects themselves identified them as different from the other objects in the scenes.

Apparatus. Eye movements were monitored using a Generation 5.5 Stanford Research Institute Dual Purkinje Image Eyetracker (Crane, 1994; Crane & Steele, 1985). The eyetracker has a resolution of 1' of arc and a linear

Figure 2. Sample scene illustrating regions used for critical objects and a scan pattern used by one of the participants on one of the trials. Full-colour versions were used in the experiment.

output over the range of the visual display used. A bite-bar and a forehead rest were used to maintain the participant's viewing position and distance. The position of the right eye was tracked, though viewing was binocular. Signals were sampled from the eyetracker using the polling mode of the Data Translations DT2802 analogue-to-digital converter, producing a sampling rate slightly faster than 1000 Hz.

Stimuli were displayed at a resolution of 800 × 600 pixels × 256 colours on an NEC Multisync P750 monitor driven by a Hercules Dynamite 128/Video graphics card. The screen refresh rate was 143 Hz. The room was dimly illuminated by an indirect, low-intensity light source.

Button-presses were collected with a button panel connected to a dedicated input–output (I/O) card. The eyetracker, display monitor, and I/O card were interfaced with a 90 MHz, Pentium-based microcomputer. The computer controlled the experiment and maintained a complete record of eye position and time values, and button press events and times, over the course of each trial.

Procedure. Upon arriving for the experimental session, participants were given a written description of the experiment along with a set of instructions. The description informed participants that their eye movements would be monitored while they viewed images of naturalistic scenes on a computer monitor. Participants were instructed to view pairs of scenes and to judge as quickly and accurately as possible whether they were the same or different. They were explicitly told that differences between scenes would be due to differences between one object in one scene and the corresponding object in the other scene and they were given examples of the types of differences that could be found. Participants were instructed to press the right button as soon as such a difference was detected or the left button as soon as they were satisfied that the two scenes were the same. Following review of the instructions, the experimenter calibrated the eyetracker. Calibration was considered accurate if the computer's estimate of the current fixation position was within +/− 5 min arc of each marker. The participant then completed the experimental session. Calibration was checked every three–four trials, and the eyetracker was recalibrated when necessary.

A trial consisted of the following events. First, a fixation screen was shown. When the participant fixated a central box in this screen (as indicated by a computer-generated display of its estimated fixation position), the experimenter initiated the trial. Each image remained visible for as long as was needed to make the decision; however, subjects were instructed to perform the task both quickly and accurately. Each participant viewed 32 scenes. There were four conditions to the experiment: Three different conditions (different-type, different-token, and deletion) and one same condition. Scenes were assigned to conditions via a Latin-square design so that each scene appeared in each condition an equal number of times across participants. The order of scene presentation (and hence the order of condition presentation) was determined randomly for each participant within each session. Each session lasted approximately 35 min.

Results and discussion

Single factor within-subjects analyses of variance (ANOVAs) were performed to evaluate the general performance measures of accuracy and reaction time prior to the analyses of primary interest. Accuracy varied marginally across the four conditions, $F(3, 33) = 2.54$, $MSE = 0.02$, $p = .07$, and was highest in the same condition (mean = 0.95). Overall performance was fairly high: The mean proportion correct was .88.

The analysis of reaction times was based on correct responses only. Reaction times varied across the four conditions, $F(3, 33) = 12.16$, $MSE = 20.57$, $p < .001$; however, there were no differences in reaction time when only the three different conditions were considered, $F < 1$. Viewing times were longer in the same condition (mean = 13.9 s) than in the three different conditions (mean = 4.8 s), $F(1, 11) = 13.20$, $MSE = 37.68$, $p < .01$. This difference is expected given that subjects would have to perform an exhaustive search in order to make a decision in the same condition but not in the others.

To generate and test predictions about the use of memory in this task, a series of analyses were performed using the eye-tracking data. First, the number of scene glances was determined to generate minimal and maximal memory models for the expected numbers of objects fixated per scene. Second, an estimate of the number of fixations that generally contribute to an object gaze was determined on the basis of the target objects and this estimate was multiplied by the expected number of objects fixated per scene to express the predictions in terms of the dependent measure—the number of fixations per scene. The observed numbers of fixations per scene were then tested against each of the predicted values. To compare these values, object- and scene-sized regions were defined for each display. Fixation data was recorded in a pixel-based coordinate system and tabulated as a function of occurrence in these regions. The analyses were based on correct trials only. Eye-tracking data from all 12 subjects went into the analyses; however, 21 trials were eliminated (approximately 5%) because of track loss.

Since the predicted number of objects fixated per scene depended on the total number of scene glances, the number of times each scene was fixated was tabulated for each of the two scene regions for every trial. The values for each of the pairs of regions were summed and the averages were entered into the analysis for each subject and condition. The total number of scene glances varied by condition, $F(3, 33) = 12.46$, $MSE = 24.93$, $p < .001$; however, there was no effect when only the three different conditions were considered, $F < 1$. There were more scene glances made in the same condition (16.5) than in the three different conditions (mean = 6.4), $F(1, 11) = 13.69$, $MSE = 44.96$, $p < .01$. Again, this

TABLE 1
Summary of the observed and predicted values in Experiment 1

	Same	*Different*
Expected objects per scene		
Minimum	1.9	1.7
Maximum	5.6	5.0
Expected fixations per scene		
Minimum	2.3	2.0
Maximum	6.8	6.0
Observed fixations (S.E.) per scene	2.73 (.11)	2.46 (.10)

would be expected given exhaustive searching in the same condition. The expected numbers of objects fixated per scene were therefore formulated for minimal and maximal memory based on 17-glance models for the same condition and 6-glance models for the mean of the different conditions (see Figure 1). The resulting values are listed in the top section of Table 1.

An estimate of the number of fixations that are generally made on an object was generated by counting the number of fixations that occurred during the first glance within regions defined for each of the target objects. The object-sized regions were 0.25 degrees larger than the smallest rectangle that would enclose each object (see Figure 2). The regions were slightly larger than the objects to reduce the number of fixations that would be inappropriately excluded from the analysis due to minor errors in calibration and saccade targeting. The mean fixation count for each of the two regions was averaged and entered into the analysis for each subject and for each condition. The number of fixations per object did not vary by condition, $F(3, 33) = 0.79$, $MSE = 0.02$, $p = .51$. On average, 1.2 fixations contributed to the first glances within the target object regions. This value was used to compute the expected number of fixations per scene for minimal and maximal use of memory: The expected numbers of objects fixated per scene derived from each of the models were multiplied by the number of fixations per object to evaluate the observed numbers of fixations per scene. Table 1 presents the values resulting from these computations.

The first measure of interest, the number of fixations per scene, was computed by dividing the total number of fixations in each of the two scene regions by the total number of scene glances for every participant and every trial. The number of fixations per scene varied by condition, $F(3, 33) = 4.08$, $MSE = 0.09$, $p < .05$. However, there was not a reliable difference in the number of fixations per scene when only the three different conditions were considered, $F < 1$. The number of fixations per scene was greater in the same condition (2.73) than in the three different conditions (mean = 2.46), $F(1, 11) = 6.99$, $MSE = 0.06$, $p < .05$. The direction of this difference is consistent with the predictions generated by the models (see Table 1).

The observed values for both the same and different conditions were compared to the respective expected numbers given minimal and maximal use of memory using one-sample *t*-tests. The observed values were greater than values predicted by minimal memory in the same condition, $t(11) = 3.77$, $p < .01$, as well as in the different conditions, $t(11) = 4.85$, $p < .01$. However, the predicted values for maximal memory were greater than the observed values in the same condition, $t(11) = 35.73$, $p < .001$, as well as in the different conditions, $t(11) = 36.84$, $p < .001$. Importantly, the means were considerably closer to those predicted by the minimal memory models. The average difference between the observed and predicted values was 0.3 fixations per scene for minimal memory versus 3.8 fixations per scene for maximal memory. Thus, although the observed values fall outside of the range of predictions made by either strategy, they clearly favour a minimal memory interpretation.

The second analysis of interest is the number of fixations that intervened between the first fixation on the first critical object fixated and the first fixation on the corresponding object in the other scene. However, since by definition this analysis includes only trials where both critical items were fixated, it is important to know how frequently this occurs. Thus, the proportion of trials where both items were fixated was examined prior to the analysis of primary interest. Both analyses were accomplished using the object-sized regions described above. The proportion of trials with both objects fixated differed reliably across the four conditions, $F(3, 33) = 6.17$, $MSE = 0.034$, $p < .01$. This effect, however, is completely driven by the deletion condition. The proportion of trials with both objects fixated did not differ when the deletion condition was excluded from the analysis, $F < 1$. The proportion of trials with both objects fixated was less in the deletion condition (.43) than the average of the other three conditions (.67), $F(1, 11) = 20.37$, $MSE = 0.018$, $p < .01$. This result is not surprising given that object presence can be detected from eccentricities of 4 degrees (Henderson, Williams, Castelhano, & Falk, 2003). Thus, excluding the deletion condition, both objects were fixated on the majority of trials.

The number of fixations that intervened between the first fixation on the first critical object fixated and the first fixation on the corresponding object in the other scene was determined for all trials where both critical objects were fixated. If the use of memory is minimized in this task, participants should most frequently direct the eyes from one object immediately to the corresponding object in the other scene. If memory use is maximized, participants should frequently fixate two objects in between glances to each of the two critical objects. The mean number of fixations per object can be used to determine the number of intervening fixations that would be expected in each of these cases. Given an average of 1.2 fixations per object, if memory is minimized in this task, the most frequent number of intervening fixations should be somewhere between zero and one. However, a mode of between two and three intervening fixations would be expected if the use of memory were maximized. The most frequent number of intervening fixations was either zero or one in all four conditions of this experiment. Additionally, in all conditions zero and one intervening fixations were more frequent than two and three intervening fixations. Figure 3 shows the distribution of intervening fixations collapsed across conditions and pooled across subjects. This analysis suggests that participants frequently directed their eyes from one item immediately to the other and provides rather strong evidence in favour of the minimal memory hypothesis.

To convey more intuitively what is going on in this task, a scan pattern used by one of the participants on one of the trials is included in Figure 2. The hypothesized one-to-one viewing strategy is evident in the frequent shifts made between scenes. Support for the minimal memory hypothesis was provided by two analyses designed to make inferences about the use of memory in this task: the number of fixations per scene and the number of intervening fixations. The numbers of fixations per scene were somewhat greater than predicted by minimal memory; however, the observed values were considerably smaller than predicted had participants been making full use of memory. The intervening fixations analysis suggests that participants frequently held a single item in memory.

Taken together, the analyses suggest a preference to minimize the use of memory in a task that is primarily visual. However, because different levels of encoding were needed to detect the differences between scenes in this experi-

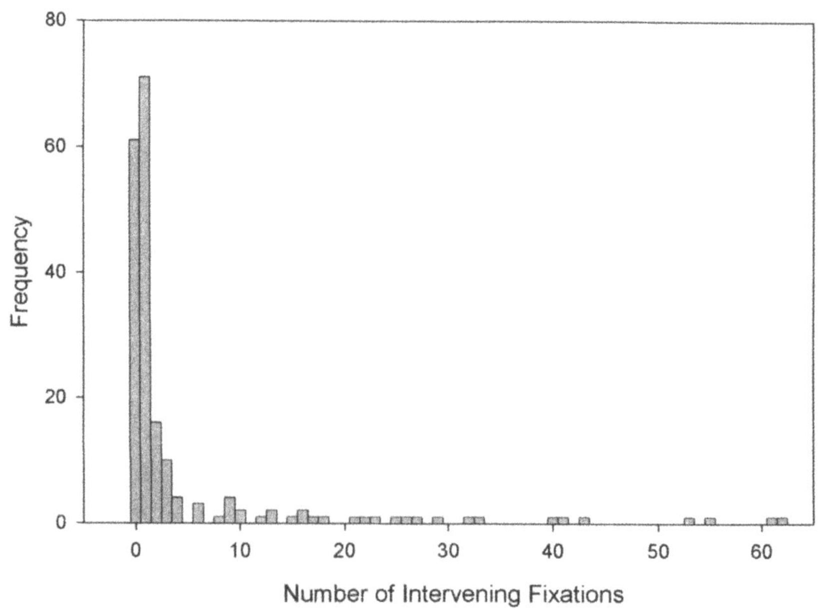

Figure 3. Histogram of the number of intervening fixations in Experiment 1.

ment, the observed viewing strategy might depend on the nature of the difference between scenes. Deletions and different-types required only the encoding of the presence and the identity of the objects. Different-tokens could only be detected on the basis of the visual details of the objects. Although we assume the items can be encoded as integrated objects, one could argue that the need to compare tokens produced a bias to encode the objects in terms of their features. If this were the case, having to detect this kind of difference could have elevated the demand on capacity and the one-to-one viewing strategy would not have reflected minimal use of memory. Given that the nature of the differences between scenes varied randomly from trial to trial, the overall viewing pattern we observed might therefore have been driven by a bias in the encoding strategy. To address this issue, in Experiment 2 we ran only the different-type and different-token conditions, and we ran them in separate blocks.

EXPERIMENT 2

The purpose of Experiment 2 was to examine the possibility that the pattern of viewing depends on the type of encoding needed to make the discrimination. To test this hypothesis, the nature of the difference between scenes was administered in blocks. By isolating the kind of difference that needed to be detected, it was assumed that the encoding strategy would be stable within each block and the generality of the minimal memory preference could be put to a more stringent test. In this experiment, only two kinds of differences between scenes were used: Different-types and different-tokens. In all other respects, Experiment 2 was identical to Experiment 1.

Method

Participants. Sixteen Michigan State University undergraduate students participated in the experiment in exchange for course credit or were paid. All participants had normal vision, were naive with respect to the hypotheses under investigation, and had not participated in Experiment 1.

Stimuli and apparatus. The stimuli and apparatus were the same as in Experiment 1, except that the deletion condition was eliminated.

Procedure. The procedure was the same as in Experiment 1, except the session was divided into two blocks. At the beginning of each block, participants were explicitly told the nature of the difference that could occur between scenes and examples were provided. As a result of the blocking procedure, the experiment was treated as a 2 × 2 factorial design with two levels of the kind of difference between scenes (different-type and different-token) and the two levels of same versus different. Scenes were assigned to conditions via a Latin-square design so that each scene appeared in each condition an equal number of times across participants. The order of blocks was counterbalanced and the order of scene presentation was determined randomly for each participant within each session. In addition, the initial point of fixation was moved from the centre to the top of the screen. Because the initial fixation was not under viewer control, this change reduced the possibility that fixations would be inappropriately assigned to one of the critical regions.

Results and discussion

In this experiment, the kind of difference that was possible between scenes varied between blocks. Although the order of blocks was counterbalanced across participants, it is possible that the strategy used by participants in the second block was systematically influenced by the strategy adopted in the first block. To insure that this was not the case, all analyses included tests for order effects. Two within-subjects factors of the kind of difference and same versus different were entered into ANOVAs along with order as a between-subjects factor. Because there were no order interactions, these tests will not be discussed below.

Performance measures of accuracy and reaction time were evaluated prior to the analyses of primary interest. Participants were again more accurate on the same conditions relative to the different conditions, $F(1, 14) = 6.71$, $MSE = 0.01$, $p < .05$; however, there was no effect of the kind of difference between scenes and no interaction, $F\text{s} < 1$. Accuracy was quite high in this experiment. The mean proportions correct were .97 and .90 in the same and different conditions, respectively.

Reaction times were included for correct responses only. Participants took longer in the same conditions relative to the different conditions, $F(1, 14) = 76.51$, $MSE = 7.92$, $p < .001$; however, there was no effect of the kind of difference between scenes and no interaction, $F\text{s} < 1$. The mean reaction times were 10.5 s and 4.4 s in the same and different conditions, respectively.

Data from all 16 subjects went into the scene- and object-level analyses; however, 38 trials were eliminated (approximately 7%) because of track loss. The mean number of scene glances was determined in order to generate predictive models. Scene glances were more frequent in the same relative to the different conditions, $F(1, 14) = 64.68$, $MSE = 11.13$, $p < .001$. However, there was no effect of the kind of difference between scenes and no interaction, $F\text{s} < 1$. The mean number of scene glances was 12.7 and 6.0 in the same and different conditions, respectively. Therefore, the expected number of objects fixated per scene was formulated for minimal and maximal memory based on 13-glance models for the same conditions and 6-glance models for the different conditions. The resulting values are listed in the top section of Table 2.

The number of fixations per object was determined by counting the number of fixations that occurred during the first glance within each of the object-sized regions as discussed in Experiment 1. There were fewer fixations per object in the same conditions relative to the different conditions, $F(1, 14) = 7.13$, $MSE =$

TABLE 2
Summary of the observed and predicted values in Experiment 2

	Same	*Different*
Expected objects per scene		
Minimum	1.8	1.7
Maximum	5.5	5.0
Expected fixations per scene		
Minimum	2.2	2.2
Maximum	6.6	6.5
Observed fixations (S.E.) per scene	2.81 (.10)	2.30 (.08)

0.027, $p < .05$. However, there was no effect of the kind of difference between scenes and no interaction, *F*s < 1. There was an average of 1.2 and 1.3 fixations per object in the same and different conditions, respectively. The expected numbers of objects fixated per scene derived from each of the models was multiplied by these values in order to evaluate the observed numbers of fixations per scene. Table 2 presents the values resulting from these computations.

The number of fixations per scene was computed by dividing the total number of fixations in each of the two scene regions by the total number of scene glances for every trial. The number of fixations per scene was greater in the same relative to the different condition, $F(1, 14) = 147.29$, $MSE = 0.029$, $p < .001$. However, there was no effect of type of difference between scenes and no interaction, *F*s < 1. The mean number of fixations per entry was 2.81 and 2.30 in the same and different conditions, respectively.

The analyses thus far demonstrate that the pattern of results was not affected by the kind of difference participants expected to find between the scenes. Because there were no effects of the kind of difference on any of the preceding measures, the observed values for both the same and different conditions were collapsed across the kind of difference and these were compared to the predicted values using one-sample *t*-tests. The observed values were greater than the predicted values for minimal memory in the same conditions, $t(15) = 5.88$, $p < .001$, but not in the different conditions, $t(15) = 1.23$, $p = .24$. The predicted values based on maximal memory were greater than the observed values in the same conditions, $t(15) = 36.34$, $p < .001$, as well as in the different conditions, $t(15) = 52.95$, $p < .001$. The average difference between the observed and predicted values was 0.1 fixations per scene for minimal memory versus 4.0 fixations per scene for maximal memory. Thus, the results of this analysis closely mirror the pattern found in Experiment 1.

The second analysis of interest is the number of fixations that intervened between the first fixation on the first critical object fixated and the first fixation on the corresponding object in the other scene given that both critical objects were fixated. In addition, the proportion of trials that both objects were fixated was determined. Both objects were fixated more frequently in the different conditions (mean = 0.89) relative to the same conditions (mean = 0.71), $F(1, 14) = 15.13$, $MSE = 0.03$, $p < .01$. This difference is not surprising given that correct detections require the inspection of both critical items. If participants were biased toward reporting that the scenes were the same, correct responses could have been made in the same conditions whether both critical items were inspected or not. There was, however, no effect of the kind of difference between scenes and no interaction, Fs < 1. In both cases both critical objects were fixated on the majority of trials. As in Experiment 1, the most frequent number of intervening fixations was either zero or one in all four conditions. Figures 4A and 4B shows the distribution of intervening fixations collapsed across the same and different conditions for the different-type and different-token conditions, respectively. In both cases the mode was zero intervening fixations and zero and one intervening fixations were better than twice as frequent as two and three intervening fixations.

In sum, the results of Experiment 2 are nearly identical to those of Experiment 1. The number of fixations per scene was slightly more than would be predicted by minimal memory in the same conditions but not in the different conditions. In addition, there were many fewer fixations per scene than predicted by maximal memory. The number of intervening fixations was even more aligned with the minimal memory view in Experiment 2. The mode of zero intervening fixations strongly suggests a preference for the one-to-one viewing strategy associated with minimized use of working memory. Importantly, however, there were no differences in the viewing patterns when participants were looking for different-types versus different-tokens. If the preferred viewing strategy observed in Experiment 1 was driven by the level of encoding needed to detect differences in the visual details of the objects, the viewing patterns should have diverged here. This was clearly not the case. The results of Experiment 2 suggest that the minimal memory preference is general—visual working memory is minimized whether it is the identity of the object or the visual details that need to be remembered. An alternative possibility, however, is that the visual details of the objects were encoded in both cases. That is, visual differences existed between the different-tokens as well as the different-types. If participants opted to use the same level of encoding in both blocks, the one-to-one viewing strategy might still be limited to the case when visual details are encoded. In order to test this possibility, a version of the task was designed that eliminated judgements based on the visual details of the objects.

EXPERIMENT 3

To ensure that the viewing strategy observed in Experiments 1 and 2 is not limited to cases when the visual details of the objects are encoded, an object array version of the task was introduced. Participants compared an array of pictures of objects to an array of the names of those objects. The arrays were the

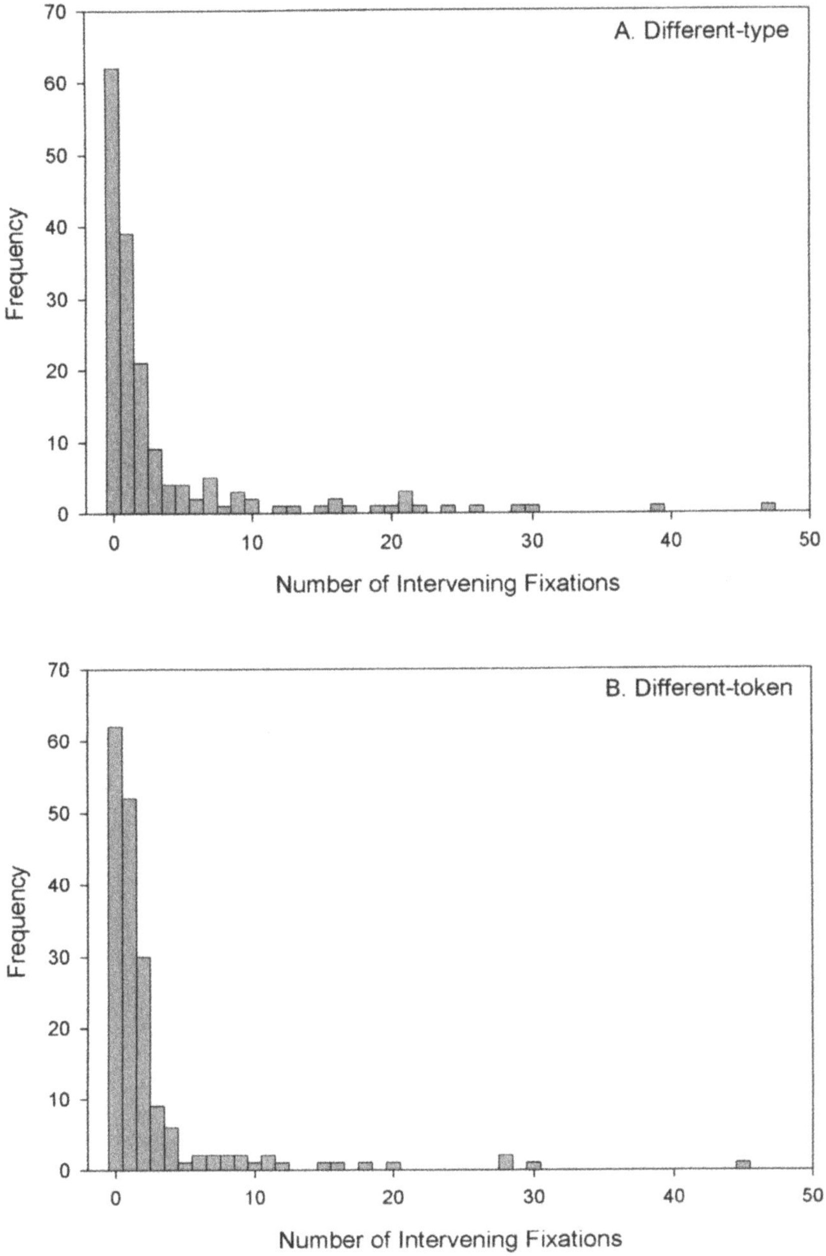

Figure 4. Histogram of the number of intervening fixations in the different-type (A) and different-token (B) conditions of Experiment 2.

same if all of the pictures and names matched, and different if one pair of pictures and names mismatched. Importantly, encoding the visual details of the objects did not assist in the execution of this task. The pictures needed to be encoded at the level of the object's identity in order to detect differences between arrays. If the minimal memory preference is limited to the case when visual details are encoded, the viewing pattern should diverge here. An additional benefit of using object arrays is that fixations can be more readily assigned to objects. As a result, the number of items fixated per array and the number of intervening items can be determined directly.

Method

Participants. Eight Michigan State University undergraduate students participated in the experiment for course credit. All participants had normal vision and were naive with respect to the hypotheses under investigation.

Stimuli. Sixteen images were constructed containing two arrays with a neutral area on top so that the initial fixation would be outside either of the arrays. Figure 5 shows a sample of an image used in this experiment. The two arrays subtended $7.7° \times 9.9°$ visual angle; the neutral area subtended $1.9° \times 15.7°$ visual angle. The three areas were demarcated by light grey borders against a dark grey background. The left-hand arrays were filled with pictures of objects and the right-hand arrays contained the words that named the objects. The object pictures were taken from the Hemera Photo-objects 50,000 Premium Image Collection and were sized so that the largest dimension, vertical or horizontal, subtended 1.9° visual angle. The font size was selected so that the largest words would fit within the same square regions. The objects and words were positioned so that there was a spatial correspondence between matching items in the two arrays. Pictured objects, such as the moose, the fan, and the bug in the sample image (Figure 5), were located in the same position in the picture array as the words that named them in the word array. A total of 184 objects were selected so that they could be named by one-syllable words. Each of the 16 images contained 11 object/word pairs. To create arrays for the different condition, one item in one of the arrays was randomly selected and replaced with one of the eight remaining items. There were no item repetitions.

Apparatus. The apparatus was the same as in Experiments 1 and 2.

Procedure. The procedure was the same as in Experiment 1, except that there were only two conditions in this experiment—same and different. Participants were provided with a visual example that illustrated the nature of

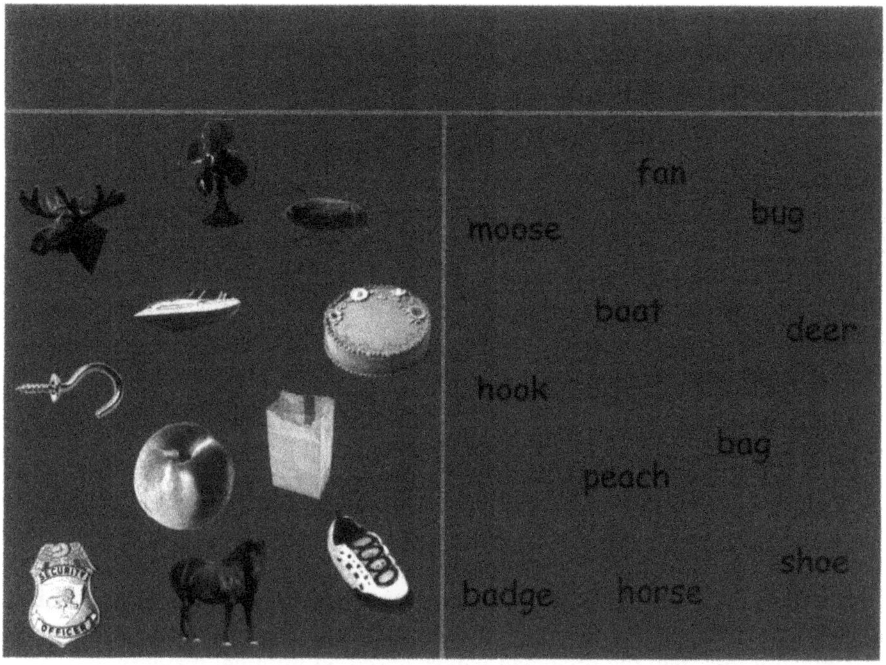

Figure 5. Sample image from the different condition of Experiment 3. Note that the location in the picture array that corresponds to the location of the word "deer" contains a picture of a cake. Full-colour displays were used in the experiment.

differences that could be found between arrays. Each session lasted approximately 20 min.

Results and discussion

Performance measures of accuracy and reaction time were evaluated prior to the analyses of primary interest. Accuracy was moderately high (mean = 0.85) and did not reliably differ between the same and the different conditions, $F(1, 7) = 0.50$, $MSE = 0.07$, $p = .50$. Reaction times were entered for correct responses only. Participants viewed the arrays for longer in the same condition (8.5 s) relative to the different condition (6.7 s), $F(1, 7) = 27.12$, $MSE = 0.71$, $p < .01$.

Data from all 16 subjects went into the scene- and item-level analyses; however, 12 trials were eliminated (approximately 9%) because of track loss. The mean number of array glances was determined in order to generate predictive models. Array glances were more frequent in the same relative to the different condition, $F(1, 7) = 18.49$, $MSE = 2.36$, $p < .01$. The mean number of array glances was 14.0 and 10.7 in the same and different conditions, respectively. Therefore, the expected number of items fixated per array was formulated for minimal and maximal memory based on 14- and 11-glance models, respectively. The resulting values are presented in Table 3.

In this experiment, due to the clean separation between items, we could directly determine the number of items fixated per array. This was accomplished by tabulating the number of item regions that receive fixations during each glance to each of the arrays. Because fixations that fall outside of these regions do not contribute to the analysis, it is important to determine the mean proportion of fixations that fall within the item-sized regions. This was accomplished by tallying the total number of fixations within the item regions and dividing by the total number of fixations falling within regions defined for each of the arrays for every trial. The mean proportion of fixations falling within the item regions was .89 and did not vary between the same and the different conditions, $F(1,7) = 0.45$, $MSE = 0.00$, $p = .52$. Thus, a large majority of the fixations were assigned to items in the arrays.

The mean number of items fixated per array glance was computed by dividing the total number of items fixated by the total number of array glances for every trial. The number of items fixated per array glance did not reliably differ between the same (1.76) and different (1.67) conditions, $F(1,7) = 1.04$, $MSE = 0.03$, $p = .34$. However, both means were tested because of the reliable differences found in the number of array glances. The observed values were numerically less than the predicted values based on minimal memory, but they did not reliably differ in the same condition, $t(7) = -1.77$, $p = .12$, nor in the different condition, $t(7) = -1.968$, $p = .09$. The predicted values based on maximal memory, however, were greater than the observed values in the same condition, $t(7) = 49.23$, $p < .001$, as well as in the different condition, $t(7) = 59.54$, $p < .001$. Thus, the results of this analysis are in agreement with the predictions based on the minimal memory hypothesis.

In this experiment we could also determine the actual number of intervening items fixated. Because there was no need to make assumptions about the number of fixations per item, predictions for this measure were more clear: minimal memory predicts a mode of zero intervening items, and maximal memory predicts a mode of two intervening items. Figure 6 shows the distribution of

TABLE 3

Summary of observed and predicted values in Experiment 3

	Same	*Different*
Expected items per array		
Minimum	1.9	1.8
Maximum	5.6	5.5
Observed items (S.E.) per array	1.76 (.08)	1.67 (.06)

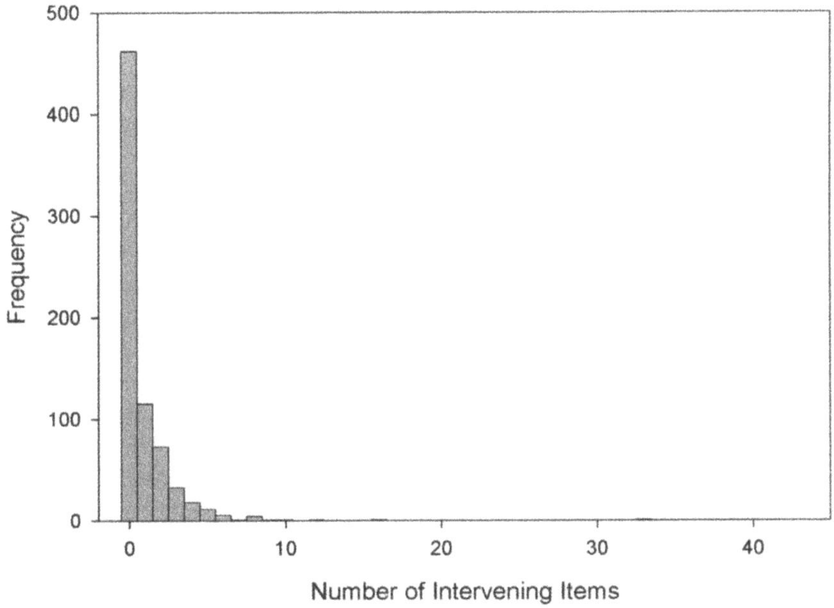

Figure 6. Histogram of the number of intervening items in Experiment 3.

intervening items collapsed across conditions. The most frequent number of intervening items was zero in both conditions.

The pattern of results found in Experiment 3 was consistent with those found in Experiments 1 and 2. In fact, these analyses most unambiguously support the preference for a one-to-one viewing strategy in this task. The number of items fixated per array was fewer than predicted by maximal memory but did not differ reliably from the values predicted by minimal memory. In addition, the intervening items analysis suggests that participants most frequently hold a single item in memory. Importantly, the persistence of this viewing pattern despite the irrelevance of the visual details suggests a minimal memory preference that is general.

GENERAL DISCUSSION

The purpose of this study was to test the generality of the minimal memory preference found by Ballard et al. (1995; Hayhoe et al., 1998). Is the preference to minimize the use of visual working memory limited to the case when visual information must be coordinated with basic motoric interactions with the environment? In order to test the minimal memory hypothesis, a scene comparison task was introduced. Participants compared side-by-side simultaneously presented pairs of scenes that were either identical or differed in terms of one object. Eye movements were examined to make inferences about the use of memory in this task. If the minimal memory strategy extends to this primarily visual task, participants would be expected to move the eyes frequently back and forth between scenes holding just one object in memory. There were two dependent measures of primary interest: The number of fixations per scene and the number of fixations between glances to corresponding objects. The observed number of fixations per scene was compared to the number of fixations per scene predicted by models generated with one and with three items occupying working memory. The observed values were reliably greater than predicted by minimal memory in the same and different conditions of Experiment 1 and in the same but not the different conditions of Experiment 2; however, these differences were always very small relative to the differences between the observed values and those predicted by maximal memory. In addition, the number of items fixated per array glance in Experiment 3 did not differ from the predictions based on minimal memory. Strong evidence in favour of the minimal memory view was obtained from the intervening fixation analyses of Experiments 1 and 2 and the intervening item analysis of Experiment 3. These data suggest that corresponding objects are most frequently fixated one after the other.

The pattern of results found in Experiment 1 suggested that the preference to minimize the use of visual working memory generalizes to tasks that are primarily visual. However, the presence of multiple kinds of differences between scenes left open the possibility that the observed viewing pattern depended on the level of encoding needed to detect the differences between scenes. Only the different-token condition required the encoding of the visual details of the objects. Although we assume the objects can be encoded as integrated units, having to detect this kind of difference might have biased participants towards a feature-based encoding strategy. If this were the case, capacity could no longer be expressed in terms of objects and the one-to-one viewing strategy would no longer reflect a minimal memory preference. To explore this possibility, different-types and different-tokens were run in separate blocks in Experiment 2. By isolating the kind of differences that were possible between scenes, it was assumed that a single encoding strategy would be adopted within each block. If the one-to-one viewing strategy observed in Experiment 1 was driven by the demands of the different-token condition, this pattern should only have been observed in the different-token condition of Experiment 2. The failure to find a difference between blocks in this experiment suggests a preference to minimize the use of memory even when only the identity of the objects need be encoded. However, the possibility remained that participants opted to encode the visual details in the different-type condition despite the fact that the identity would have been sufficient. Experiment 3 tested this possibility by introducing an object array version of the task. Participants compared arrays of object pictures to arrays of words that named the objects. In this task, items needed to be processed to the level of the object's identity to make the comparisons, but participants would have no need to encode the visual details of the objects. The persistent observation of the one-to-one viewing strategy in Experiments 2 and 3 suggests that the minimal memory preference generalizes to the case when only the identity of the object need be encoded.

Because one of the primary goals of this study was to gain a better understanding of the use of visual working memory in scene perception, we chose to begin by using naturalistic scenes as stimuli. Although hypotheses need to be tested using stimuli that approach the complexity of normal viewing situations, using scenes as stimuli introduces challenges to experimentation. In order to overcome these challenges, inferences were made on the basis of one pair of objects in each of the pairs of scenes. The number of fixations on critical objects was used as an index of the number of fixations that generally contribute to a gaze on an object, and this number was used to express the predicted number of objects fixated per scene in terms of the number of fixations per scene. The consistency of the results found in the scene and array versions of the task confirm the validity of this approach. Although the purpose of Experiment 3 was to test for differences that might have resulted from the level of encoding, an additional benefit of using object arrays is that fixations on all the objects could be tabulated. As a result, the actual number of items fixated per array glance and the actual number of intervening items could be determined. Interestingly, the observed number of items fixated per scene and intervening items fit even more closely to the predictions based on minimal memory. This suggests that the reduced level of support coming from the fixations per scene analyses of Experiments 1 and 2 might be the result of saccade targeting errors. However, even given a preference to minimize the use of memory, some fixations would be expected to reflect processes other than the explicit coding of one object to be compared to the other. These fixations could be associated with other aspects of processing such as searching for items to check as well as keeping track of which items have already been inspected.

What does minimal memory in a scene comparison task tell us about memory in scene perception? An issue that has received considerable attention by those studying scene perception is the nature of the information that accumulates in memory as a result of visually exploring a scene. Based on capacity estimates for visual working memory of three–four objects (Irwin, 1992; Irwin & Andrews, 1996; Luck & Vogel, 1997; Vogel et al., 2001), theorists have proposed that the active representation includes the three–four most recently attended items (e.g., Hollingworth & Henderson, 2002; Irwin & Andrews, 1996; Irwin & Zelinsky, 2002). The present study suggests a functional capacity that is more limited. Consistent with the ''just in time'' processing strategy favoured by proponents of the active vision framework, the visual system appears to have a bias towards minimal online representation. The viewing behaviour observed in this study suggests that it is easier to sample the environment with eye movements than to keep representations active in working memory, but it says nothing about the storage or the retrievability of these representations with the passage of time. In this sense, eye movements are cheaper than memory.

REFERENCES

Aloimonos, J., Weiss, I., & Bandopadhay, A. (1987). Active vision. *International Journal of Computer Vision, 1*, 333–356.

Ballard, D. H., Hayhoe, M. M., & Pelz, J. B. (1995). Memory representations in natural tasks. *Journal of Cognitive Neuroscience, 7*, 66–80.

Ballard, D. H., Hayhoe, M. M., Pook, P. K., & Rao, R. P. N. (1997). Deictic codes for the embodiment of cognition. *Behavioral and Brain Sciences, 20*, 723–767.

Crane, H. D. (1994). The Purkinje image eyetracker, image stabilization, and related forms of stimulus manipulation. In D. H. Kelley (Ed.), *Visual science and engineering: Models and applications* (pp. 15–89). New York: Marcel Dekker.

Crane, H. D., & Steele, C. M. (1985). Generation-V dual-Purkinje-image eyetracker. *Applied Optics, 24*, 527–537.

Hayhoe, M. M., Bensinger, D. G., & Ballard, D. H. (1998). Task constraints in visual working memory. *Vision Research, 38*, 125–137.

Henderson, J. M., Weeks, P. A., Jr., & Hollingworth, A. (1999). The effects of semantic consistency on eye movements during complex scene viewing. *Journal of Experimental Psychology: Human Perception and Performance, 25*, 210–228.

Henderson, J. M., Williams, C. C., Castelhano, M. S., & Falk, R. J. (2003). Eye movements and picture processing during recognition. *Perception and Psychophysics, 65*, 725–734.

Hollingworth, A., & Henderson, J. M. (2002) Accurate visual memory for previously attended objects in natural scenes. *Journal of Experimental Psychology: Human Perception and Performance, 28*, 113–136.

Irwin, D. E. (1992). Memory for position and identity across eye movements. *Journal of Experimental Psychology: Learning, Memory, and Cognition, 18*, 307–317.

Irwin, D. E., & Andrews, R. (1996). Integration and accumulation of information across saccadic eye movements. In T. Inui & J. L. McClelland (Eds.), *Attention and performance XVI: Information integration in perception and communication* (pp. 125–155). Cambridge, MA: MIT Press.

Irwin, D. E., & Zelinsky, G. J. (2002). Eye movements and scene perception: Memory for things observed. *Perception and Psychophysics, 64*, 882–895.

Luck, L. J., & Vogel, E. K. (1997). The capacity of visual working memory for features and conjunctions. *Nature, 390*, 279–281.

Nelson, W. W., & Loftus, G. R. (1980). The functional visual field during picture viewing. *Journal of Experimental Psychology: Human Learning and Memory, 6*, 391–399.

Rensink, R. A. (2000). The dynamic representation of scenes. *Visual Cognition, 7*, 17–42.

Vogel, E. K., Woodman, G. F., & Luck, L. J. (2001). Storage of features, conjunctions, and objects in visual working memory. *Journal of Experimental Psychology: Human Perception and Performance, 27*, 92–114.

Wheeler, M. E., & Treisman, A. M. (2002). Binding in short-term visual memory. *Journal of Experimental Psychology: General, 131*, 48–64.

VISUAL COGNITION, 2005, 12 (6), 1003–1016

Memory for object position in natural scenes

Andrew Hollingworth
Department of Psychology, The University of Iowa, Iowa City, USA

Memory for the positions of objects in natural scenes was investigated. Participants viewed an image of a real-world scene (preview scene), followed by a target object in isolation (target probe), followed by a blank screen with a mouse cursor. Participants estimated the position of the target using the mouse. Three conditions were compared. In the *target present preview* condition, the target object was present in the scene preview. In the *target absent preview* condition, the target object not present in the scene preview. In the *no preview* condition, no preview scene was displayed. Localization accuracy in the target present preview condition was reliably higher than that in the target absent preview condition, which was reliably higher than localization accuracy in the no preview condition. These data demonstrate that participants can remember both the spatial context of a scene and the specific positions of local objects.

Successful behaviour within an environment depends on knowing the nature of the objects present and their locations. Vision plays the primary role in extracting this information. Vision allows us to perceive the shape of objects and to classify these objects into meaningful categories. Vision also allows us to accurately perceive the spatial location of objects, relative to ourselves and to other objects. If one can perceive an object's visual form, know what kind of object it is, and know where it is located, then one can typically interact with it intelligently. For example, reaching for a cup of coffee depends on perceiving the shape of an object and recognizing it as belonging to the category of cups that contain coffee; perceiving the perceptual details of the particular cup in view, its orientation, the shape of the handle, and so on, to support grasping; and perceiving the position of the cup in the visual field, to support accurate reaching.

The environments we typically inhabit contain many, often hundreds, of individual objects. The visual system cannot process the details of all these objects simultaneously due to resource limitations and due to the fact that high-

Please address all correspondence to: Andrew Hollingworth, The University of Iowa, Department of Psychology, 11 Seashore Hall E, Iowa City, IA, 52242-1407, USA.
Email: andrew-hollingworth@uiowa.edu

This research was support by NIMH grant R03 MH65456.

DOI:10.1080/13506280444000625

resolution sensory information is available for only a very small portion of the visual field (corresponding to the foveal region of the retina). These limitations lead to the sequential selection of objects, with the eyes shifted via saccadic eye movements to fixate individual objects and obtain high-resolution information (see Henderson & Hollingworth, 1998). As a result, visual scene perception is extended over time and space as the eyes are oriented from object to object. To construct a scene representation, memory is necessary to integrate information acquired from different eye fixations on local objects. Thus, our ability to represent which objects are where in an environment depends critically on visual memory.

The phenomenon of change blindness, in which participants can be surprisingly insensitive to scene changes when detection requires memory, has been taken by many authors to indicate that scene representations are visually impoverished, with little or no accumulation of information as the eyes and attention are oriented from object to object within a scene (Becker & Pashler, 2002; Irwin & Andrews, 1996; Irwin & Zelinsky, 2002; O'Regan, 1992; Rensink, 2000; Rensink, O'Regan, & Clark, 1997; Scholl, 2000; Simons, 1996; Simons & Levin, 1997; Wolfe, 1999). In contrast to this view, Hollingworth and Henderson (Hollingworth, 2003a, 2004, in press; Hollingworth & Henderson, 2002; Hollingworth, Williams, & Henderson, 2001) have demonstrated significant accumulation of visual information from many individual objects within a scene (see Castelhano & Henderson, this issue, for evidence that such accumulation can occur incidentally). These studies examined memory for the visual form of individual objects. The typical method involved initial presentation of a complex natural scene, containing many individual objects, for viewing. Memory for a single object was then tested in either a change detection or two-alternative forced choice test. The changed version or distractor was either a different object from the same basic-level category or the same object rotated 90° in depth. Participants could accurately perform these tasks even when as many as 10 objects intervened between target fixation and test (Hollingworth, 2004; Hollingworth & Henderson, 2002), and memory for object form was still well above chance across many intervening scenes (Hollingworth, in press; Hollingworth & Henderson, 2002) and across more than 400 intervening objects (Hollingworth, 2004).

These studies demonstrate that people can accurately remember and accumulate information about the visual form of many individual objects. But how accurately can participants remember the locations of individual objects in scenes? Simons (1996) showed that participants can detect changes to the global spatial configuration (or layout) of objects, but that study did not require memory for the binding of a particular object to a particular location. In fact, Simons argued that changes were detected based on the retention of spatial layout independently of information about local object properties (i.e., as a configuration of abstract locations). Irwin and Zelinsky (2002) displayed an array of seven objects, followed by a blank screen and a location probe. Participants were able to accurately identify which object had appeared at the probed location, at least for objects that had been fixated recently. This study provides evidence that visual memory can retain the binding of a particular object to a scene location. However, the Irwin and Zelinsky study used the same set of seven discrete locations on every trial and thus can only provide limited evidence regarding memory for the positions of objects in real-world scenes.

The present experiment sought to examine memory for the positions of individual objects in natural scenes. There were three conditions. In the *target present preview* condition, an image of a realistic, natural scene was presented for 20 s (the preview scene). This was followed by an image of a single target object in the centre of the screen (the target probe image). Participants did not know which object in the preview scene would be the target. Next, the target probe was removed, and a blank screen was displayed with a mouse cursor. Participants moved the cursor to indicate the position at which the target object had appeared in the preview scene. The *target absent preview* condition was identical to the target present preview condition, except that the target object was not present in the preview scene. Participants were instructed to place the mouse cursor at the position where the target object would likely have appeared if it had been present in the preview scene. In this condition, participants could use memory for the layout of contextual objects and surfaces—combined with information about target identity, size, and other cues—to estimate placement. Such knowledge of where an object is likely to appear in a scene has been demonstrated to facilitate object localization during search (Henderson, Weeks, & Hollingworth, 1999; Hollingworth, 2003b; Oliva, Torralba, Castelhano, & Henderson, 2003). The critical comparison in the present experiment was between position accuracy in the target present and target absent preview conditions. If position estimates were found to be more accurate in the former condition, this would provide strong evidence that participants remembered the specific position of the target object when it was present in the preview.

Finally, in the *no preview* condition, no preview scene was displayed. Participants were instructed to imagine a scene that could contain the target object and choose a position where that object might likely appear. Even with no preview scene, target information could guide placement: A small airplane target object would more plausibly be found in the upper portions of a scene than in the lower portions. A second important comparison was between performance in the no preview and target absent preview conditions. If position estimates in the target absent preview condition were more accurate than in the no preview condition, this would demonstrate that participants remembered information about the spatial context of the scene, such as the arrangement of surfaces and objects.

METHOD

Participants

Eighteen participants from the University of Iowa community, 18–30 years' old, completed the experiment. All participants reported 20/20 uncorrected or corrected vision.

Stimuli

Scene images were created from a set of 48 different 3-D models. Each model depicted a complex, real-world environment. A unique target object was chosen within each model, with the constraint that the target could not appear in the centre of the image. To produce the target present preview images, each model was rendered with the target object present in the scene. To produce the target absent preview images, the scene was rendered after the target object had been removed from the model. Sample target present and absent preview images are displayed in panels A and B of Figure 1. Target probe images (panel C of Figure 1) presented the target object in isolation in the centre of the screen. For each scene, the same target probe was used in all three preview conditions. The target probe images were created by applying a transparent texture to all objects except the target object. Though not visible in the rendered image, background objects still reflected light and cast shadows within the model, ensuring that the target object appearance was identical to that in the preview images. The target object was then moved to the centre of the image. The target probe background was set to a uniform olive green, chosen because none of the target objects contained this colour and thus would not blend into the background.

Scene images subtended $16.9^\circ \times 22.8^\circ$ visual angle at a viewing distance of 80 cm. Target objects subtended 3.33° on average along the longest dimension in the picture plane.

Apparatus

The stimuli were displayed at a resolution of 800×600 pixels $\times$ 24-bit colour on a 17 inch monitor at a refresh rate of 100 Hz. Position responses were collected using a mouse. The presentation of stimuli and collection of responses was controlled by E-Prime software running on a Pentium IV-based computer. Viewing distance was maintained at 80 cm by a forehead rest. The room was dimly illuminated by a low-intensity light source.

Procedure

Participants were tested individually. Each participant was given a written description of the experiment along with a set of instructions. Participants were informed that on some trials they would see an image of a scene for 20 s. On

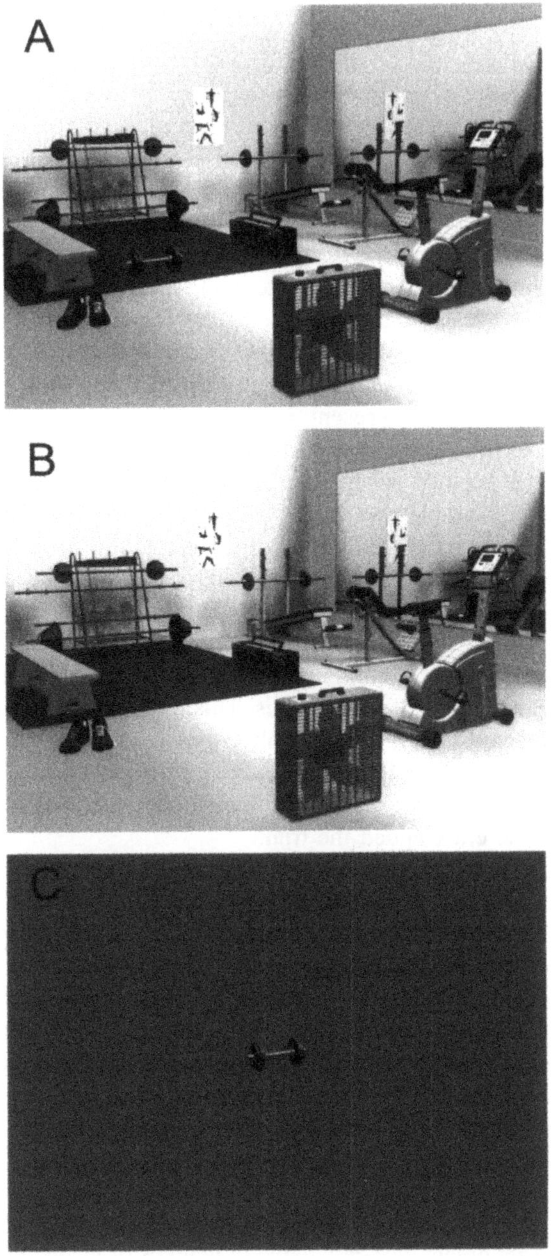

Figure 1. Principal stimuli for one of the 48 scene items. A shows the preview scene with the target object present (barbell). B shows the preview scene with the target object absent. C shows the target probe image.

other trials no scene would be displayed. Next, a single target object would be presented in the centre of the screen (the target probe image). On some trials, the target object had been present in the preview scene; on other trials it had not. A label appearing between the scene and the probe image would inform them whether the target object had been present in the scene, had been absent from the scene, or no preview had been displayed. Participants were instructed that after the target probe image, a mouse cursor would appear on a blank screen. For the target present condition, they were to move the mouse cursor to the position on the blank screen corresponding to the centre of the object as it appeared in the scene. For the target absent condition, they were instructed to place the mouse cursor at a position where such an object was likely to have appeared in the preview scene if it had been present. For the no preview condition, participants were instructed to imagine a scene that might contain that object and move the mouse cursor to a position where such an object might likely appear. For this final condition, an example was used in which a fire hydrant probe object might appear in a street scene and would likely be found in the lower half of the scene.

Participants pressed a mouse button to initiate each trial. Then, a blank, olive-green screen was displayed for 1000 ms. This was followed by the preview scene for 20 s, except in the case of the no preview condition. Next, an olive-green screen was presented with either "present", "absent", or "no preview" displayed in large, white text for 1200 ms. This was followed by the target probe image for 2200 ms. Finally, a blank, olive-green screen was presented with a white, plus-sign cursor initially in the centre. Participants moved the cursor with the mouse to their chosen position on the screen and clicked to register the response. A mouse click ended the trial.

Participants first completed a practice session of six trials, two in each of the three preview conditions. Practice scenes were not used in the experimental session. The practice trials were followed by 48 experimental trials, 16 in each of the three preview conditions. Each scene item was displayed once; there was no scene repetition. Trials from the three preview conditions were randomly intermixed. Across participants, each of the 48 scene items appeared in each condition an equal number of times. The entire experiment lasted approximately 40 min.

RESULTS AND DISCUSSION

On each trial, the Euclidean distance between the response position and the centre of the target object was calculated in degrees of visual angle, the *position error*. For each scene item, the same target position was used in all three preview conditions. Mean position error in each of the preview conditions is displayed in the top panel of Figure 2. Mean position error is also illustrated in the bottom panel of Figure 2, superimposed over the sample scene. There was a reliable main effect of preview condition, $F(2, 34) = 179.52$, $p < .001$. Mean

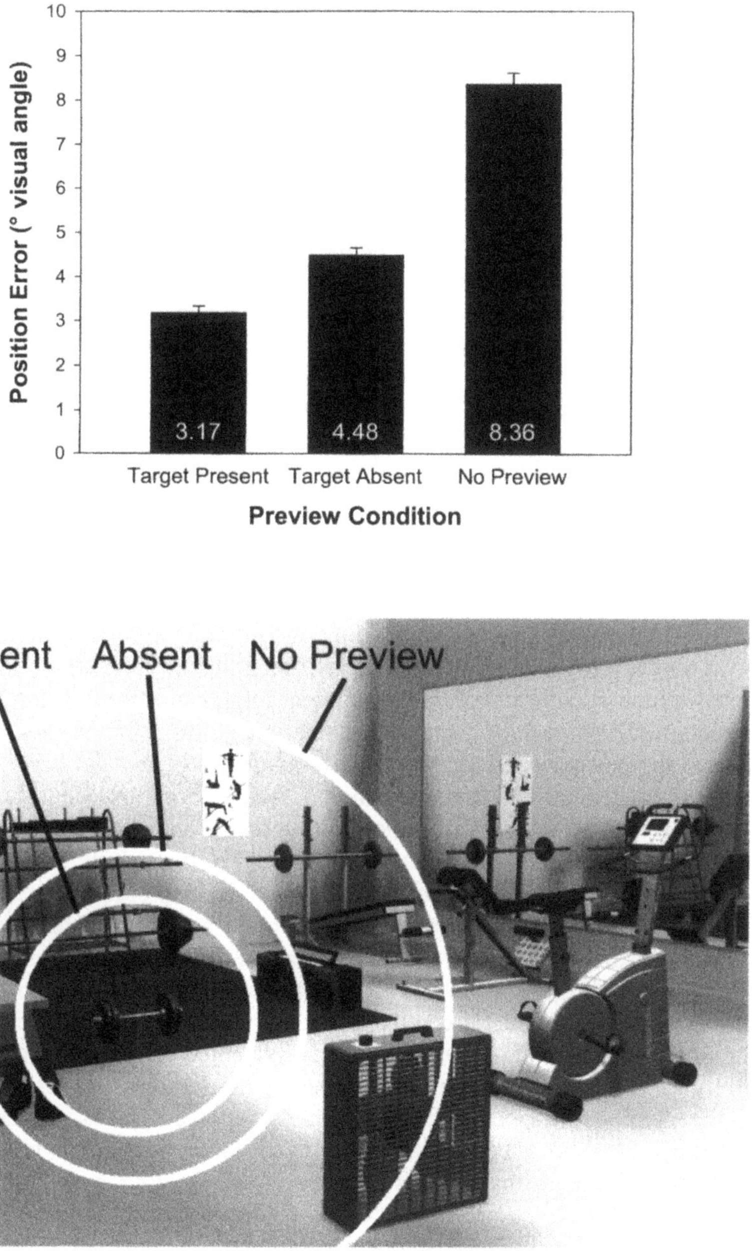

Figure 2. Top panel: Mean position error in degrees of visual angle for the three preview conditions. Error bars are standard errors of the means. Bottom panel: Mean position error (white circles) superimposed over a sample scene and target object (barbell) to illustrate position error relative to the size of the scenes. Note that the means are collapsed across all 48 scene items.

position error was reliably smaller in the target present preview condition (3.17°) than in the target absent preview condition (4.48°), $F(1, 17) = 31.57, p < .001$. And mean position error was reliably smaller in the target absent preview condition than in the no preview condition (8.36°), $F(1, 17) = 151.32, p < .001$. This pattern of results has been replicated using a modified set of scenes and a preview duration of 10 s (Hollingworth, 2003b).

These data demonstrate that scene representations maintain both information about the general spatial context of the scene and information about the specific locations of individual objects. The large difference in position error for the target absent preview and no preview conditions demonstrates that participants can remember the spatial structure of a scene. These two conditions were controlled for reasoning based on target object features (such as identity or size), since the same target objects were used in both conditions. The only difference was the presentation of the scene context prior to position estimation. The contextual preview clearly provided a great deal of information. Despite the absence of the target object from the preview scene, participants could quite accurately estimate the target object position. This implies they were able to remember the global spatial structure of the scene, such as the arrangement of surfaces and objects. In addition, it seems unlikely that accurate placement in the target absent preview condition could have been supported solely by memory for global spatial structure. Accurate placement implies that participants were able to combine knowledge of contextual features (e.g., that kitchen counters tend to support appliances) with target object information (e.g., that the target object is a toaster). In general, performance in the target absent preview condition is consistent with previous work demonstrating that people can generate fairly rich and accurate representations of natural scenes (Hollingworth, 2003a, 2004, in press; Hollingworth & Henderson, 2002; Hollingworth et al., 2001)

Although localization performance was quite accurate in the target absent preview condition, it was even more accurate in the target present preview condition. The target present and absent preview conditions were controlled for reasoning based on general contextual memory and target object features. The only difference was the presence of the target object in the preview. Thus, the advantage for the target present preview condition demonstrates that participants could remember the specific position of the target object (at least on a significant proportion of trials). Mean position error in the target present preview condition was 3.17° within an image that subtended 16.9° × 22.8°. Note that error was calculated from the centre of the target object, and target objects subtended approximately 3° on average, so position estimates in the target present preview condition were typically less than 2° from the nearest contour of the object. This relatively accurate localization was observed despite the fact that participants did not know which object in the preview scene would be the target. As is evident from the sample scene in Figure 1, there were multiple objects in each scene. Accurate position estimates therefore imply that participants remembered the positions of multiple objects within the scene. The precise number of objects is difficult to estimate from these data. One interesting question for future research is whether memory for object position in natural scenes exceeds the three or four object capacity of VSTM (Hollingworth, 2004; Irwin & Zelinksy, 2002).

Evidence of memory for target location and spatial context has implications for the recent debate over the role of memory in visual search (e.g., Shore & Klein, 2000; Wolfe, 1999). If participants can remember target location and the spatial context of a scene, it seems reasonable that such memory representations could guide visual search. Hollingworth (2003b) tested this possibility in a search task. Participants were shown a scene preview for 10 s. As in the present study, the target object was either present in the preview or absent from the preview. A no preview control condition provided baseline search data. After the scene preview, participants saw the target probe in isolation. This was followed by a search scene, which was the same basic scene as the preview. The target object was always present in the search scene, and it was either the same as the target probe or mirror reversed. Participants' task was to find the target in the search scene and determine whether it was in the same orientation as the target probe or not. Hollingworth found a similar pattern of results as in the present experiment. Search speed (measured both as overall response time and elapsed time until target fixation) was fastest in the target present preview condition, next fastest in the target absent preview condition, and slowest in the no preview condition. These data indicate that the types of spatial memory observed in the present study can be employed dynamically to guide visual search in natural scenes. In addition, preview benefits were still observed when the preview duration was reduced to 2 s and to 500 ms, demonstrating that spatial information sufficient to guide search can be acquired relatively rapidly from a scene.

In other studies of object position memory, memory for object position displays systematic biases. For example, for an object hidden in a rectangular sand box, older children show a systematic position bias toward the centre of the left or right half of the rectangle (Huttenlocher, Newcombe, & Sandberg, 1994). Are systematic biases present in memory for the positions of real-world objects in natural scenes? One difficulty in using the present data to answer this question is that on some trials, participants may simply have failed to remember the position of the target object, leading them to guess a position. Guessing itself could lead to what appears to be systematic bias. If an object is on the right-hand side of the screen, the mean position estimate based on guessing should be near the centre of the scene, which could appear to be a bias in position estimate toward the centre of the image.

To avoid this problem, only relatively accurate position estimates were used to examine possible position biases in the present data. Position estimates in the target present preview condition were examined, with the constraint that the estimate must be less than 2° from the centre of the target. Figure 3 illustrates

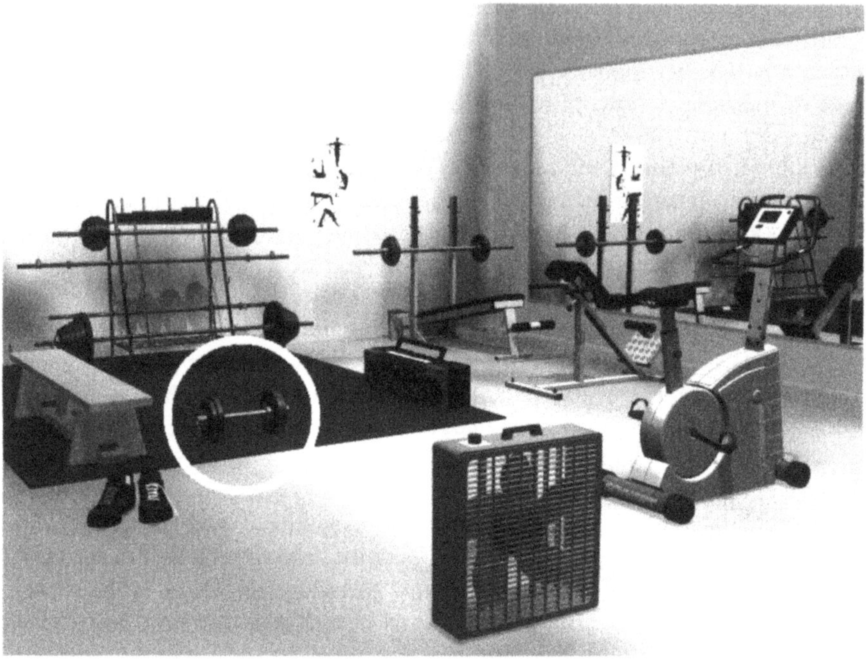

Figure 3. The size of the response region used to examine position biases.

the size of this 2° constraint using the sample scene. Of the total 288 position estimates in the target present preview condition, 110 met this criterion. These position estimates were divided based on the position of the target object in the scene, left- versus right-hand side of the screen and upper versus lower half of the screen. Each position estimate is plotted in Figures 4 and 5 as a function of the horizontal and vertical distance from the centre of the target object, which is plotted at the intersection of the dotted lines. Note that each position estimate appears twice, once for the horizontal division and once for the vertical division.

For target objects appearing on the left-hand side of the screen, there was no clear horizontal bias in position estimate. Thirty-one of sixty position estimates (51.7%) were to the right of target centre, which was not reliably different from chance of 50%, $\chi^2 = 0.02$, $p = .90$. For the target objects appearing on the right-hand side of the screen, there was a clear bias to the left. Thirty-seven of the forty-nine position estimates (75.5%) were to the left of target centre, which was reliably higher than chance, $\chi^2 = 12.76$, $p < .001$. (One position estimate on the left/right dimension was at the precise horizontal centre of the target and was dropped from the analysis.) For target objects appearing in the upper half of the screen, there was a clear position bias lower. Nineteen of the twenty-three position estimates (82.6%) were below target centre, which was reliably greater

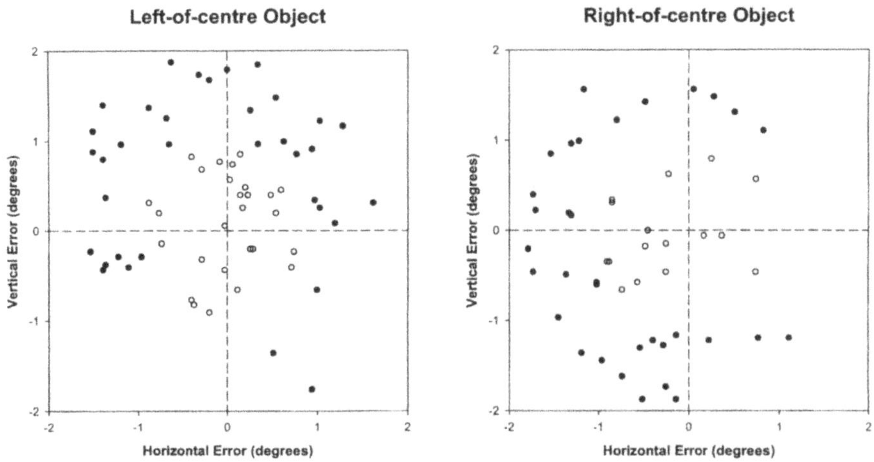

Figure 4. Individual position estimates in the target present preview condition for objects appearing on the left and right sides of the screen. Each estimate is plotted relative to the centre of the target object on that trial (the intersection of the dotted lines). Only estimates within 2° of target centre are displayed. Open circles depict estimates within 1° of target centre.

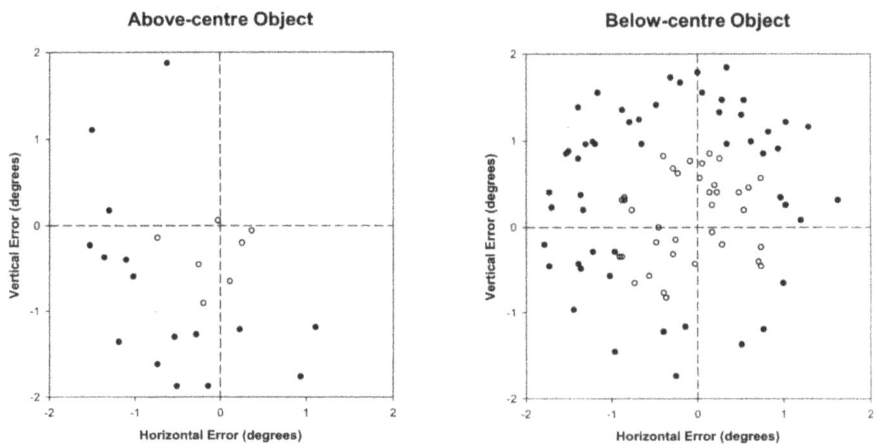

Figure 5. Individual position estimates in the target present preview condition for objects appearing on the top and bottom halves of the screen. Each estimate is plotted relative to the centre of the target object on that trial (the intersection of the dotted lines). Only estimates within 2° of target centre are displayed. Open circles depict estimates within 1° of target centre.

than chance, $\chi^2 = 9.78$, $p < .005$. Finally, for target objects appearing in the lower half of the screen, there was a clear position bias higher. Fifty-seven of the eighty-seven position estimates (65.5%) were above target centre, which was reliably greater than chance, $\chi^2 = 8.38$, $p < .005$. With the exception of target objects on the left-hand side of the screen, estimates were consistently biased toward the centre of the image.

The same trends were evident even when considering only estimates within 1° of target centre (open circles in Figures 4 and 5). For target objects on the left-hand side of the screen, 15 of 27 estimates were to the right of target centre (55.6%). For target object on the right-hand side of the screen, 11 of 16 estimates were to the left of target centre (68.8%). For target objects in the upper half of the screen, 6 of 7 estimates were below target centre (85.7%). For target objects in the lower half of the screen, 20 of 35 estimates were above target centre (57.1%).

This position bias toward the centre of the scene is consistent with studies examining spatial biases in memory for groups of objects in two-dimensional configurations (Hund & Plumert, 2002, 2003; Plumert & Hund, 2001). Plumert and Hund (2001) presented children and adults with a box divided into four quadrants. A set of five objects was placed in each quadrant in a random configuration. The objects were removed, and participants then attempted to place the objects back in the original positions. Plumert and Hund found that position estimates were consistently biased toward the centre of each quadrant. The present study shows that such spatial biases toward the centre of a geometric region generalize to adult position memory for objects in natural scenes. In addition, these biases were observed despite the fact that natural scenes contain significant spatial contextual information to constrain object position memory.

Finally, position memory biases toward scene centre may provide an explanation for the scene memory phenomenon of boundary extension. Intraub and colleagues (see Intraub, 1997 for a review) have found that when participants are asked to draw a complex picture from memory, they tend to expand the scope of the picture, suggesting that scene memory is expanded to include information beyond the boundaries of the original image. If, as suggested by the present results, participants remember the locations of objects as closer to scene centre than they actually were, one would expect precisely the boundary extension effect observed by Intraub: When drawing a scene from memory, participants should compress objects toward the centre of the drawing, resulting in a depiction that extends beyond the original scene boundaries.

CONCLUSION

After viewing a preview of a natural scene, participants were able to fairly accurately estimate the position at which a target object would have appeared in the scene if it had been present (target absent preview condition). This result demonstrates memory for the general spatial and contextual features of a scene. In addition, position estimates were even more accurate when the target object was actually present in the preview scene (target present preview condition), demonstrating memory for the positions of individual objects. These findings inform our basic understanding of the nature of scene representations. Contrary to claims that scene representations are visually impoverished (e.g., O'Regan, 1992; Rensink et al., 1997), the present data are consistent with the claim that visual memory supports scene representations containing information about the visual form and locations of multiple individual objects (Hollingworth & Henderson, 2002). In addition, these data suggest that memory for the positions of objects in a scene could support important visual tasks, such as visual search.

REFERENCES

Becker, M. W. & Pashler, H. (2002). Volatile visual representations: Failing to detect changes in recently processed information. *Psychonomic Bulletin and Review*, *9*, 744–750.

Castelhano, M. S., & Henderson, J. M. (this issue). Incidental visual memory for objects in scenes. *Visual Cognition*, *12*, 1017–1040.

Henderson, J. M., & Hollingworth, A. (1998). Eye movements during scene viewing: An overview. In G. Underwood (Ed.), *Eye guidance in reading and scene perception* (pp. 269–283). Oxford, UK: Elsevier.

Henderson, J. M., Weeks, P. A., Jr., & Hollingworth, A. (1999). The effects of semantic consistency on eye movements during complex scene viewing. *Journal of Experimental Psychology: Human Perception and Performance*, *25*, 210–228.

Hollingworth, A. (2003a). Failures of retrieval and comparison constrain change detection in natural scenes. *Journal of Experimental Psychology: Human Perception and Performance*, *29*, 388–403.

Hollingworth, A. (2003b). *Visual memory and the online representation of complex scenes*. Paper presented at the Munich Visual Search symposium, Munich, Germany.

Hollingworth, A. (2004). Constructing visual representations of natural scenes: The roles of short- and long-term visual memory. *Journal of Experimental Psychology: Human Perception and Performance*, *30*, 519–537.

Hollingworth, A. (in press). The relationship between online visual representation of a scene and long-term scene memory. *Journal of Experimental Psychology: Learning, Memory, and Cognition.*

Hollingworth, A., & Henderson, J. M. (2002). Accurate visual memory for previously attended objects in natural scenes. *Journal of Experimental Psychology: Human Perception and Performance*, *28*, 113–136.

Hollingworth, A., Williams, C. C., & Henderson, J. M. (2001). To see and remember: Visually specific information is retained in memory from previously attended objects in natural scenes. *Psychonomic Bulletin and Review*, *8*, 761–768.

Hund, A. M., & Plumert, J. M. (2002). Delay-induced bias in children's memory for location. *Child Development*, *73*, 829–840.

Hund, A. M., & Plumert, J. M. (2003). Does information about what things are influence children's memory for where things are? *Developmental Psychology*, *39*, 939–948.

Huttenlocher, J., Newcombe, N., & Sandberg, E. H. (1994). The coding of spatial location in young children. *Cognitive Psychology*, *27*, 115–147.

Intraub, H. (1997). The representation of visual scenes. *Trends in Cognitive Sciences*, *1*, 217–222.
Irwin, D. E., & Andrews, R. (1996). Integration and accumulation of information across saccadic eye movements. In T. Inui & J. L. McClelland (Eds.), *Attention and performance XVI: Information integration in perception and communication* (pp. 125–155). Cambridge, MA: MIT Press.
Irwin, D. E., & Zelinsky, G. J. (2002). Eye movements and scene perception: Memory for things observed. *Perception and Psychophysics*, *64*, 882–895.
Oliva, A., Torralba, A., Castelhano, M. S., & Henderson, J. M. (2003). Top down control of visual attention in object detection. In *IEEE proceedings of the international conference on Image Processing* (Vol. I, pp. 253–256).
O'Regan, J. K. (1992). Solving the ''real'' mysteries of visual perception: The world as an outside memory. *Canadian Journal of Psychology*, *46*, 461–488.
Plumert, J. M., & Hund, A. M. (2001). The development of memory for location: What role to spatial prototypes play? *Child Development*, *72*, 370–384.
Rensink, R. A. (2000). The dynamic representation of scenes. *Visual Cognition*, *7*, 17–42.
Rensink, R. A., O'Regan, J. K., & Clark, J. J. (1997). To see or not to see: The need for attention to perceive changes in scenes. *Psychological Science*, *8*, 368–373.
Scholl, B. J. (2000). Attenuated change blindness for exogenously attended items in a flicker paradigm. *Visual Cognition*, *7*, 377–396.
Shore, D. I., & Klein, R. M. (2000). On the manifestations of memory in visual search. *Spatial Vision*, *14*, 59–75.
Simons, D. J. (1996). In sight, out of mind: When object representations fail. *Psychological Science*, *7*, 301–305.
Simons, D. J., & Levin, D. T. (1997). Change blindness. *Trends in Cognitive Sciences*, *1*, 261–267.
Wolfe, J. M. (1999). Inattentional amnesia. In V. Coltheart (Ed.), *Fleeting memories* (pp. 71–94). Cambridge, MA: MIT Press.

VISUAL COGNITION, 2005, 12 (6), 1017–1040

Incidental visual memory for objects in scenes

Monica S. Castelhano and John M. Henderson
Department of Psychology and Cognitive Science Program, Michigan State University, East Lansing, MI, USA

Three experiments were conducted to investigate the existence of incidentally acquired, long-term, detailed visual memory for objects embedded in previously viewed scenes. Participants performed intentional memorization and incidental visual search learning tasks while viewing photographs of real-world scenes. A visual memory test for previously viewed objects from these scenes then followed. Participants were not aware that they would be tested on the scenes following incidental learning in the visual search task. In two types of memory tests for visually specific object information (token discrimination and mirror-image discrimination), performance following both the memorization and visual search conditions was reliably above chance. These results indicate that recent demonstrations of good visual memory during scene viewing are not due to intentional scene memorization. Instead, long-term visual representations are incidentally generated as a natural product of scene perception.

What is the nature of the representation that is created during ongoing, natural scene perception? Intuitively, it seems that the visual system generates a complete and highly detailed model of the external world. The perceptual experience of a stable and detailed visual world has led many vision researchers in the past to conclude that the visual representation formed for a scene is veridical and complete (McConkie & Rayner, 1976; Neisser, 1967). Such a detailed visual

---

Please address all correspondence to: Monica S. Castelhano, Department of Psychology, 129 Psychology Research Building, Michigan State University, East Lansing, MI 48824, USA. Email: monica@eyelab.msu.edu

This research was submitted in partial fulfilment of the requirements for the Masters of Arts degree by Monica Castelhano, who was supported by the MSU graduate school on behalf of the NSF IGERT program in Cognitive Science. The authors wish to thank Jeremy Athy, for his help with data collection, and the members of the masters committee, Erik Altmann, Thomas Carr, and Fred Dyer, for their guidance and comments. This work was also supported by the National Science Foundation (BCS-0094433 and ECS-9873531) and the Army Research Office (DAAD19-00-1-0519; the opinions expressed in this paper are those of the authors and do not necessarily represent the views of the Department of the Army or any other governmental organization. Reference to or citations of trade or corporate names do not constitute explicit or implied endorsement of those entities or their products by the author or the Department of the Army).

DOI:10.1080/13506280444000634

representation could serve as the basis for computations involving motor interaction with the environment, could underlie perceptual learning and priming, and might also serve as the basis for our visual phenomenology of a panoramic, full-colour, and visually detailed experience of the world. Such a representation, if it exists, would need to be created across multiple eye fixations so that the fine detail and colour information provided by the fovea during each fixation could be included in the composite global scene representation. Many of these theories were based on research in scene memory, which showed that observers could recognize a vast number of previously viewed photographs with high accuracy (Nickerson, 1965; Shepard, 1967; Standing, 1973; Standing, Conezio, & Haber, 1970).

Despite its intuitive appeal, the *composite scene representation* hypothesis has been challenged over the years by a number of investigators coming from a variety of empirical and theoretical perspectives (e.g., Ballard, 1992; Brooks, 1987; Churchland, Ramachandran, & Sejnowski, 1994; O'Regan, 1992; Wolfe, 1999). Studies of transsaccadic memory have shown that visual information is not fused across saccades (Bridgeman, Hendry, & Stark, 1975; Irwin, 1991; Irwin, Yantis, & Jonides, 1983; McConkie & Zola, 1979; Rayner & Pollatsek, 1983). The failure to fuse images across saccades has been replicated with simple patterns (Irwin et al., 1983), visual features of words (O'Regan & Levy-Schoen, 1983), contours of objects (Henderson, 1997), and features of scenes (Henderson & Hollingworth, 2003a). Without the ability to fuse visual information from one fixation to the next, it is improbable that the visual system creates and uses a composite point-by-point visual representation across fixations.

Studies on change blindness have also contested the existence of a complete, veridical visual representation of the outside world. Research on change detection has found that large discrepancies in a scene can go unnoticed, if changes are made during a saccade (Bridgeman et al., 1975; Grimes, 1996; Henderson & Hollingworth, 1999; Hollingworth & Henderson, 2002) or some other disruption, like a mudsplash (O'Regan, Rensink, & Clark, 1999), insertion of a blank screen (Rensink, O'Regan, & Clark, 1997), movie cut (Levin & Simons, 1997), or a temporary physical occlusion (Simons & Levin, 1998). In other words, participants show change blindness when the change is not accompanied by the usual transient motion cues. In many of these studies, participants showed change blindness even when they were actively looking for a change (e.g., Grimes, 1996; Henderson & Hollingworth, 1999; Hollingworth & Henderson, 2002; Rensink et al., 1997). Change blindness has thus been taken as evidence for the general absence of a composite scene representation (O'Regan, 1992; O'Regan & Noë, 2001; Rensink, 2000a, 2000b, 2002), and instead for a *localist-minimalist* approach to scene representation.

As a class, localist-minimalist theories assume that the visual representation is temporally and spatially limited to the object that is currently the focus of attention (Rensink, 2000a, 2000b, 2002; Wolfe, 1999), with additional representation in visual short-term memory (VSTM) of up to three or four recently attended objects (Irwin, 1992a, 1992b; Irwin & Andrews, 1996). In this view, once an object is no longer attended or preserved in visual short-term memory, visual information about that object is no longer retained; instead, information about the object is stored as an abstract, semantically-based, nonvisual representation in long-term memory (LTM).

Although localist-minimalist theories of scene representation provided an initial account of the basic change blindness phenomenon, it makes several predictions about the preservation of visual detail that have not been supported by direct empirical testing (see Henderson & Hollingworth, 2003c). First, this class of theory predicts that changes to objects in scenes should always be very difficult to detect across saccades. Contrary to this prediction, good change detection is observed when care is taken to ensure that the changed scene region is fixated (and therefore attended) both before and after the change (Henderson & Hollingworth, 1999; Hollingworth & Henderson, 2002). Second, this view predicts that other behavioural measures of visual representation beyond explicit change detection should also show no evidence for the preservation of visual representations. Counter to this prediction, fixation duration, a covert behavioural measure of change detection, is elevated following unreported scene changes, suggesting that the information needed to detect the change has been generated and preserved (Henderson & Hollingworth, 2003c; Hollingworth & Henderson, 2002; Hollingworth, Williams, & Henderson, 2001). Third, localist-minimalist theories predict the absence of memory for the visual attributes of an object that is no longer held in VSTM during scene perception. In an online memory test for objects that had been fixated (and therefore attended) in scenes several seconds and multiple fixations before the test, Hollingworth and Henderson (2002) showed that participants were able to discriminate between a previously fixated object and a visual distractor from the same basic-level category (e.g., two kinds of radios). Exceptional memory performance was also observed when participants had to discriminate between the original view of the critical object and a 90° rotated view of that same object. Hollingworth and Henderson also tested long-term visual memory with type and token discrimination memory tests administered following presentation of the entire set of scenes. Participants were able to discriminate a viewed token from a distractor token differing only in visual detail with high accuracy despite multiple intervening scenes between study and test.

Contrary to localist-minimalist theories, then, recent change detection and visual memory results suggest that visual information acquired from an object during scene perception is represented in a relatively stable form in memory once attention has been withdrawn from that object and deployed elsewhere (Henderson & Hollingworth, 2003a; Hollingworth & Henderson, 2002; see also Henderson & Hollingworth, 2003c; Hollingworth et al., 2001). To account for change blindness under some conditions and not others, as well as the overwhelming evidence that iconic representations are not fused across saccades (Irwin, 1992a, 1992b; Irwin & Andrews, 1996; Pollatsek & Rayner, 1992), Hollingworth and Henderson (2002) proposed a *visual memory theory* of scene representation (see also Henderson & Hollingworth, 2003c). Like the localist-minimalist theories described above, visual memory theory assumes that attention to (or fixation on) an object is necessary to create and encode into VSTM a local visual representation. Unlike localist-minimalist theories, however, this theory posits that these visual representations leave a lingering long-term visual trace in memory. In this view, change blindness is caused not by the lack of visual representation, but rather by a failure of one (or more) of three memory-related processes: Initial encoding of a prechange representation, encoding of a postchange representation, and retrieval of the prechange representation following the change. When care is taken to ensure that a viewer has fixated and therefore has had an opportunity to encode a scene region prior to and after a change, as well as to retrieve the encoded prechange information following the change, then change detection (i.e., absence of change blindness) is observed (Henderson & Hollingworth, 1999; Hollingworth & Henderson, 2002; see also Henderson & Hollingworth, 2003b; Hollingworth, 2003; Hollingworth et al., 2001).

Results from the scene memory literature support the notion that visual details of objects are encoded in memory (Bahrick & Boucher, 1968; Mandler & Parker, 1976; Mandler & Ritchey, 1977). Although relatively simpler scenes were used in these earlier studies (line drawings of between six to nine objects in each), participants were able to distinguish between the target object and similar distractors of the same basic-level category (Bahrick & Boucher, 1968) and were able to recall different types of visual details (Mandler & Ritchey, 1977). Several studies also demonstrated that participants were able to recognize object type (Friedman, 1979; Goodman, 1980; Hock, Romanski, Galie, & Williams, 1978), visual details (Friedman, 1979; Mandler & Parker, 1976; Pezdek, Maki, Valencia-Laver, Whetstone, Stoeckert, & Dougherty, 1988; Pezdek, Whetstone, Reynolds, Askari, & Dougherty, 1989), and verbally recall and recognize object descriptions (Brewer & Treyens, 1981). More recent studies have found that the context of a target object is not only encoded incidentally over the course of training, but that this information is retained in LTM and used to improve performance in subsequent visual searches (Chun & Jiang, 1998, 2003).

The studies demonstrating visual memory appear to provide compelling evidence against localist-minimalist theories and for detailed long-term visual scene representation. These data, however, have been largely ignored or dismissed in the scene perception literature. A lingering concern is whether these results are due to memory processes normally recruited in the service of scene perception, or instead are the result of aberrant scene processing brought about by the use of viewing instructions that stress scene memorization. In prior studies demonstrating visual scene memory, it is possible that participants employed strategies that called into play processes and mechanisms not typical of normal scene viewing (e.g., storing visual information as verbal descriptions). It could be that these long-term representations can be created when viewers engage in intentional memory encoding, but that these representations are not typically generated during natural scene perception tasks that do not require intentional memorization. For example, viewers are unlikely to engage in intentional scene memorization when locomoting through an environment, or searching through a scene. In this view, although natural dynamic scene perception does not involve the retention of detailed visual representations, atypical viewing strategies involving memorization can be engaged during scene perception when necessary. If this view is correct, then the evidence for good visual memory performance obtained in prior studies can be dismissed as artifactual and irrelevant to normal visual processing. Instead, the conclusion might be that localist-minimalist theories are correct and visual representations are in fact minimal, local, and transient during natural scene perception.

If the strategic memory encoding explanation for good visual memory is true, then evidence for long-term memory for the visual details of previously viewed objects should only be observed in intentional memorization tasks. Viewing tasks for which intentional memory encoding is unnecessary (i.e., an incidental memory task) should produce no lingering visual memory. Such a result would be consistent with localist-minimalist theories. On the other hand, if visual memory theory is correct and long-term visual representations are generated and stored as a natural product of visual scene perception, then evidence for the long-term preservation of visual detail in memory should be observed in both intentional and incidental memorization conditions. The present study was designed to test these competing predictions.

In the current study, the nature of the visual memory representation generated during scene perception was investigated by examining memory performance for visual information obtained from objects either intentionally or incidentally during scene viewing. In each of three experiments, participants viewed scenes while engaged in an intentional-learning memorization task or an incidental-learning visual search task. A memory test for a critical object in each scene was administered after both viewing tasks had been completed, although for the visual search task no memory test was anticipated by the participant during initial viewing. In the memorization task, participants were instructed to view the scenes in preparation for a difficult memory test that would require knowledge of details of specific objects. These instructions are similar to those given in other recent tests of visual memory (Hollingworth & Henderson, 2002). In the visual search task, participants were instructed to search for a specified target object in each scene, and were not told that they would receive a memory test for the objects in these scenes. In both cases, the memory test (given after all scenes had been viewed) always focused on the visual properties of a specific critical object drawn from each scene. The critical objects for the search scenes were distractor objects in the scenes and were never the search targets. In Experiments 1 and 2, all participants took part in both the memorization and visual search tasks. In Experiment 1, the memory test involved discriminating between a previously seen critical object drawn from each scene and another foil object that was a different token of the same basic-level category type. In Experiment 2, participants had to discriminate between the previously viewed orientation of the critical object and a mirror-reversal distractor version of the same object. In Experiment 3, a between-participant version of Experiment 1 was conducted in which each participant was given only one of the two study conditions to ensure that there was no contamination from memorization to visual search. The main question in the three experiments was whether long-term visual memory would be observed for objects that were incidentally encoded during scene viewing.

EXPERIMENT 1

The purpose of Experiment 1 was to determine whether the visual properties of objects are stored incidentally in long-term memory as a natural consequence of scene viewing, or whether such representations are only stored when participants are intentionally memorizing a scene. Participants viewed digitized photographs of real-world scenes in each of two tasks: memorization and visual search. In the memorization task, participants were asked to examine each scene in preparation for a difficult memory test that would require memory for specific objects; in the visual search task, participants were asked to search for and locate a prespecified target object in each scene. After viewing all of the scenes in both tasks, a memory test was given. The memory test was expected for the memorization scenes but not for the search scenes. The memory test consisted of forced-choice token discrimination involving a critical test object taken from each viewed scene and a distractor object of the same basic-level category taken from a similar scene not viewed during the study session (see Figure 1). In the case of the search scenes, the critical test object was never the search target, but instead was another (distractor) object from the scene. Neither the scene from which the critical object was taken nor the task in which that scene had appeared were cued at the time of the memory test, although a small patch of scene appeared around the target and distractor objects. Thus, during the test the participant was presented with pairs of objects with minimal contextual aid.

If localist-minimalist theories of scene representation are correct in their claim that visual memory for scene information is an artifact of intentional memorization, then token discrimination should be above chance for critical objects taken from the memorization scenes and be at chance for critical objects taken from search scenes. If instead visual memory theory is correct and the generation of a visual representation is a natural consequence of scene

Figure 1. Example of the scene and test screens used in Experiments 1 and 2. (A) Example scene studied in one of two task instructions used in Experiments 1 and 2. For the visual search task, participants were asked to search for a swimsuit in this scene. (B) Example of test screen used in Experiment 1. The distractor was always a different object taken from the same basic-level category. (C) Test screen from Experiment 2 that depicts the target and the distractor, made by horizontally flipping the selected section of the scene.

perception, then above-chance memory performance should be observed for objects viewed in both the memorization and visual search tasks.

Method

Participants. Twenty Michigan State University students with normal vision participated in this experiment in exchange for credit in an introductory psychology course.

Stimuli. The critical stimuli were 30 photographed scenes of indoor and outdoor environments. Fifteen scenes were shown in each of the memorization and search tasks, with assignment of scene to task counterbalanced across participants. In the search task, the object that was the target of the search was never included in the scenes; instead, five noncritical filler scenes containing search targets were presented. These filler scenes allowed participants to successfully find a target on some trials, but were challenging enough (determined in a pilot study) to prevent them from guessing that most of the search scenes contained no target. Each participant thus viewed 35 scenes in total, 15 memorization scenes and 20 search scenes (15 critical scenes that contained no target and 5 filler scenes that contained targets). The search target differed for each search scene. A word naming the search target was presented before the onset of each search scene. The word was printed in a 72-point Times New Roman font in black, also centred on a uniformly grey background. The words averaged $3.96^\circ \times 1.24^\circ$, with a range from 8.57° to 1.7° in width and 1.51° to 0.92° in height. In both study conditions, each trial began with a black fixation cross centred in a uniformly grey screen.

The memory test images were created by cutting out a square region containing the test object from each scene. Each test object was then matched with a foil object in its basic-level conceptual category (i.e., conceptual type), size, and orientation. Foils were taken from a separate picture set not viewed during the study session. The two objects (test object and foil) were placed on a uniformly grey background (see Figure 1B). Thirty memory test images were created this way in total, one for each critical scene viewed in the two tasks. The correct response was on the left for half the test images and on the right for the others. The test objects and foils were the same for the memorization and visual search conditions, although a given participant saw each test object and foil only once.

Apparatus. Participants' eye movements were monitored during scene viewing using a Generation 5.5 Stanford Research Institute Dual Purkinje Image Eyetracker (Crane, 1994; Crane & Steele, 1985), which has a resolution of $1'$ of arc and a linear output over the range of the visual display used. A bite-bar and forehead rests were used to maintain the participant's viewing position and distance. The position of the right eye was tracked, although viewing was binocular. Signals were sampled from the eyetracker using the polling mode of the Data Translations DT2802 analogue-to-digital converter, producing a sampling rate of better than 1000 Hz.

The monitor was placed 1.13 m from the participant. The screen subtended 15.20° horizontally and 11.93° vertically. The resolution was set at 800×600 pixels in 16-bit colour and the refresh rate was 100 Hz. The eyetracker and display monitor were interfaced with a microcomputer running a 90 MHz Pentium processor. The computer controlled the experiment and maintained a complete record of time and eye position values and participant responses over the course of each trial.

Procedure. Participants took part in two sessions, a learning session and a test session. In the learning session, the participants performed two tasks (memorization and visual search) while viewing the scenes. The tasks were presented in blocks and block order was counterbalanced across participants. In the test session, participants' memory for objects in scenes was tested using forced-choice recognition.

At the beginning of the experiment, the two tasks were described to participants as two separate experiments to discourage them from guessing that the visual search scenes would be included in the memory test. A bite-bar was prepared and the experimenter explained the eyetracker requirements. The first task was then explained in further detail and the eyetracker was calibrated. Each trial in both tasks began with an eyetracker calibration check; the eyetracker was recalibrated whenever the computer's estimate of fixation position (as shown on the screen by a white dot) was off by approximately +/− 8 pixels. Participants began each trial in both tasks by fixating a central fixation cross; once the cross was fixated, the experimenter initiated the trial.

In the memorization task, participants were instructed to view each scene in preparation for a memory test, to be administered at the end of the session that would examine memory for specific objects in the scenes. Participants were given three practice trials before commencing the experimental trials. Each participant then viewed each of the 15 scenes for 10 s.

In the visual search task, participants were instructed to indicate when they had located the target object by holding fixation on the object and pressing the response button. At the beginning of a trial, a word naming the search target for the immediately following scene was presented in the centre of the computer screen for 2 s. The scene was then displayed for 10 s or until the participant pressed the response button, whichever came first. To encourage participants to exhaustively search the critical scenes for the entire 10 s presentation duration, the search target was never present in these scenes. The search target was present in five filler scenes. The upcoming memory test for the distractor objects in the search scenes was not mentioned. Each viewing task was explained separately just before it was to be performed, allowing for each participant to receive a 2–3 min break between the two tasks.

After both learning tasks were completed, participants were given the memory test. Participants were told that they would be receiving a memory test for objects that had appeared in the scenes that they had studied in the memorization task. These instructions coincided with the previous description of the experiment and were given to prevent any last minute strategic encoding of the previously viewed information (especially when the visual search task was administered in the second block). Participants were instructed that they would see a series of trials in which two objects would appear together, one on either side of a grey screen. They were informed that one of these objects had been cut out from one of the scenes that they had studied previously, while the other was from a photograph that they had not seen. They were instructed to respond by pressing the button that corresponded to the side of the screen on which the previously seen object appeared (left or right). If they were not sure which object they had seen before, they were told to guess. Participants were informed that they could take as long as they desired to make their decisions. Instructions for the memory test required about 2–3 min, ensuring that there was some delay between presentation of the last scene in the second block of the learning session, and the first trial in the memory test.

At the end of the experiment, participants were debriefed. They were told that the test had included objects from the scenes in the visual search task, and were asked whether they had tried to memorize the scenes while looking for the target objects in the visual search task.

Results

Responses to debriefing question at the end of the experiment. All participants responded that they had concentrated on finding the target objects during the search task. No participant reported having known, guessed, or suspected that that their memory for the visual search scenes would be tested.

Percentage of critical objects fixated. Memory performance on both immediate and long-term memory tests is dependent on fixation of the critical object at the time of memory encoding (Henderson & Hollingworth, 1999; Hollingworth & Henderson, 2002; Nelson & Loftus, 1980). Therefore, in the present experiment, memory accuracy scores in the memory test were calculated only for trials in which the critical object was fixated at least once during the learning session. Fixation on the critical object was defined as follows: At least one fixation with a minimal duration of 90 ms within a region defined as the smallest rectangle encompassing the bounding contour of the critical object. On this definition, the critical objects were fixated on 84% of the memorization trials and 66% of the visual search trials, $t(19) = 4.54, p < .01$. The average total fixation time on the critical objects that were fixated (sum of the durations of all fixations) was 929 ms during memorization and 860 ms during visual search, $t(19) = 0.91, p > .1$. Figure 2 shows all fixations across all participants in each task on an example scene.

Memory accuracy by task. Table 1 shows the mean accuracy for both tasks. One sample *t*-tests revealed that memory accuracy in both task conditions was significantly above chance (50%): memorization task, $t(19) = 13.652, p < .05$; visual search task, $t(19) = 5.998, p < .05$. To determine if there was any difference in the means between tasks, a paired sample, two-tailed *t*-test was computed and showed that performance in the memorization task was marginally higher than in the visual search task, $t(19) = 1.973, p = .063$.

Memory accuracy by task and block. The learning tasks were administered in blocks, with task order counterbalanced across participants. It is possible that task order had an effect on memory accuracy. First, memory accuracy could be affected by a recency effect, with test items studied in the second block remembered better than those studied in the first. Second, and more critically,

Figure 2. All fixations across all participants in Experiment 1 are displayed on an example scene (A) for the memorization task and (B) for the visual search task. The participants were searching for an apple when performing the visual search task on this scene.

TABLE 1
Mean memory accuracy percentages as a function of block and task for Experiment 1

Task	*Block 1*	*Block 2*	*Task means*
Memory	80.3%	81.6%	81.0%
Visual search	64.8%	78.9%	71.8%
Block means	72.6%	80.2%	

performing the memorization task first could have placed the participants in a memorization mind-set for the following visual search task. This possibility would undermine the use of the visual search task as an incidental learning task.

To investigate the possibility that memory accuracy in the visual search task was contaminated by exposure to the prior memorization task, memory performance was analysed as a function of task and block in a mixed-design repeated-measures ANOVA. If memory for objects in search scenes was a consequence of contamination, memory performance for the objects from the search scenes should be above chance only when search was completed in the second block. For this analysis, block was treated as a within-participants factor, while task order (memory followed by visual search versus visual search followed by memory) was treated as a between-participants factor (see Table 1 for memory accuracy means by block and task). The analysis revealed that there were no main effects of block, $F(1, 18) = 3.023$, $MSE = 0.014$, $p > .1$, or task order, $F(1, 18) = 2.977$, $MSE = 0.0193$, $p > .1$, although there was a marginally significant interaction, $F(1, 18) = 4.308$, $MSE = 0.0193$, $p = .053$. The trend toward an interaction is consistent with the possibility that the above-chance memory performance for the visual search task was an artifact of contamination by the prior memorization task. To test this hypothesis directly, we examined whether memory performance for the visual search task differed from chance when visual search was presented in the first block. The results showed that in both blocks, memory accuracy was significantly above chance: Block 1, $t(9) = 4.504$, $p < .01$; Block 2, $t(9) = 4.942$, $p < .01$. Thus, above-chance memory performance for objects viewed during the visual search task was not the result of having engaged in a memorization task first.

Discussion

Localist-minimalist theories of scene perception posit that once attention has been withdrawn from a viewed object, visual information about that object is not retained in memory. Therefore, this class of theory predicts that memory performance for the visual attributes of previously viewed objects should be at chance. In this view, prior demonstrations of good visual memory for scene information are dismissed as the product of unusual strategic memory encoding strategies. In the present study, visual search was used as an incidental memory task. The results clearly demonstrated that the visual detail needed to discriminate between two objects of the same basic-level category was stored in memory during scene viewing under these incidental-learning conditions. This visual memory was established in less than 1 s of total fixation time on the tested objects, suggesting that even a relatively short direct glimpse of an object is sufficient to incidentally establish a visual representation that endures over a relatively long period of time.

Experiment 1 demonstrated that participants were able to distinguish between an object incidentally encoded into memory during a visual search task and another perceptual token of the same basic-level category. Experiment 2 was designed to provide a further test of incidental long-term scene memory for visual information. In Experiment 2, the foils in the forced-choice memory test were created by mirror reflecting the critical objects seen in the learning session; in this way, they contained the identical visual and semantic information as the critical objects. Accurate memory performance in this case requires that object orientation be encoded in memory.

EXPERIMENT 2

Experiment 2 used an orientation discrimination task to explore further the nature of the visual information stored in long-term memory during extended scene viewing. Hollingworth and Henderson (2002) have shown in both immediate and long-term memory tests that object orientation is acquired and stored in memory during scene viewing when memory encoding is required for the task. If object orientation is encoded into memory as a natural consequence of scene perception, then accuracy in an orientation-discrimination task should be above chance following an incidental-learning scene viewing task. If instead the generation of detailed visual representations requires strategic encoding, then accuracy for orientation discrimination should be above chance for scenes viewed under memorization instructions but at chance for scenes viewed under visual search instructions. The orientation discrimination also controls for the scene patch surrounding the object. For instance, in Figure 1, the blue background behind the green baseball cap may have provided additional cues for the token discrimination test in Experiment 1. These cues are not available in the orientation discrimination test of Experiment 2.

Method

Participants. Twenty Michigan State University undergraduate students with normal vision received credit in an introductory psychology course or received $7 for their participation.

Stimuli. The stimuli were the same as those used in Experiment 1, with the exception of the test images. Again, a square region containing the test object was taken from each critical scene. The region was copied and mirror-reversed using commercial graphics software. This manipulation maintained all visual features of the critical object (as well as the surrounding scene patch) except for orientation. The two alternatives were placed to the right and left of the centre of the screen, as was done in Experiment 1, with position of the target and foil counterbalanced across scenes. Figure 1C shows an example of a test image used in Experiment 2. Again, the same target objects and foils were used in the memorization and visual search conditions, although a given participant saw each only once.

Procedure. With the exception of the foils used in the memory test, the procedure for Experiment 2 was identical to Experiment 1.

Results

Responses to debriefing question at the end of the experiment. As in Experiment 1, no participant reported having known, guessed, or suspected at the time of the search task that their memory for the visual search scenes would be tested.

Percentage of critical objects fixated. As in Experiment 1, the memory accuracy analyses were based on trials in which the critical objects were fixated. The critical objects were fixated (as defined in Experiment 1) on 76% of the trials during memorization and 69% of the trials during visual search, $t(19) = 1.421, p > .1$. The average total fixation time on the critical objects was 896 ms in the memorization task and 764 ms in the visual search task, $t(19) = 1.625$, $p > .1$.

Memory accuracy by task. Memory accuracy as a function of task is shown in Table 2. Memory performance following both the memorization and visual search conditions was significantly above chance, $t(19) = 3.242$, $p < .05$ and $t(19) = 4.376$, $p < .05$, respectively. Performance following the memorization task was not significantly better than following the visual search task, $t(19) = 0.141, p > .1$, although as found in Experiment 1, there seemed to be a tendency for performance to be slightly higher for the memorization task.

Memory accuracy by task and block. Memory accuracy as a function of block and task is shown in Table 2. There were no significant main effects of task order, $F(1, 18) = 3.434$, $MSE = 0.0345$, $p = .08$, or block, $F(1, 18) = 3.049$, $MSE = 0.0189$, $p = .098$, and no interaction, $F(1, 18) = 0.784$, $MSE = 0.0189$, $p > .1$. Thus, in contrast to Experiment 1, there was no evidence in the current

TABLE 2
Mean memory accuracy percentages as a function of block and task for Experiment 2

Task	*Block 1*	*Block 2*	*Task means*
Memory	55.6%	72.2%	63.9%
Visual search	60.8%	59.4%	60.1%
Block means	58.2%	65.8%	

experiment that memory accuracy for objects viewed during visual search was better when the visual search task followed the memorization task. As in Experiment 1, memory performance for objects viewed during visual search was significantly above chance in both blocks: Block 1, $t(9) = 6.877, p < .01$; Block 2, $t(9) = 2.551, p < .05$. Thus, above-chance memory performance for the visual search scenes was not the result of having engaged in a memorization task first.

Discussion

Although the accuracy scores for Experiment 2 were lower overall than in Experiment 1, the main results were replicated. Participants were able to perform a difficult visual memory test, in this case discriminating the orientation of a viewed object from its mirror reflection, both when they had intentionally tried to memorize the objects while viewing scenes and when they had searched through the scenes for a different (nonpresent) target object. As found in Experiment 1, no participant reported suspecting that they would be given a memory test for the visual search scenes. Furthermore, memory accuracy for objects that had appeared as distractors in the visual search task was equivalent whether this task was performed in the first or second block. Therefore, the possibility that memory performance in the visual search task was due to strategic encoding of visual information is remote.

EXPERIMENT 3

The purpose of Experiment 3 was to provide a final test of the strategic encoding hypothesis. Specifically, the experiment was designed to rule out the possibility that the memory results observed following visual search in the first two experiments was due to contamination of the visual search task by the memorization task. In Experiment 3, each participant performed only one of the two viewing tasks; the unassigned task for a given participant was never mentioned. Following presentation of all 30 critical scenes in the assigned memorization or search task, each participant took part in the token discrimination memory test used in Experiment 1. If the above-chance memory performance observed for the search task in Experiments 1 and 2 was due to contamination of the search task by the memorization task, then in Experiment 3, visual object memory should be above chance under intentional encoding (memorization) instructions and should be at chance under the incidental encoding (visual search) instructions. If, however, detailed visual memory for objects results from both intentional and incidental memory encoding, then memory for the critical test objects should be above chance following both viewing conditions.

Method

Participants. Twenty Michigan State University undergraduate students with normal vision received $7 for their participation.

Stimuli. The stimuli were identical to those used in Experiment 1.

Procedure. The procedure for Experiment 3 was identical to Experiment 1 with the following exceptions. Each participant took part in only the memorization or the visual search task. Participants assigned to the memorization task were informed that they would be memorizing the scenes, and participants in the visual search task were informed that they would be searching for target objects in the scenes. The specific instructions for each task were the same as those given in Experiments 1 and 2. Unlike Experiments 1 and 2, each participant viewed all 30 critical scenes in the assigned task (participants in the search task also saw the 5 filler scenes that contained a target). They also received the exact same set of targets and foils in the memory and search conditions. The order of scene presentation was randomized for each participant. For those participants assigned to the visual search task, the memory test was not mentioned until all search scenes had been viewed. The memory test instructions were the same for both groups and were those given in Experiment 1.

Results

Responses to debriefing question at the end of the experiment. As found in Experiments 1 and 2, no participant reported having known, guessed, or suspected at the time of the search task that their memory for the visual search scenes would be tested.

Percentage of critical objects fixated. The memory accuracy analyses were again based only on trials in which the critical object was fixated during initial scene viewing. The critical objects were fixated on 75% of the memorization trials and 76% of the visual search trials, $t(18) = -0.73$, $p > .1$. The average total fixation time on the critical objects was 939 ms during memorization and 796 ms during visual search, $t(18) = 1.718$, $p > .1$.

Memory accuracy by task. Memory accuracy following both the memorization task (78%) and the visual search task (75%) was significantly above chance, $t(9) = 10.582$, $p < .05$ and $t(9) = 9.643$, $p < .05$, respectively. Memory performance following the two viewing tasks did not differ reliably from each other, $t(18) = 0.818$, $p > .1$.

Discussion

The purpose of Experiment 3 was to determine whether the visual memory performance observed in Experiments 1 and 2 for distractor objects fixated during visual search was due to contamination from exposure to the memorization task. In Experiment 3, viewing task in the learning session was manipulated between participants so that those individuals assigned to the visual search task would have no knowledge that the study included a memory component. Nevertheless, participants assigned to the visual search task clearly demonstrated memory for the visual properties of the distractor objects they had viewed.

GENERAL DISCUSSION

The purpose of the present study was to test competing predictions generated by two theoretical perspectives on the nature of visual memory and scene representation. According to localist-minimalist theories, sustained visual scene representation is nonexistent (O'Regan & Noë, 2001; Rensink, 2000a, 2000b, 2002), or at best, is limited to high-level semantic representations of scene gist, spatial layout, and object identity (Irwin, 1992a, 1992b; Irwin & Andrews, 1996). In contrast, according to visual memory theory (Henderson & Hollingworth, 2003a; Hollingworth & Henderson, 2002), a relatively detailed visual scene representation for attended scene information accumulates over time in long-term memory as a natural consequence of scene viewing. Recent investigations of scene memory have suggested that the visual attributes of objects are retained over both the short and long term, and that these preserved visual attributes can support both overt and covert change detection as well as direct memory test (Hollingworth & Henderson, 2002; Hollingworth et al., 2001), supporting visual memory theory. A possible criticism of these studies, however, is that participants may have engaged in intentional memory encoding strategies that are not characteristic of typical scene perception in order to complete the experimental task.

The present study contrasted the localist-minimalist and visual memory perspectives by investigating the degree to which detailed visual information is incidentally stored in memory during a scene viewing task that does not require intentional encoding. In three experiments, participants' memory for object detail was compared following an intentional scene memorization task and an incidental-learning visual search task. In the memorization task, participants intentionally attempted to encode object detail. In the search task, participants searched for target objects and were unaware that their memory for distractor objects in the search scenes would later be tested. After completing the tasks, participants were given a difficult two-alternative forced-choice memory test involving previously viewed objects and foil objects distinguished from the viewed objects only by visual detail. Localist-minimalist theories predict either (1) that memory performance for objects appearing in both viewing tasks should be at chance because visual representations simply cannot be formed during scene viewing, or (2) that memory performance for objects appearing in the memorization task should be above chance due to strategic memory encoding, but that visual memory performance for objects appearing as distractors in the visual search task should be at chance because memorization strategies will not have been engaged. Visual memory theory, in contrast, predicts above-chance performance for objects viewed in both intentional and incidental learning tasks, because this theory holds that incidental storage of visual information from attended scene elements is a natural consequence of scene viewing.

In Experiment 1, two objects from the same basic-level category were presented in the memory test, one of which had been viewed in a scene during the study session. Accuracy was above chance for objects whether they had appeared in the visual search task or the memorization task. In Experiment 2, the same paradigm was used, except that the memory test involved distinguishing between a previously viewed object and its mirror reflection. Again, above chance performance was found whether the objects were viewed during the visual search task or the memorization task. In Experiment 3, viewing task was manipulated between participants to ensure that memory in the search task was not contaminated by exposure to the memorization instructions or task. The memory test was the same as in Experiment 1. Performance was again found to be above chance for both the intentional memorization and the incidental visual search conditions. The results from all three experiments are inconsistent with the hypothesis that visual memory for scene information is only established during intentional memorization of scene details. The evidence presented in this study thus demonstrates that visual information is encoded and stored in memory during scene viewing whether or not the viewer is intentionally trying to memorize that information. The results also suggest that prior demonstrations of good visual memory for scene information (Henderson & Hollingworth, 1999, 2003b; Hollingworth & Henderson, 2002; Hollingworth et al., 2001) were not the result of intentional memory encoding strategies.

The present study provides a relatively stringent test of visual memory for objects encoded during scene viewing, and hence should place a lower boundary on the degree of visual memory generated from viewed scenes. Some of the factors likely to contribute to low memory performance include the following. First, memory performance for the critical objects was based on a total of only 10 s of viewing time per scene and an average of less than 1 s of foveal fixation time per tested object. Thus, the amount of time available for memory encoding was quite limited. Second, for both the memorization and search tasks, test objects were presented in the memory session without the contextual support that would be provided by the entire scene. Thus, the memory cues available for aiding retrieval were severely limited during test. Third, the retention interval in this study was relatively long (approximately 4–20 min) and the number of intervening objects relatively large between initial scene viewing and the object memory test, ruling out the use of short-term memory as the system supporting memory performance. All of these factors should have conspired to depress memory performance.

Finally, the total number of objects likely to have been encoded across the scenes was relatively large. A conservative estimate of the number of objects encoded per scene can be generated by considering the presentation durations of the scenes (10 s) and the average total fixation time per object, which can be estimated from the average fixation time on the critical objects. Assuming that only fixated objects were encoded (Hollingworth & Henderson, 2002; Nelson & Loftus, 1980), and that all objects were fixated for the same amount of time on average as the critical objects (929, 896, 939 ms in the memorization task of Experiments 1–3, respectively), we can estimate that about 10.7 objects were fixated per scene on average across experiments in the memorization task. In the search task, implicit encoding would have taken place for even more objects (12.4 per scene), because average fixation time per object was lower (860, 764, and 796 ms in Experiments 1–3, respectively). Using average fixation times for each experiment, and the fact that 15 scenes were presented in the memorization task and 20 (15 critical and 5 filler) in the search task, we estimate that 394 objects were fixated in Experiment 1 and 429 objects were fixated in Experiment 2. In Experiment 3, because participants saw all 30 scenes in only one condition or the other, we get two estimates: 373 objects for participants who memorized and 440 objects for participants who searched (including the five filler scenes that were not tested later). Given that our choice of critical test objects constituted a random sampling of all the objects in the scenes, memory performance for the test objects can be taken as an indication of the level of memory performance that untested objects would have shown if tested. We would therefore estimate that memory performance for token-level detail and left–right orientation for any of the hundreds of objects any given participant fixated in the scenes would have been at the levels we observed. Another way to think about this is that participants remembered x% of the objects viewed, with x estimated by the observed forced-choice performance y corrected for guessing ($x = [y - .5]/.5$). In terms of number of objects, in Experiment 1, we estimate that participants remembered the details of 244 objects (~16 objects per scene) in the memory condition and 173 in the search condition (~9 objects per scene). In Experiment 2, they remembered the orientation of 120 objects (~8 objects per scene) in the memory condition and 87 objects (~4 objects per scene) in the search condition. Finally, in Experiment 3, they remembered the details of 178 objects (~6 objects per scene) in the memory condition and 219 (~6 objects per scene) in the search condition.

Given these considerations, the demonstrated existence of visual object memory poses a serious challenge to localist-minimalist theories, which assume that no visual representation is stored in memory from a viewed object once attention is allocated elsewhere in the scene.

Intentional versus incidental memory for objects in scenes

It is tempting to compare directly the memory results from the intentional and incidental learning conditions. In all three experiments, memory performance was statistically equivalent across the learning conditions, and although performance was numerically better in the intentional condition, the difference was not great, varying from a high of 9% in Experiment 1 to 4% in Experiment 2 and 1% in Experiment 3. The lack of large memory differences across conditions is striking given that participants were motivated to try to encode the critical objects into memory in the memorization condition, but simply had to reject the critical objects as targets in the search task. Furthermore, as noted above, fixation time on the critical objects was greater on average for the memorization than the search conditions in all three experiments, providing more time for memory encoding in the former case. Despite these advantages, performance for the critical objects was statistically equivalent following the memorization and search tasks. This result must be treated with some caution because the experiments were not designed to investigate this difference directly, and the result itself is based on a null effect. Nevertheless, the result is intriguing, and at the least suggests that memory performance in the implicit learning condition was approaching the ceiling level of performance set by the factors that limited performance in the intentional memorization condition.

Visual search does have memory

The results of the present study are also relevant to the issue of memory for distractor items in visual search. Horowitz and Wolfe (1998) presented a series of experiments apparently demonstrating that participants did not retain in memory information about previously viewed distractor items in a visual search task. In these studies, a target letter that had been a distractor in previous search trials was not found more rapidly than a target letter that was new to the search array. The lack of improvement in search efficiency suggested to Wolfe (1999) that attending to a distractor item on prior search trials did not establish a memory representation that could later be used to facilitate search for that item. In contrast to that conclusion, the data from the present study demonstrate that participants do have visual memory for previously viewed distractors in a visual search task. These results are consistent with more recent results from the array search literature, which have shown in various ways that visual search does produce memory (Gibson, Li, Skow, Brown, & Cooke, 2000; Kristjánson, 2000; Peterson, Kramer, Wang, Irwin, & McCarley, 2001: Wolfe, Kemplen, & Dahlen, 2000). The present study contributes to this body of work by demonstrating that attention to a distractor object in search through a natural scene similarly leaves a lingering memory trace of the visual details of that object.

Level of analysis: Objects and scene patches

We have emphasized memory for objects because objects are the targets of fixations (Buswell, 1935; Yarbus, 1967), and in the memorization and visual search tasks used in this study the object was the task-relevant level of analysis. However, it could be argued that our tests of object memory were contaminated by the inclusion of a small amount of the surrounding scene in the memory test. We would argue that the main conclusions hold whether the level of analysis is objects or scene patches containing objects. In either case, visual details must be encoded to discriminate the target patches from the foils; in the case of orientation discrimination, the visual information in the target and foil patches was identical except for orientation, so discrimination required memory for the specific orientation of that patch. At both of these levels of analyses, the conclusion is that relatively detailed visual representations have been formed and retained in both intentional and incidental learning viewing tasks.

CONCLUSION

The present study was designed to test competing predictions from localist-minimalist and visual memory theories of scene representation. The differences between these views reflect a fundamental contrast in perspectives on the functional operation of the visual system. In the former view, the concept of visual representation is unnecessary at best and misguided at worst. In the latter view, visual representation is a necessary and natural consequence of vision. The results of the present study strongly suggest that visual information acquired from attended objects during scene viewing is stored in memory whether or not the viewer intends to remember that information, consistent with the latter view.

REFERENCES

Bahrick, H. P., & Boucher, B. (1968). Retention of visual and verbal codes of the same stimuli. *Journal of Experimental Psychology*, *78*, 417–422.

Ballard, D. (1996). On the function of visual representation. In K. Akins (Ed.), *Vancouver studies in cognitive science: Vol. 5. Perception* (pp. 111–131). New York: Oxford University Press.

Brewer, W. F., & Treyens, J. C. (1981). Role of schemata in memory for places. *Cognitive Psychology*, *13*, 207–230.

Bridgeman, B., Hendry, D., & Stark, L. (1975). Failure to detect displacement of the visual world during saccadic eye movements. *Vision Research*, *15*, 719–722.

Brooks, R. A. (1987). Intelligence without representation. *Artificial Intelligence*, *47*, 139–159.

Buswell, G. T. (1935). *How people look at picture: A study of the psychology of perception in art.* Chicago: University of Chicago Press.

Chun, M. M., & Jiang, Y. H. (1998). Contextual cueing: Implicit learning and memory of visual context guides spatial attention. *Cognitive Psychology*, *36*, 28–71.

Chun, M. M., & Jiang, Y. H. (2003). Implicit, long-term spatial contextual memory. *Journal of Experimental Psychology: Learning, Memory, and Cognition*, *29*(2), 224–234.

Churchland, P. S., Ramachandran, V. S., & Sejnowski, T. J. (1994). A critique of pure vision. In C. Koch & J. L. Davis (Eds.), *Large-scale neuronal theories of the brain.* Cambridge, MA: MIT Press.

Crane, H. D. (1994). The Purkinje image eyetracker, image stabilization, and related forms of stimulus manipulation. In D. H. Kelley (Ed.), *Visual science and engineering: Models and applications* (pp. 15–89). New York: Macel Dekker.

Crane, H. D., & Steele, C. M. (1985). Generation-V dual-Purkinje-image eyetracker. *Applied Optics*, *24*, 527–537.

Friedman, A. (1979). Framing pictures: The role of knowledge in automatized encoding and memory for gist. *Journal of Experimental Psychology: General*, *108*, 316–355.

Gibson, B. S., Li, L., Skow, E., Brown, K., & Cooke, L. (2000). Searching for one versus two identical targets: When visual search has memory. *Psychological Science*, *11*, 324–327.

Goodman, G. (1980). Picture memory: How the action schema affects retention. *Cognitive Psychology*, *12*, 473–495.

Grimes, J. (1996). On the failure to detect changes in scenes across saccades. In K. Akins (Ed.), *Vancouver studies in cognitive science: Vol. 5. Perception* (pp. 89–110). New York: Oxford University Press.

Henderson, J. M. (1997). Transsaccadic memory and integration during real-world object identification. *Psychological Science*, *8*, 51–55.

Henderson, J. M., & Hollingworth, A. (1999). The role of fixation position in detecting scene changes across saccades. *Psychological Science*, *10*, 438–443.

Henderson, J. M., & Hollingworth, A. (2003a). Global transsaccadic change blindness during scene perception. *Psychological Science*, *14*, 493–497.

Henderson, J. M., & Hollingworth, A. (2003b). Eye movements and visual memory: Detecting changes to saccade targets in scenes. *Perception and Psychophysics*, *65*, 58–71.

Henderson, J. M., & Hollingworth, A. (2003c). Eye movements, visual memory and scene representation. In M. A. Peterson & G. Rhodes (Eds.), *Perception of faces, objects and scenes: Analytic and holistic processes.* New York: Oxford University Press.

Hock, H. S., Romanski, L., Galie, A., & Williams, C. S. (1978). Real-world schemata and scene recognition in adults and children. *Memory and Cognition*, *6*, 423–431.

Hollingworth, A. (2003). Failures of retrieval and comparison constrain change detection in natural scenes perception. *Journal of Experimental Psychology: Human Perception and Performance*, *29*, 388–403.

Hollingworth, A., & Henderson, J. M. (2002). Accurate visual memory for previously attended objects in natural scenes. *Journal of Experimental Psychology: Human Perception and Performance*, *28*, 113–136.

Hollingworth, A., Williams, C., & Henderson, J. M. (2001). To see and remember: Visually specific information is retained in memory from previously attended objects in natural scenes. *Psychonomic Bulletin and Review*, *8*, 761–768.

Horowitz, T. S., & Wolfe, J. M. (1998). Visual search has no memory. *Nature*, *394*, 575–577.

Irwin, D. E. (1991). Information integration across saccadic eye movements. *Cognitive Psychology*, *23*, 420–453.

Irwin, D. E. (1992a). Memory for position and identity across eye movements. *Journal of Experimental Psychology: Learning, Memory, and Cognition, 18*, 307–317.

Irwin, D. E. (1992b). Visual memory within and across fixations. In K. Rayner (Ed.), *Eye movements and visual cognition: Scene perception and reading* (pp. 146–165). New York: Springer-Verlag.

Irwin, D. E., & Andrews, R. V. (1996). Integration and accumulation of information across saccadic eye movements. In T. Inui & J. L. McClelland (Eds.), *Attention and performance XVI: Information integration in perception and communication*. Cambridge, MA: MIT Press.

Irwin, D. E., Yantis, S., & Jonides, J. (1983). Evidence against visual integration of information across saccadic eye movements. *Perception and Psychophysics, 34*, 35–46.

Kristjánson, A. (2000). In search of remembrance: Evidence for memory in visual search. *Psychological Science, 11*, 328–332.

Levin, D. T., & Simons, D. J. (1997). Failure to detect changes to attended abject in motion pictures. *Psychonomic Bulletin and Review, 4*, 501–506.

Mandler, J. M., & Parker, R. E. (1976). Memory for descriptive and spatial information in complex pictures. *Journal of Experimental Psychology: Human Learning and Memory, 2*, 38–48.

Mandler, J. M., & Ritchey, G. H. (1977). Long-term memory for pictures. *Journal of Experimental Psychology: Human Learning and Memory, 3*, 386–396.

McConkie, G. W., & Rayner, K. (1976). Identifying the span of the effective stimulus in reading: Literature review and theories of reading. In H. Singer & R. B. Ruddell (Eds.), *Theoretical models and processes in reading* (pp. 137–162). Newark, DE: International Reading.

McConkie, G. W., & Zola, D. (1979). Is visual information integrated across successive fixations in reading? *Perception and Psychophysics, 25*, 221–224.

Neisser, U. (1967). *Cognitive psychology*. Appleton: Century-Crofts.

Nelson, W. W., & Loftus, G. R. (1980). The functional visual field during picture viewing. *Journal of Experimental Psychology: Human Learning and Memory, 6*, 391–399.

Nickerson, R. S. (1965). Short-term conceptual memory for complex meaningful visual configurations: A demonstration of capacity. *Canadian Journal of Psychology, 19*, 155–160.

O'Regan, J. K. (1992). Solving the "real" mysteries of visual perception: The world as an outside memory. *Canadian Journal of Psychology, 46*, 461–488.

O'Regan, J. K., & Levy-Shoen, A. (1983). Integrating visual information from successive fixations: Does trans-saccadic fusion exist? *Vision Research, 23*, 765–768.

O'Regan, J. K., & Noë, A. (2001). A sensorimotor theory of vision and visual consciousness. *Behavioral and Brain Sciences, 24*, 939–1031.

O'Regan, J. K., Rensink, R. A., & Clark, J. J. (1999). Change blindness as a result of "mudsplashes". *Nature, 398*, 34.

Peterson, M. S., Kramer, A. F., Wang, R. F., Irwin, D. E., & McCarley, J. S. (2001). Visual search has memory. *Psychological Science, 12*, 287–292.

Pezdek, K., Maki, R., Valencia-Laver, D., Whetsone, T., Stoeckert, J., & Dougherty, T. (1988). Picture memory: Recognizing added and deleted details. *Journal of Experimental Psychology: Learning, Memory, and Cognition, 14*, 468–476.

Pezdek, K., Whetstone, T., Reynolds, K., Askari, N., & Dougherty, T. (1989). Memory for real-world scenes: The role of consistency with schema expectation. *Journal of Experimental Psychology: Learning, Memory, and Cognition, 15*, 587–595.

Pollatsek, A., & Rayner, K. (1992). What is integrated across fixations? In K. Rayner (Ed.), *Eye movements and visual cognition: Scene perception and reading* (pp. 166–191). New York: Springer-Verlag.

Rayner, K., & Pollatsek, A. (1983). Is visual information integrated across saccades? *Perception and Psychophysics, 34*, 39–48.

Rensink, R. A. (2000a). Seeing, sensing, and scrutinizing. *Vision Research, 40*, 1469–1487.

Rensink, R. A. (2000b). The dynamic representation of scenes. *Visual Cognition, 7*(1–3), 17–42.

Rensink, R. A. (2002). Change detection. *Annual Review of Psychology, 53*, 245–277.

Rensink, R. A., O'Regan, J. K., & Clark, J. J. (1997). To see or not to see: The need for attention to perceive changes in scenes. *Psychological Science*, *8*, 368–373.

Shepard, R. N. (1967). Recognition memory for words, sentences and pictures. *Journal of Verbal Behavior*, *6*, 156–163.

Simons, D. J., & Levin, D. T. (1998). Failure to detect changes to people in real-world interaction. *Psychonomic Bulletin and Review*, *5*, 644–649.

Standing, L. (1973). Learning 10,000 pictures. *Quarterly Journal of Experimental Psychology*, *25*, 207–222.

Standing, L., Conezio, J., & Haber, R. N. (1970). Perception and memory for pictures: Single-trial learning of 2500 visual stimuli. *Psychonomic Science*, *19*(2), 73–74.

Wolfe, J. M. (1999). Inattentional amnesia. In V. Coltheart (Ed.), *Fleeting memories* (pp. 71–94). Cambridge, MA: MIT Press.

Wolfe, J. M., Klempen, N., & Dahlen, K. (2000). Postattentive vision. *Journal of Experimental Psychology: Human Perception and Performance*, *26*, 693–716.

Yarbus, A. I. (1967). *Eye movements and vision*. New York: Plenum Press.

VISUAL COGNITION, 2005, 12 (6), 1041–1056

Are boundary extension and representational momentum related?

Margaret P. Munger, T. Ryan Owens, and John E. Conway
Department of Psychology, Davidson College, NC, USA

Boundary extension (BE) is misremembering more of a particular scene than actually presented. Representational momentum (RM) is misremembering the final position of a moving object as further along its implied trajectory. There is a sense in which both BE and RM are predictions, and the current experiments explore the possibility that these are related phenomena. We showed participants single photographs to measure baseline BE, and then approach sequences and measured BE and RM. We found significant BE and RM, and a significant interaction between baseline BE and BE following an approach sequence, but no interaction with RM. The initial evidence suggests that BE and RM are separate processes, and that BE (the establishing of spatial layout) occurs before RM (the continuation of movement) within a scene.

We experience a stable visual world, and can make reasonably accurate predictions about the continuing layout of a scene and where moving objects are heading. Boundary extension (BE) occurs when we remember a little more of a picture than we actually saw (Intraub, Bender, & Mangels, 1992; Intraub & Bodamer, 1993; Intraub, Gottesman, Willey, & Zuk, 1996; Intraub, Gottesman, & Bills, 1998), in a way predicting what's just outside our view (Intraub, 2002). Representational momentum (RM) occurs when we remember the final location of a moving object as a little further along its path of motion (Freyd & Finke, 1984; Hubbard, 1995; Munger, Solberg, Horrocks, & Preston, 1999c). Both of these distortions involve prediction—what's just out of view, and where an object is heading—and therefore might be related (Hubbard, 1995; Intraub, 2002).

Boundary extension experiments typically ask participants to look at photographs of scenes involving a central object with surrounding texture (e.g., a

Please address all correspondence to: Margaret P. Munger, Department of Psychology, Davidson College, Box 7001, Davidson, NC 28035-7001, USA. Email: mamunger@davidson.edu

We thank Tim Hubbard, Cathy Reed, and Ian Thornton for helpful comments, Dave Munger for the photographs, Ben Covington for the pyramids, and the students of Psy 301, Fall 2001, for assistance with data collection. This research was presented at the American Psychological Society, June 2003.

DOI:10.1080/13506280444000643

teddy bear sitting on the steps of a building; Intraub et al., 1996), with instructions to remember the entire scene, including the background. When memory is later tested, participants remember seeing more of the picture than they were initially shown. When BE is tested using a rating system to indicate if the current view is closer or farther away, participants rate the same view as closer, indicating they are remembering more of the background (e.g., Gottesman & Intraub, 2002). When BE is tested by having participants draw the scene, again more background is presented, and participants will even complete background objects that were initially cropped (for review see Intraub, 2002). BE occurs over a variety of conditions, including rather simple scenes, but the key appears to be that the individual has considered the presented objects as being part of a scene (Gottesman & Intraub, 2002; Intraub et al., 1996, 1998).

Intraub, Gottesman, and Bills (1998) found BE when participants were asked to imagine the background surrounding an object, but not when they were asked to imagine the object's colour. The critical difference was the surface layout included in the imagined background. Gottesman and Intraub (2002) found that directed imagery was not necessary, but that simply showing a photograph cutout of an object mounted on a card could be enough to trigger an understanding that the object was on a surface. However, when the separation between the cutout and black card was highlighted, by physically placing the cutout onto the card, no BE occurred—now it was obvious that the object was quite separate from its background. Gottesman and Intraub propose that any time pictures are understood to be fragments of a continuous world, amodal continuation of that world will occur. It is not a perceptual illusion that the world continues outside your immediate view, but a top-down understanding of how the world looks.

There is no question that BE occurs over a variety of scenes, but what will happen when the views of the scene are moving? DeLucia (2001) showed participants approaching and receding movies and measured BE for the final view. BE was observed, but the direction of motion mattered, with more BE for approach than for recede conditions. An increase in BE for an approach is opposite what RM would suggest, leading DeLucia to suggest that judgements of boundary are different from judgements of position (personal communication). We would like to take a closer look at how an approach sequence will impact BE, using a within-participant design. The approach is particularly interesting because if any continuation of the movement were to occur (e.g., RM), this would imply a restricting of the boundaries, the opposite of boundary extension.

In contrast with the photographs or line drawings used in BE experiments, RM stimuli are typically much simpler, often using small squares (e.g., Hubbard, 1995). Line drawings of various objects have been used, and the particular connotations or characteristics of an object have an impact on RM (Reed & Vinson, 1996; Vinson & Reed, 2002). Some RM work has been done that depicts multiple objects (a "scene"), and there can be interactions between the objects (Hubbard, 1995; Hubbard & Ruppel, 1999, 2000; Munger & Owens,

2004). Halpern and Kelly (1993) presented line drawings of animals and vehicles in a simple scene, and RM still occurred, so the mere presence of a textured background does not eliminate RM (see also, Munger, Covington, Minchew, & Starr, 1999a). A related phenomenon to RM is an asymmetry in reaction times to frozen action photographs (e.g., a man jumping down off a low wall). Freyd (1983) presented a single photograph and asked participants to determine if a subsequent photograph was identical. It took longer to judge frames further along in the sequence (e.g., the man closer to the ground) than to judge frames behind the original (e.g., the man higher up). Futterweit and Beilin (1994) replicated this effect with children, and together these results suggest that RM should occur in more elaborate scenes. In addition, Kourtzi and Kanwisher (2000) found increased activation in brain regions involved with motion when participants viewed photographs that implied motion. The current experiment also uses more detailed scenes to explore the role of additional spatial layout on RM.

Most RM work has depicted movement in the picture plane, though RM does occur for rotations in depth (Munger, Solberg, & Horrocks, 1999b; Munger et al., 1999c; Munger & Minchew, 2002). Hubbard (1996) looked at RM for translations in depth by depicting a square changing size and instructing participants to think of this as movement in depth. He found more RM for receding movement (the square getting smaller) than for approaching movement (the square getting bigger). In this context, Hubbard observed that BE, remembering more of the background, is geometrically equivalent to the central object moving away from the viewer, potentially linking movement in depth to BE. For motion receding from the observer, BE and RM would both lead to accepting smaller central objects (remembering more background *and* continuing the receding motion). Motion approaching the observer, however, would have RM continuing to an even closer view, and BE still expanding the background, perhaps cancelling each other out. Smaller RM distortions have also been found with approaching motion using stereoscopic displays (Hayes, Sacher, Thornton, Sereno, & Freyd, 1996; Nagai, Kazai, & Yagi, 2002). In these experiments (Hubbard, 1996; Nagai et al., 2002), the moving object was a simple square with no surrounding texture or context. RM distortions for depth translations have also been observed using more complex scenes, including smooth animation along a virtual road with landmarks (Thornton & Hayes, 2004) and implied movement using perspective drawings (DeLucia, 2001). Thornton and Hayes (2004) also found RM distortions for complex crowd scenes, involving the movement of many people in the scene.

Instead of thinking of BE as the central object moving away, Intraub (2002) suggests that BE reflects an active exploration of a scene, an engagement with the environment. She points out that RM experiments feature an observer watching an object move, and that BE is more like the observer moving through the scene. One way to compare BE and RM tasks would be to show participants an approach, as if they are moving into the scene (see Figure 1). It is with an approach sequence that BE and RM distortions oppose one another. BE results tell us that we make predictions about what is just out of view, and so should remember seeing more of the background. RM results show that we make predictions about movement continuing, and so in an approach sequence participants should misremember the scene as even closer. The current experiment depicted approach sequences and used both the ratings measure of BE (is this picture closer or farther away?), and the same/different decision of RM (is this picture exactly the same as the previous one?). In addition to the approach sequences, we measured the baseline BE for each participant. We are particularly curious about the role individual differences in BE ratings might play in using approach information. Are the individuals who initially include relatively more spatial layout information in their perceptual schemas (high BE), more or less sensitive to the inclusion of a dynamic approach? What will the additional information provided about the background in an approach sequence allow individuals with relatively less BE to remember? The approach sequence literally shows you the surrounding area of the final close-up—does this eliminate or encourage BE? The approach itself should lead to a closer view of the final position if RM is playing a role, but will the richer context of the scene enhance RM, or will the additional information from the surrounding context eliminate the distortion?

Participants were asked to judge entire scenes in three different conditions: A standard BE rating task using single, close-up photographs to give us a baseline measure of individual BE; a BE rating task following a sequence of photographs that implies an approach; and finally, a standard RM task depicting an implied approach (see Figure 2).

The first block presents a slightly modified BE task: Participants are shown a single close-up photograph of a scene and asked to indicate immediately if a second view is closer or further away using the rating scale of previous BE research (e.g., Intraub, 2002). Performance on this task will be used to separate participants into groups for subsequent data analysis to investigate how approach sequences impact BE.

The second block presents a sequence of photographs that implies an approach, ending with a final close-up photograph. Participants perform the same BE rating task as block one on this close-up. The approach sequence is similar to the implied events used to study RM (e.g., Munger et al., 1999c) and will allow comparisons of BE ratings where motion has been implied to ratings resulting from single photographs (Block 1). Approaching a scene is the critical comparison condition for BE and RM—BE should lead to accepting scenes in which the central object is smaller (from having extended the boundaries), but RM should lead to accepting scenes in which the central object is larger (from continuing the implied, looming approach). Previous studies with simple objects approaching or receding (either implied or animated) have found

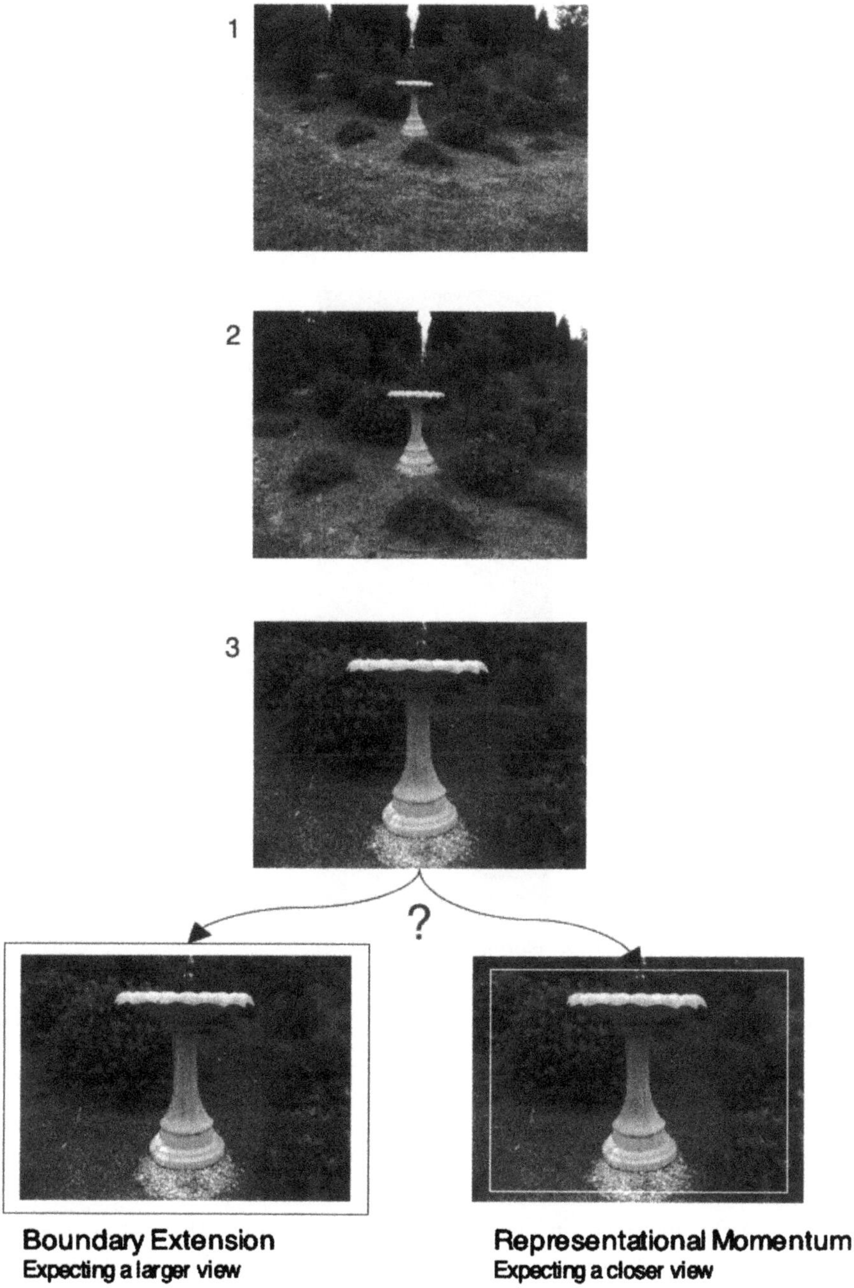

Figure 1. Boundary extension (BE) and representational momentum (RM) predict opposite reactions following the viewing of an approach sequence.

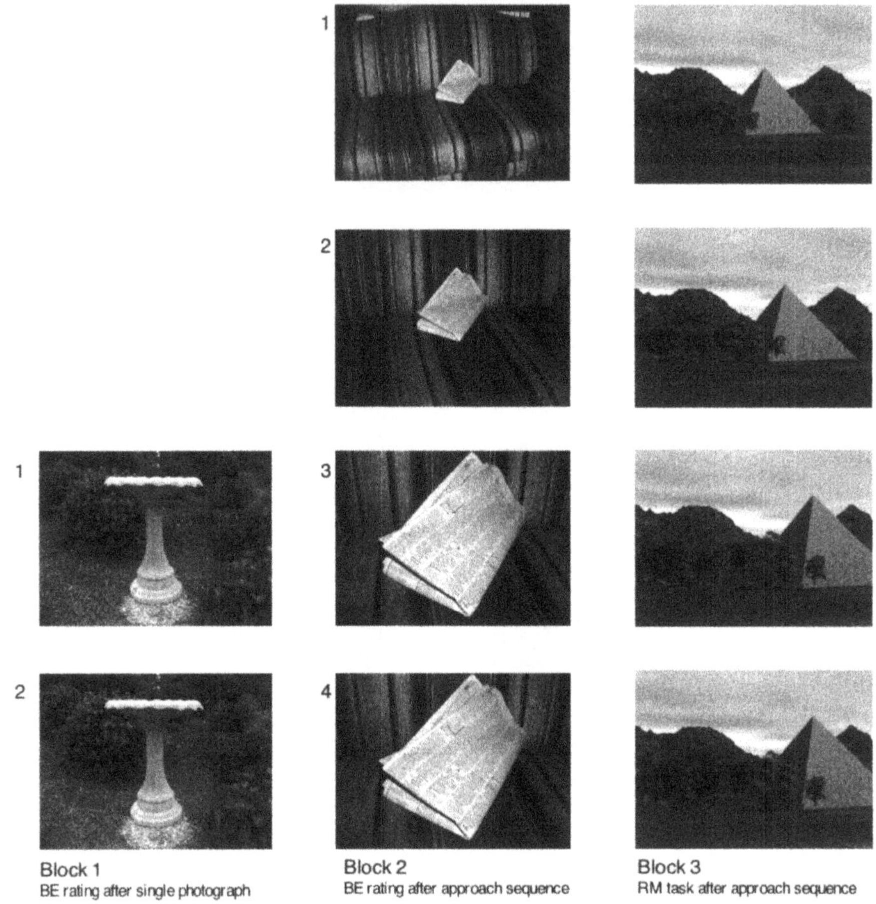

Figure 2. Example stimuli, although the actual images were in colour. The left column depicts a single photograph followed by BE rating task (Block 1). The central column depicts an implied approach, with the first frame at the top and subsequent frames descending; following the third frame, a fourth frame appeared for a BE rating (Block 2). The right column depicts the ''Mexico'' scene in a similar manner, with an RM task occurring at Frame 4 (Block 3).

smaller RM for approach, in keeping with some level of BE even without a surrounding spatial layout (Hubbard, 1995; Nagai et al., 2002). When asked to rate the distance of the redisplayed scene under the approach-motion condition (Block 2), smaller BE ratings, or even a shift to positive ratings indicating boundary restriction (BR; the object appears too far away, the remembered boundaries are smaller) would indicate that RM was occurring within the BE rating measurement.

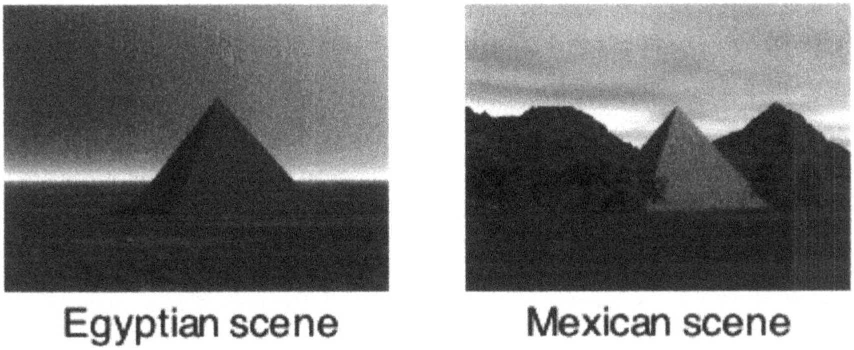

Figure 3. The pyramid scenes used in the RM task (Block 3). Note that there is surface texture in both scenes, and additional objects in the ''Mexico'' scene.

The final block presents two photorealistic landscape scenes generated using a computer graphics program in which an approaching motion is implied, and participants are asked to judge whether the third and fourth scenes are identical, an RM task. The computer-generated scenes allow a relatively easy manipulation of the scenic details and construction of the test positions necessary for the RM task. The same pyramid was presented in two contexts (see Figure 3) to begin to see if background impacts RM. For example, the more detailed scene may provide more cues to the viewer and reduce forward distortions. Given how readily individuals will show BE to photographic cutouts (Gottesman & Intraub, 2002), there is also the possibility that this surface information will lead only to BE, with no forward RM distortion present at all. The addition of a standard RM task will also allow an explicit comparison between BE rating data (Block 1) and RM performance (e.g., will high BE raters show more or less RM?).

Method

Participants. The participants were 43 introductory psychology students (13 men and 30 women) at Davidson College who received class credit for participating in the half-hour experiment.

Apparatus. The stimuli were displayed and collected by a Power Macintosh G4 Cube computer. Subjects sat a comfortable viewing distance (approximately 50 cm) from a 17-inch Apple Multiscan CRT monitor.

Stimuli. Blocks 1 and 2 used digital colour photographs of common objects in natural settings (e.g., fire hydrant with grass in the background; see Appendix for stimulus descriptions). The final photograph was always a close-up, but the central object was never cropped. The photographs measured 400 pixels wide

(14 cm, approximately 15° of visual angle) and 300 pixels high (10.5 cm, approximately 12° of visual angle). In Block 1, a single close-up photograph was presented for 250 ms, followed by a 250 ms ISI and a second presentation of the close-up photograph which remained on-screen until participants responded.

Block 2 presented a sequence of three photographs depicting an approach, with each photograph displayed for 250 ms followed by a 250 ms ISI. Photographs for the approach were made by keeping the focal length of the camera fixed and literally pacing out the distance to close-up object; so the photographs depict an actual approach, not simply a zoom-lens manipulation. The third photograph in every sequence was a close-up, and was presented a second time, following a 250 ms ISI, as in Block 1. Each trial of Block 2 presented a total of four photographs: The first three depicted an approach, and the fourth photograph was identical to the third and remained on-screen until participants responded.

Block 3 used computer-rendered pyramids presented in two different scenes (''Egypt'' and ''Mexico'', developed in Bryce 3D—a computer graphics program). The Egyptian scene depicted the pyramid in a desert, and the Mexican scene depicted the pyramid with a forest and mountain background (see Figure 3). The pyramid was positioned on the right of the screen in approximately the same location for both scenes. In the Egyptian scene, the pyramid was 182 pixels wide (6.4 cm, approximately 7.3° of visual angle) and 148 pixels in height (5.2 cm, approximately 6.0° of visual angle). In the Mexican scene, the pyramid was 174 pixels wide (6.1 cm, approximately 7.0° of visual angle) and 127 pixels in height (4.5 cm, approximately 5.1° of visual angle). Both backgrounds for the scenes were 320 pixels wide (11 cm, approximately 13° of visual angle) and 240 pixels in height (8.5 cm, approximately 9.7° of visual angle). The scenes were rendered using a single light source simulating sunlight. The Egyptian pyramid was shaded dark sand on the left side (on an Apple RGB scale of 0–100: red 24, green 26, and blue 15) and a lighter sand on the right side (red 44, green 40, and blue 24). The Mexican pyramid was similarly shaded with dark sand on the left (red 34, green 33, and blue 24) and a lighter right side (red 60, green 55, and blue 38). The surrounding scenery was developed using additional objects and textures from Bryce 3D.

In Block 3, the presentation of the stimuli was similar to Block 2, with an approach into a scene being presented in three brief, static views (constituting the inducing display) followed by a fourth scene (probe). Each scene in the inducing display lasted for 250 ms with an ISI of 250 ms resulting in an inducing display lasting 1.25 s. The probe scene was displayed until the participant responded. The approach depicted the centre of the scene approaching the observer, resulting in slight rightward movement of the pyramid. The location of the probe test positions varied about the final position with equal probability. The change in pixels for each test position from the final position of the pyramid was (horizontal shift, vertical shift): (left 20, up 6), (left 15, up 5), (left 10, up 3), (left 5, up 2), 0, (right 5, down 2), (right 10, down 3), (right 15, down 5), and (right 20, down 7). Notice that there are very similar amounts of change for positive and negative test positions, so any asymmetry in participant's responses should reflect RM, not overall discriminability of the test probes.

Procedure. All participants viewed all three blocks in the same order (BE rating following single photograph, BE rating following approach, RM task; see Figure 2). We fixed the order of the blocks for two reasons: To obtain a baseline BE score for each participant before any approaches were viewed, and to simplify the instructions. One set of instructions were provided for both BE tasks, and then new instructions were provided for the final RM task. Block 1 (BE rating following a single photograph) presented 10 experimental trials, in a unique random order for each participant. Block 2 (BE rating following approach) also presented 10 experimental trials in a unique random order for each participant. Finally, Block 3 (RM task) presented 72 experimental trails (2 scenes × positions × 4 replications) in a unique random order for each participant. For the BE rating tasks (Blocks 1 and 2), picture Set 1 was used in the single photograph condition (Block 1), followed by Set 2 in the approach condition (Block 2) for half the participants. The picture sets were switched for the other half. Within the RM task, all participants viewed both scenes (Egypt and Mexico).

A trial was initiated when the participants pressed the "return" key on the keyboard. For Blocks 1 and 2, participants rated the relative distance of the final photograph compared to the previous display (a BE rating task). They were instructed to determine whether the final scene was identical (0), much closer (−2), a little closer (−1), a little farther away (1), or much farther away (2) than the previous view. Participants indicated their rating by pressing labelled keys corresponding to each possible response. Prior to beginning the task, participants received examples on the computer of the degree of each possible response (i.e., the difference between "a little closer" and "much closer"). The rating scale is similar to those used in previous BE experiments (e.g., Intraub et al., 1998), though our participants entered ratings using the keyboard. For Block 3, participants were instructed to watch the complete presentation carefully and judge if the third and fourth scenes were the same or different, the typical RM task (e.g., Munger et al., 1999c). At the start of Block 3, nine representative practice trials with feedback were presented in a different random order for each participant prior to the experimental trials. Experimental trials were performed without feedback and in a unique random order for each participant. The instructions emphasized responding to what the display looked like, and the participants were not told of the unequal probabilities of "same" and "different" test positions. Following the experiment, participants were informed of the background and purpose of the study.

Results

The overall BE ratings for Blocks 1 and 2 were calculated, with negative ratings indicating boundary extension and positive ratings indicating boundary restriction. With this and all other analyses, $p < .05$ unless otherwise noted. Overall, significant BE was observed for both Block 1, BE rating following single photograph, mean rating = −0.21, $t(42) = 4.29$, and Block 2, BE rating following approach, mean rating = −0.26, $t(42) = 4.60$. The magnitude of our observed BE is slightly smaller than those typically observed; using similar timing (250 ms exposure) Intraub et al. (1996) observed BE of −0.74 for close-up photographs. Our next step was to examine individual performance in Block 1, the typical BE task. Using individual scores from Block 1, three groups emerged: individuals who show BE (BE group, $n = 29$), individuals who have accurate memory (no distortion, $n = 8$), and individuals who show boundary restriction (BR), which is remembering less of the picture than was presented (BR group, $n = 6$). An ANOVA with block (single photo, approach) as a within-subject variable and group (BE group, no distortion, and BR group) and picture set (Sets 1 and 2, counterbalanced between participants) as a between-subject variables found a significant main effect of block, $F(37) = 8.05$, a significant interaction between block and group, $F(37) = 9.83$, no interaction between block and picture set, and a significant three-way interaction between block, group, and picture set, $F(37) = 4.03$. Group means from Blocks 1 and 2 are presented in Figure 4 and the striking interaction is revealed. Individuals who show BE to single photographs (most of our participants), show somewhat less BE following an approach, as if they were beginning to react to the dynamic information of the approach. Significant BE ratings were observed for both blocks for the BE group: Single photograph, mean rating = −0.37, $t(28) = 8.11$; approach, mean rating = −0.23, $t(28) = 3.56$; and there is a marginally significant difference, $t(28) = 1.73$, $p < .10$.

The two other groups (no distortion and BR group) actually do the opposite when presented with an approach. The additional information presented by the initial views in the approach is not leading to RM-like continuation of the event, which would show as positive ratings (BR). Both groups actually show BE following the approach, even the group that was initially remembering less of the single photographs (BR group). For the no distortion group (individuals who literally had scores of 0 from the first block), significant BE is observed, mean rating = −0.43, $t(7) = 2.55$, which is a significant change from their Block 1 (single photograph) performance. A similar downward shift in BE ratings occurred for the BR group, starting with significant BR, mean rating = 0.25, $t(5) = 3.47$, and dropping to a nonsignificant BE rating, mean rating = −0.18, with a marginally significant difference between Blocks 1 and 2, $t(5) = 2.33$, $p < .10$.

Both picture sets led to BE (see Figure 5) and the ANOVA revealed no interaction between block and picture set. More of the individuals in the no

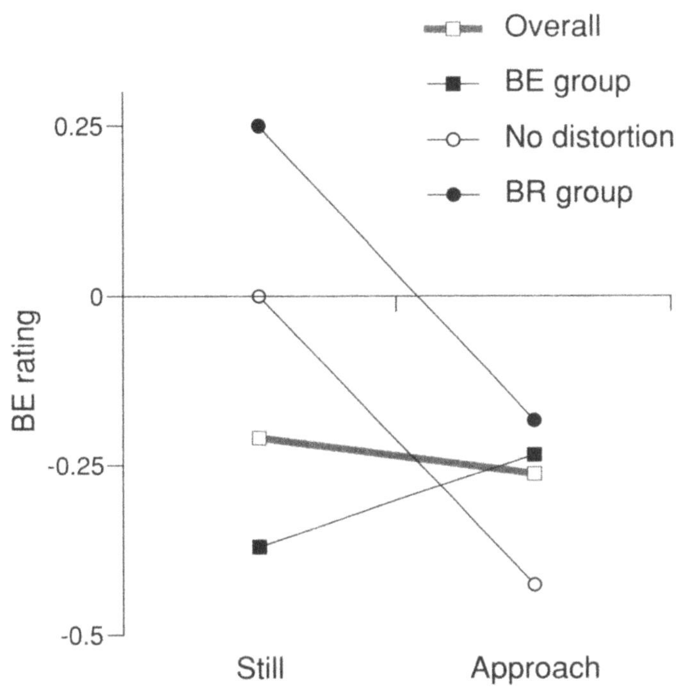

Figure 4. BE ratings, ratings from Block 1 (single photograph) were used to identify three groups: BE group, no distortion group, and BR group.

distortion (6 of 8) and BR groups (4 of 6) had Set 2 for Block 1, which probably explains the three-way interaction.

For the RM task of Block 3, percentage "same" responses as a function of test position are presented in Figure 6. To estimate the size and direction of the distortion, weighted means were calculated by summing the products of the test positions and percentage "same" responses and dividing by the total percentage "same" responses (e.g., Freyd & Jones, 1994; Hubbard, 1996). This weighted mean is positive when more "same" responses occur for test positions further along the implied event. Significant RM was observed for both the Egypt, weighted mean = 2.62, $t(42) = 3.69$, and Mexico, weighted mean = 2.80, $t(42) = 4.45$, scenes. To see if initial BE rating is related to RM scores, we ran a correlation between Block 1 performance (BE rating following single photograph) to the RM task (Block 3) and found no correlation (Pearson's correlation = .07). We also ran a correlation between BE rating following and approach (Block 2) and RM scores, again finding no correlation (Pearson's correlation = .15). To compare the experimental conditions, a 2 (scene) within-subject ANOVA with a between-variable of BE/no-distortion/BR grouping was

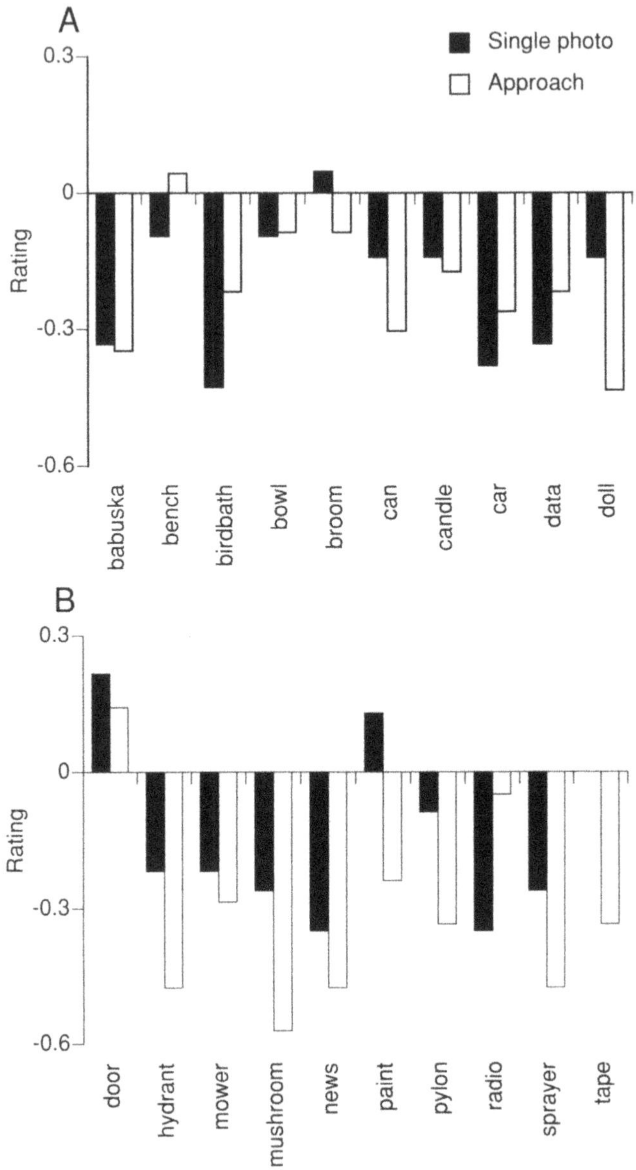

Figure 5. Mean BE ratings for single photo (Block 1) and approach (Block 2) for (A) Picture set 1 and (B) Picture set 2. Picture sets were counterbalanced, so participants who saw Set 1 in Block 1, then saw approach sequences using Set 2 in Block 2.

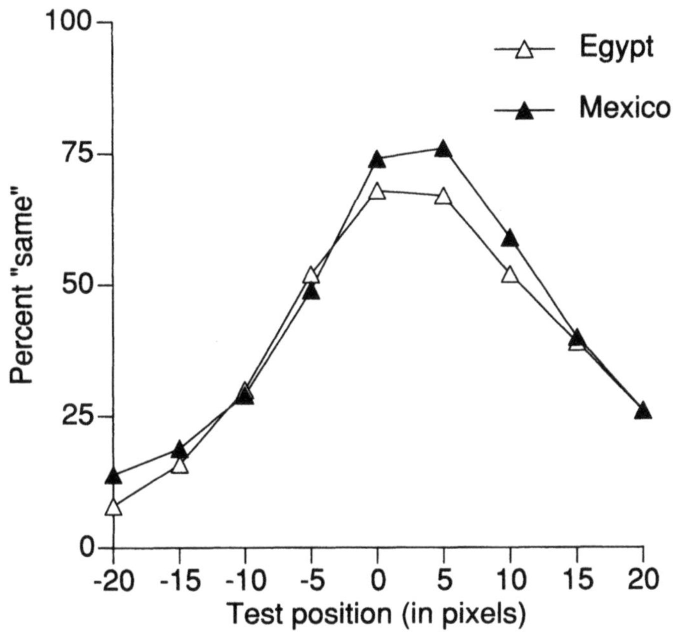

Figure 6. Percentage "same" responses for each scene in RM task (Block 3). Test positions are in pixels, and positive positions indicate test positions further along the approach.

computed on the weighted mean measure. There was not a significant main effect of scene, and no significant interactions. All participants show equivalent RM, despite differences in the BE task.

Discussion

Boundary extension is a prediction about what is just around the corner of your vision, and representational momentum is a prediction about the position of a moving object. While both distortions reveal slight inaccuracies within the perceptual system, both actually provide useful hints about what is likely to happen in the immediate future. The current experiment used a variety of scenes with both BE and RM tasks to examine whether or not these small predictions are related. We found both BE and RM, and the overall pattern suggests a separation of these distortions, not an overlap.

We found that how readily participants extended the boundaries on a single photograph had a profound impact on how they responded to subsequent approach conditions in a BE task, but not for an RM task. The majority of our participants showed BE to single photographs, and continued to show BE following an approach sequence. An interesting trend towards less BE was observed, as if they were beginning to include the approach sequence into their representation of the scene, in keeping with RM. In contrast, participants who either had no distortion for the single photographs or showed significant BR had more BE following the approach—the opposite of RM. Apparently, these individuals need to witness the surround before they are able to include additional layout information in their representation of the scene. Having been shown an approach, they readily extend the boundaries of the final, close-up photograph; they do not continue the approach. Despite this interaction between baseline BE and BE following an approach, there was no correlation or interaction with RM performance in Block 3. Significant RM was observed for both scenes, but it appears to be independent of BE. In the current experiment, the order of BE and RM tasks was fixed, treating BE to a single photograph as a baseline measure of the scene layout. It would be interesting to see if repeated exposure to approach sequences would impact later BE performance, either to the same scenes or new ones. There is evidence that expectations regarding moving objects impacts RM (Hubbard, 1995), so might expectations about movement impact BE?

The interaction between baseline BE and BE following an approach suggests that BE may precede RM-like inclusion of the movement into the representation of the scene. For participants who did not generate additional spatial layout information based on the single photograph (the No distortion and BR groups), the approach condition led them to BE. This is particularly intriguing for the BR group, since RM (continuing the approach) would fit with their initial reaction to the still photograph. In contrast, this is the opposite of what happened for the BE group—instead of having even larger BE, which the performance of the no distortion and BR groups might suggest—the BE group's ratings stayed the same (with a trend towards less BE). This group difference suggests that spatial layout information, as revealed by BE ratings, occurs before movement is included in the representation of a scene. In other words, before the no distortion and BR groups could include the dynamic approach information, the layout of the scene had to be extended, as revealed by their increased boundary extension following the approach. The BE group, however, who were able to establish the layout of a scene from a single photograph, showed a slight sensitivity to the new dynamic information in the approach.

Boundary extension continues when viewers have actually been presented with a wider context of a scene. In fact, presentation of the surround actually increases BE for individuals who initially do not enhance their perceptual schema with predictions about the continuing layout. Representational momentum occurs within scenes, but it does not interact with boundary extension. Even the tendency of some our participants to restrict the boundaries from a single photograph does not lead to disproportionately more RM following an approach, even though their initial representation of the single photograph was closer. BE does change with an approach sequence, but the nature of the change depends on the individual perceiver. How additional views and motion interacts with boundary extension, and what aspects of spatial layout, if any, are contributing to representational momentum for movement in a scene are questions for future research.

REFERENCES

DeLucia, P. R. (2001). *Visual memory for moving scenes: Boundary extension or representational momentum?* Poster presented at the 42nd annual meeting of the Psychonomics Society, Orlando, FL.

Freyd, J. J. (1983). The mental representation of movement when static stimuli are viewed. *Perception and Psychophysics, 33*, 575–581.

Freyd, J. J., & Finke, R. A. (1984). Representational momentum. *Journal of Experimental Psychology: Learning, Memory, and Cognition, 10*, 126–132.

Freyd, J. J., & Jones, K. T. (1994). Representational momentum for a spiral path. *Journal of Experimental Psychology: Learning, Memory, and Cognition, 20*, 968–976.

Futterweit, L. R., & Beilin, H. (1994). Recognition memory for movement in photographs: A developmental study. *Journal of Experimental Child Psychology, 57*, 163–179.

Gottesman, C. V., & Intraub, H. (2002). Surface construal and the mental representation of scenes. *Journal of Experimental Psychology: Human Perception and Performance, 28*, 589–599.

Halpern, A. R., & Kelly, M. H. (1993). Memory biases in left versus right implied motion. *Journal of Experimental Psychology: Learning, Memory, and Cognition, 19*, 471–484.

Hayes, A. E., Sacher, G., Thornton, I. M., Sereno, M. E., & Freyd, J. J. (1996). Representational momentum in depth using stereopsis. *Investigative Ophthalmology and Visual Science, 37*(Suppl. 3), S467.

Hubbard, T. L. (1995). Environmental invariants in the representation of motion: Implied dynamics and representational momentum, gravity, friction, and centripetal force. *Psychonomic Bulletin and Review, 2*, 322–338.

Hubbard, T. L. (1996). Displacement in depth: Representational momentum and boundary extension. *Psychological Research, 59*, 33–47.

Hubbard, T. L., & Ruppel, S. E. (1999). Representational momentum and the landmark attraction effect. *Canadian Journal of Experimental Psychology, 53*, 242–255.

Hubbard, T. L., & Ruppel, S. E. (2000). Spatial memory averaging, the landmark attraction effect, and representational gravity. *Psychological Research, 64*, 41–55.

Intraub, H. (2002). Anticipatory spatial representation of natural scenes: Momentum without movement? *Visual Cognition, 9*, 93–119.

Intraub, H., & Bodamer, J. L. (1993). Boundary extension: Fundamental aspect of pictorial representation or encoding artifact. *Journal of Experimental Psychology: Learning, Memory, and Cognition, 19*, 1387–1397.

Intraub, H., Gottesman, C. V., Willey, E. V., & Zuk, I. J. (1996). Boundary extension for briefly glimpsed photographs: Do common perceptual processes result in unexpected memory distortions? *Journal of Memory and Language, 35*, 118–134.

Intraub, H., Gottesman, C. V., & Bills, A. (1998). Effects of perceiving and imaging scenes on memory for pictures. *Journal of Experimental Psychology: Learning, Memory, and Cognition, 24*, 186–201.

Intraub, H. B., Bender, R. S., & Mangels, J. A. (1992). Looking at pictures but remembering scenes. *Journal of Experimental Psychology: Learning, Memory, and Cognition, 18*(1), 180–191.

Kourtzi, Z., & Kanwisher, N. (2000). Activation in human MT/MST by static images with implied motion. *Journal of Cognitive Neuroscience, 12*(1), 48–55.

Munger, M. P., Covington, B., Minchew, J., & Starr, J. A. (1999a, November). *Representational momentum in scenes.* Poster presented at Psychonomic Society, Los Angeles, CA, USA.

Munger, M. P., & Minchew, J. H. (2002). Parallels between remembering and predicting an object's location. *Visual Cognition*, *9*, 177–194.
Munger, M. P., & Owens, T. R. (2004). Representational momentum and the flash-lag effect. *Visual Cognition*, *11*, 81–103.
Munger, M. P., Solberg, J. L., & Horrocks, K. K. (1999b). The relationship between mental rotation and representational momentum. *Journal of Experimental Psychology: Learning, Memory, and Cognition*, *25*, 1557–1568.
Munger, M. P., Solberg, J. L., Horrocks, K. K., & Preston, A. S. (1999c). Representational momentum for rotations in depth: Effects of shading and axis. *Journal of Experimental Psychology: Learning, Memory, and Cognition*, *25*, 157–171.
Nagai, M., Kazai, K., & Yagi, A. (2002). Larger forward memory displacement in the direction of gravity. *Visual Cognition*, *9*, 28–40.
Reed, C. L., & Vinson, N. G. (1996). Conceptual effects on representational momentum. *Journal of Experimental Psychology: Human Perception and Performance*, *22*, 839–850.
Thornton, I. M., & Hayes, A. E. (2004). Anticipating action in complex scenes. *Visual Cognition*, *11*(2/3), 341–370.
Vinson, N. G., & Reed, C. L. (2002). Sources of object-specific effects in representational momentum. *Visual Cognition*, *9*, 41–65.

APPENDIX: PHOTOGRAPHS USED IN BLOCKS 1 AND 2

Set 1

Black tape dispenser on dark, wooden desk
Cleaning sprayer on steps
Closed wooden door in a carpeted hallway
Dave's push lawn mower at the edge of lawn and driveway
Folded newspaper on boldly striped couch
Grey and black radio on patterned tile floor
Orange traffic pylon on paved road
Paint can on paving stones in front of brick wall
White mushroom in grass
Yellow fire hydrant on grass with curb

Set 2

Broom leaning against grey house
Cement birdbath in park
Covered pottery bowl on deck
''Data'' action figure on multi-coloured carpet
Jimmy's Lego remote control car on block-quilted bed spread
Metal watering can on grass
Nora's red-haired doll in brown dress sitting on beige carpet
Russian nested doll on light, wooden table
Votive candle, with stand, on dark wood table
Wrought iron bench on paving stones in Greta's herb garden

VISUAL COGNITION, 2005, 12 (6), 1057–1092

The limits of visual resolution in natural scene viewing

Lester C. Loschky
Kansas State University, Manhattan, KS, USA

George W. McConkie
University of Illinois at Urbana-Champaign, IL, USA

Jian Yang and Michael E. Miller
Eastman Kodak Company, Rochester, NY, USA

We examined the limits of visual resolution in natural scene viewing, using a gaze-contingent multiresolutional display having a gaze-centred area-of-interest and decreasing resolution with eccentricity. Twelve participants viewed high-resolution scenes in which gaze-contingent multiresolutional versions occasionally appeared for single fixations. Both detection of image degradation (five filtering levels plus a no-area-of-interest control) in the gaze-contingent multiresolutional display, and eye fixation durations, were well predicted by a model of eccentricity-dependent contrast sensitivity. The results also illuminate the time course of detecting image filtering. Detection did not occur for fixations below 100 ms, and reached asymptote for fixations above 200 ms. Detectable filtering lengthened fixation durations by 160 ms, and interference from an imminent manual response occurred by 400–450 ms, often lengthening the next fixation. We provide an estimate of the limits of visual resolution in natural scene viewing useful for theories of scene perception, and help bridge the literature on spatial vision and eye movement control.

SCENE PERCEPTION AND RESOLUTION ACROSS THE VISUAL FIELD

Although the eyes tend to be directed toward objects of current interest, parafoveal and peripheral visual information is generally believed to play several important roles in the free viewing of natural scenes: Aiding in perceiving the gist of the image, perceiving the layout of the image, recognizing objects, and helping select the object to which the eyes are sent next. It is, therefore, important to understand the nature of information available

Please address all correspondence to: Lester Loschky, Assistant Professor, Department of Psychology, 471 Bluemont Hall, 1101 Mid-Campus Drive, Manhattan, KS 66506-5302, USA. Email: loschky@ksu.edu

DOI:10.1080/13506280444000652

from extrafoveal regions and its absolute limits (cf. similar ideas in reading Rayner, 1998).

Parafoveal and peripheral information appears to exert a large influence on the gist viewers acquire from a scene. Scene gist appears to be acquired from information across the entire image, not only in the foveal region (Boyce & Pollatsek, 1992; Boyce, Pollatsek, & Rayner, 1989; Li, VanRullen, Koch, & Perona, 2002; Schyns & Oliva, 1994). Thus, the nature of the information used to acquire gist becomes important. Some evidence suggests that the spatial frequency information for gist may be mostly low frequency, such as oriented blobs, but occasionally, it is also higher frequency, such as outlines and edges (Oliva & Schyns, 2000; Schyns & Oliva, 1994). Low frequency information may be especially useful for conveying the layout of a scene, which may be important not only for acquiring its gist, but also for understanding the spatial relations among its constituent objects (Sanocki, 2003; Sanocki & Epstein, 1997). Yet, so far, no accurate characterization of the spatial frequency information available from each part of the visual field during extended free viewing of natural scenes has been provided.

Parafoveal and peripheral vision is also useful for object recognition (Henderson, Pollatsek, & Rayner, 1987, 1989; Henderson, McClure, Pierce, & Schrock, 1997). This raises the question of how far into the visual periphery objects can be recognized. In fact, no simple distance measure is likely to provide an adequate answer since object recognition in the parafovea and periphery varies with task difficulty, image size, contrast, and eccentricity. In demanding tasks, such as using long-term memory to detect a change between two scenes involving a single critical object, good task performance is only possible when the eyes fixate within roughly 2° of the critical object (Nelson & Loftus, 1980; see also Henderson, Williams, Castelhano, & Falk, 2003). However, in simple tasks, such as determining whether a large natural scene contains an animal or not, performance is quite good at even 50° eccentricity (Thorpe, Gegenfurtner, Fabre-Thorpe, & Buelthoff, 2001). Importantly, much research has shown that peripheral visual performance can be equated with the fovea by scaling the size and contrast of stimuli such as gratings, and also natural objects such as faces (Maekelae, Naesaenen, Rovamo, & Melmoth, 2001). Since recent studies have proposed an ideal spatial frequency range for foveal object recognition (Braje, Tjan, & Legge, 1995; Gold, Bennett, & Sekuler, 1999; Olds & Engel, 1998), the limits of visual resolution across the visual field must critically influence peripheral object recognition.

Parafoveal and peripheral information is essential in guiding the eyes during free viewing. As the eyes explore a scene, they move several times per second, with attention guiding the eyes to different scene regions (Deubel & Schneider, 1996; Hoffman & Subramaniam, 1995; Irwin & Gordon, 1998; Kowler, Anderson, Dosher, & Blaser, 1995; for review see Henderson & Hollingworth, 1998). Attentional selection involves preattentively processing parafoveal and peripheral information (Levin, Takarae, Miner, & Keil, 2001; Treisman & Gelade, 1980; Wolfe, 1994). But what information is available from parafoveal and peripheral regions to guide attention? A few studies have varied the information content in scenes available on each eye fixation, in ways that can be related to variation in visual sensitivity across the visual field, and shown effects on eye guidance (Loschky & McConkie, 2002; Shioiri & Ikeda, 1989; van Diepen & Wampers, 1998). However, none of these studies varied image content to test models of visual sensitivity across the visual field. Thus, the relationship between the basic limits of parafoveal and peripheral sensitivity and eye guidance in scene viewing is still largely unknown.

All of the above-mentioned aspects of scene perception require parafoveal and peripheral visual information, sensitivity to which varies with retinal eccentricity. Thus, understanding these processes in scene perception first requires understanding the limits of visual resolution across the visual field. While there is a long history of psychophysical research on the sensitivity of the visual system as a function of distance from the centre of vision (for reviews, see Levi, 1999; Wilson, Levi, Maffei, Rovamo, & DeValois, 1990), only recently has such research been extended to natural scene stimuli and free viewing.

VARIABLE RESOLUTION OF THE VISUAL SYSTEM AND IMAGE FILTERING

At present, the most theoretically motivated measure of visual resolution is in terms of contrast sensitivity (Campbell, 1983). The resolution of any visual stimulus, from the simple alternating bars of a grating to a complex natural scene, is analysable in terms of its spatial frequency content, with fine detail (e.g., edges) comprised of high spatial frequencies, and coarse information (e.g., blobs) comprised of lower spatial frequencies. The range of visible spatial frequencies, measured by sensitivity to contrast at each frequency, is described by a contrast sensitivity function (CSF; Campbell & Robson, 1968; DeValois & DeValois, 1988), which varies as a function of distance from the fovea, or *retinal eccentricity* (Anderson, Mullen, & Hess, 1991; Banks, Sekuler, & Anderson, 1991; Cannon, 1985; Peli, Yang, & Goldstein, 1991; Pointer & Hess, 1989; Robson & Graham, 1981). Contrast thresholds increase as a function of retinal eccentricity much faster for higher spatial frequencies than for lower spatial frequencies.

A model of eccentricity-dependent contrast sensitivity

Based on much previous research using grating stimuli in simple detection or discrimination tasks, it is possible to mathematically model the drop-off of visual sensitivity across the visual field. Peli et al. (1991) described the contrast threshold for detecting a grating patch of spatial frequency f at an eccentricity E as:

$$C_t(r, f) = C_t(0, f) \exp(k f E) \quad (1)$$

where $C_t(0, f)$ is the contrast threshold at the fovea, and k determines how fast sensitivity drops off with eccentricity. Peli et al. (1991) fit this equation to data from six experiments investigating the contrast thresholds of monochromatic gratings. These fits showed that the k value ranged from 0.030 to 0.057. In Equation 1, contrast thresholds increase rapidly with eccentricity for high spatial frequencies, suggesting that this information is only useful in central vision. Since the foveal contrast threshold was not defined in Equation 1, Yang et al. (2001) used an equation to estimate it based on previous research (Yang, Qi, & Makous, 1995; Yang & Stevenson, 1998):

$$C_t(0, f) = [N + \eta^2/(f^2 + \sigma^2)] \exp(\alpha f) \quad (2)$$

where, N, η, σ, and α are parameters (for more detail, see Yang et al., 1995). Using Equations 1 and 2, we can estimate the limits of resolution across the visual field in terms of the *spatial frequency cutoff* for perception at any given eccentricity. For a set of nominal k, N, and α values being 0.045, 0.0024, and 0.14°, respectively, the cutoff spatial frequency f_c can be approximated as:

$$f_c = 43.1 * E_2/(E_2 + E) \quad (3)$$

with E_2 = 3.11°. E_2 is the retinal eccentricity at which the spatial frequency cutoff drops to half its foveal maximum (from 43.1 cpd to 21.55 cpd). (In other psychophysical studies, E_2 has often been defined as the retinal eccentricity at which foveal thresholds double, e.g., Beard, Levi, & Klein, 1997). Note that the cutoff frequency is based on the assumption of 100% contrast at each frequency. Thus, at lower contrast (e.g., in natural images) the cutoffs would drop to lower spatial frequencies.

Testing the limits of visual resolution using static multiresolutional displays

A key question is whether the above model scales up to natural images and viewing conditions. To test the model with natural images, we filter a scene's image resolution to match the resolution of the human visual system at each retinal eccentricity. If image filtering removes, at each eccentricity, only imperceptible information (e.g., frequencies higher than the threshold), the image should produce a normal perceptual experience for a viewer. On the other hand, if filtering removes perceptible information (e.g., frequencies lower than the threshold) at each eccentricity, viewers may perceive image degradation.

Several recent studies using briefly flashed (150–300 ms) multiresolutional photographic images have tested the limits of visual resolution in natural scenes (Peli & Geri, 2001; Sere, Marendaz, & Herault, 2000; Yang et al., 2001). Peli and Geri (2001) and Yang et al. (2001) tested models of eccentricity-dependent contrast sensitivity by varying parameter values producing different filtering frequency cutoffs, and then measuring viewers' perceived image degradation. In both studies, the effects of the resolution drop-off parameters on perceived image quality were both robust and close to predictions. Yet, in both studies, viewers were somewhat less sensitive to image filtering than predicted based on previous studies using isolated gratings or Gabor patches (see also Sere et al., 2000). Peli and Geri explain the latter unexpected finding in terms of masking by natural scene detail.

Importantly, because these studies used briefly flashed images, their results may not apply to dynamic, multifixation, free-viewing conditions. Attention focus may begin broad and become narrow during extended scene viewing, as suggested by viewers' initial long saccades (eye movements) with short fixations (still periods) and, later, short saccades with long fixations (Antes, 1974). Since attention focus affects visual resolution (Yeshurun & Carrasco, 1999), a rigorous test of the limits of visual resolution across the visual field in scene viewing should allow free eye movements.

Testing the limits of visual resolution using gaze-contingent multiresolutional displays

We propose to find the limits of visual resolution in dynamic natural scene viewing with a gaze-contingent display-change method (e.g., McConkie & Rayner, 1975; Rayner, 1975) in which image resolution varies with distance from the point of gaze, and the centre of highest resolution is aligned with gaze. We will use this gaze-contingent multiresolutional display (Parkhurst & Niebur, 2002; Reingold, Loschky, McConkie, & Stampe, 2003) to test predictions of the model of eccentricity-dependent contrast sensitivity (Yang et al., 2001) described in Equations 1–3, by measuring the perception of image degradation. The model predicts that image filtering matching the high frequency cutoff function in Equation 3 will be the degradation detection threshold; less filtering should be undetectably degraded, and more filtering should be more detectably degraded. We will test these predictions by varying the E_2 parameter in Equation 3 to produce a range of filtering levels, and finding the degradation threshold E_2 value. Besides testing the model's predicted threshold, the frequency cutoff function of the empirical threshold filtering level will provide an estimate of the limits of visual resolution in dynamic natural scene viewing.

To date, we know of one published study that has attempted to test the limits of visual resolution across the visual field, using natural images, under dynamic free-viewing conditions (Sere et al., 2000, Exp. 2). However, that experiment's only dependent measure was participants' reported ''subjective distinctness of the image'' at the end of each trial (p. 1410), and its only reported results were (1) that ''13 out of 14 observers reported a subjective visual change of the image,'' when filtered and unfiltered images were presented on alternating fixations, although (2) ''none of the observers reported that the images were filtered'' (p. 1410). Thus, there is clearly room for more investigation of this question.

The above study raises a critical methodological question. What is a good way to measure viewers' detection of image degradation when using a gaze-contingent multiresolutional display? If the point of highest resolution is updated on *every* fixation, and one asks viewers to report detection of image filtering only at the end of a trial, one does not know *when* the filtering was detected during the trial. In fact, a viewer could report detection when the filtering was actually only perceived on the last fixation of a trial. A more informative measure is to ask viewers to detect instances of image filtering that only occur *occasionally*, for single critical fixations, several times per trial, with a normal constant high-resolution image present at all other times (see Figure 1). This occasional gaze-contingent multiresolutional display method measures how frequently removal of specified spatial frequencies is detected, using the single fixation, rather than the entire trial, as the test unit.

The effects of above-threshold filtering on eye movements. We will also examine the effects of above-threshold filtering on eye behaviour. Several studies have shown that image degradation increases fixation durations (reduced luminance: Loftus, 1985; Loftus, Kaufman, Nishimoto, & Ruthruff, 1992; low-pass filtering: Loschky & McConkie, 2002; Mannan, Ruddock, & Wooding, 1995; Parkhurst, Culurciello, & Niebur, 2000; high-pass filtering: Mannan et al., 1995). This effect has been interpreted as reflecting disrupted processing (Loschky & McConkie, 2002), in particular, increased time needed to reach an information threshold (Loftus, 1985; Loftus et al., 1992).

Importantly, no studies have investigated the effects of removing *above-threshold* frequencies on fixation durations. None of the previous studies directly investigating resolution limits across the visual field (Peli & Geri, 2001; Sere et al., 2000; Yang et al., 2001) measured eye movements. Conversely, of those studies investigating the effects of image degradation on fixation durations (Loftus, 1985; Loftus et al., 1992; Loschky & McConkie, 2002; Mannan et al., 1995; Parkhurst et al., 2000), only Loschky and McConkie first assessed the perceptibility of the degradation. However, they did not use an image filtering method based on a model of eccentricity-dependent contrast sensitivity, so we cannot use their observed effects to predict the limits of resolution across the visual field. Thus, we will relate such limits, as measured by image filtering detection, to the effects of such filtering on eye movements, creating a bridge between the literatures on spatial vision and eye movement control.

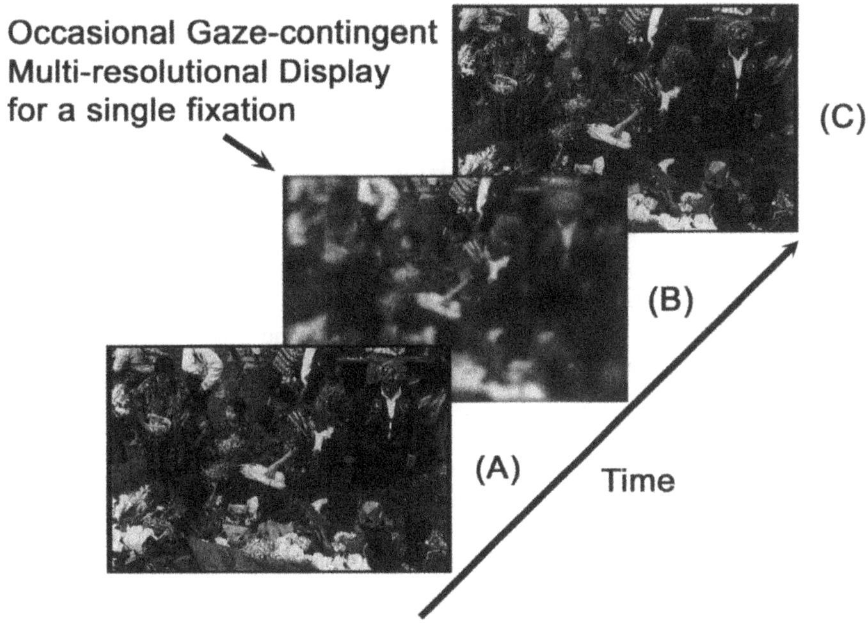

Figure 1. A schematic of an occasional GCMRD over time. (A) The viewer sees a constant high-resolution image for a prespecified number of fixations (unknown to the viewer). (B) For a single fixation, a gaze-contingent multiresolutional image is presented with the centre of high-resolution at the viewer's point of gaze. The viewer is requested to press a button whenever he/she detects that this has happened. (C) At the start of the next fixation, the constant high-resolution image is again put on the screen for another prespecified number of fixations.

The time course of processes in detecting image filtering. We will also study the time course of perception of image filtering by investigating (1) the effects of fixation durations on the perceptibility of image filtering, (2) the effects of the perceptibility of image filtering on fixation durations, and (3) the effects, if any, of making a manual response (e.g., a button press) on fixation durations. First, we will determine if longer critical fixations, and thus, longer filtered stimulus durations, increase the likelihood of detecting image filtering, as suggested by studies of stimulus duration effects on perception and memory (e.g., Bundesen & Harms, 1999; Loftus & Ruthruff, 1994). Second, we will use the temporal structure of fixation duration distributions to elucidate the time course of processes involved in extracting scene information at or below the limits of visual resolution, using similar methods to those used to study the time course of perceptual processes in reading (Yang, 2002; Yang & McConkie, 2001, 2004). Third, a complication of our occasional gaze-contingent multiresolutional display detection paradigm is that, since the observer must make a motor response in the detection task (e.g., a button press), the motor response may itself affect eye behaviour (e.g., Fischer, 1999). Thus, we will also investigate the nature of the effect, if any, of making a manual response on fixation durations.

METHOD

Participants

There were 12 paid participants (6 female, 6 male), ages 19–34 ($M = 21.6$), all but one of whom had 20/20 uncorrected near vision in both eyes. The other participant had 20/30 uncorrected near vision in the right eye and 20/20 uncorrected near vision in the left.

Stimuli

The stimuli consisted of a set of 18 monochrome, 8-bit resolution photographs shown on a computer monitor (described below). The subject matter of the images was varied (from street scenes to building interiors), and contained a large amount of visual detail (see Figure 1). Images measured 768 pixels × 512 lines and subtended 18° × 12° visual angle.

The images were filtered using an algorithm developed at Eastman Kodak Company, which is a modified version of Geisler and Perry's (1998) foveated multiresolution pyramid. In the Geisler and Perry approach, different levels of the pyramid are circularly truncated, based on the estimated spatial frequency cutoffs of the visual system at different eccentricities. The reconstructed image from the zone-limited pyramid contains fine structure at the point of gaze, and becomes more blurred with eccentricity. In the modified method, we compute a contrast threshold map based on the peak frequency of the pyramid level, and use the contrast threshold map to threshold (i.e., reduce to zero) the image content. Whenever the image contrast of a specific frequency band is below the corresponding visual contrast threshold/cutoff, the image contrast is thresholded. All other image content is kept unchanged.

We used five levels of image filtering (see Figure 2) produced by varying a parameter in the filtering algorithm corresponding to E_2 for grating resolution (see Equation 3). The filtering E_2 values are 6.22, 3.11, 1.55, 0.78, and 0.39°. Importantly, the filtering cutoff for an E_2 value of 3.11° was predicted to be at contrast threshold at each retinal eccentricity. The largest E_2 value, 6.22°, twice the predicted threshold value, removed only higher spatial frequencies than the predicted contrast threshold at each eccentricity (i.e., below-threshold frequencies), and was, therefore, predicted to produce imperceptible image filtering. Conversely, the smaller filtering E_2 values, 1.55°, 0.78°, and 0.39°, removed incrementally lower spatial frequencies than the predicted contrast threshold (i.e., above-threshold frequencies), and were, therefore, predicted to

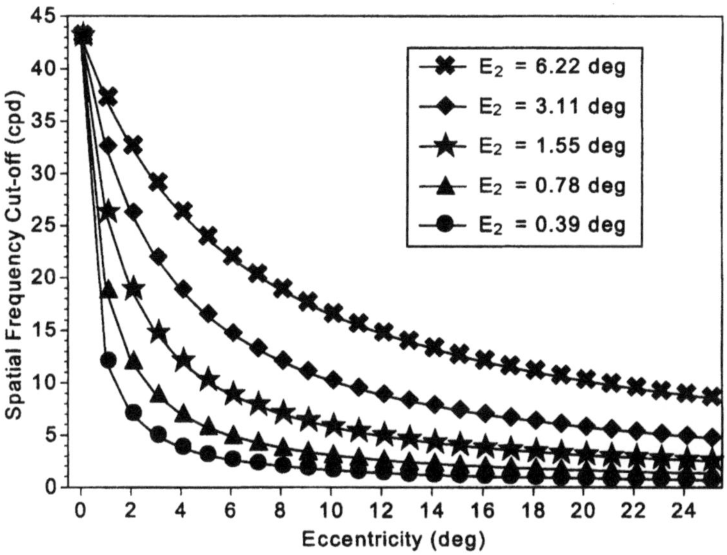

Figure 2. Spatial frequency cutoff as a function of retinal Eccentricity and filtering level for the five filtering conditions in the study. Contrast is set to 100%. cpd = cycles per degree. deg = visual angle in degrees.

produce perceptible image degradation. Example images in two filtering conditions and the control condition are shown in Figure 3.

There was a total of eighteen experimental images (three sets of six images each). Each set had an image in each of the five filtering conditions and one in a constant high-resolution, no-area-of-interest condition (i.e., a normal, high-resolution image). The six conditions were counterbalanced across images and participants.

Apparatus

Spatial alignment of the area-of-interest. In a static presentation paradigm, if the viewer carefully fixates the fixation cross, the point of highest resolution, or *area of interest*, will be accurately aligned with the centre of vision. In a dynamic presentation paradigm, imprecisely aligning the area-of-interest may result in degraded imagery being put at the fovea, even if the image resolution drop-off rate matches that of the retina. To avoid this problem, we used an eyetracker with a high level of spatial accuracy (a Dual Purkinje Image Generation 5 tracker with less than 10′ error; Crane & Steele, 1978), and an array of area-of-interest locations with relatively fine spatial resolution. The array had 330 area-of-interest locations based on a 22 × 15 imaginary grid over

Figure 3. A set of three example images. (A) A constant high-resolution control image; (B) the same image filtered at a level that removes spatial frequencies predicted to be at or below contrast threshold (E_2 = 3.11°); (C) the same image filtered at a level that removes spatial frequencies predicted to be 2 octaves above contrast threshold (E_2 = 0.39°).

the 18° × 12° image, with ~0.82° between grid points. We created 330 versions of each image, one area-of-interest version per grid point. Thus, wherever viewers looked, there was an image version whose centre of high-resolution was within 0.41° (horizontally or vertically) of their centre of gaze.

Speed of moving the area-of-interest. Unlike with a static presentation, in a gaze-contingent multiresolutional display, the area-of-interest must be quickly updated with each gaze movement. If this updating occurs too slowly, the viewer may notice degradation at the new gaze location prior to the update, or motion caused by the change in resolution across the image. Thus, we designed a system to minimize image-update delays. First, eye position was sampled at 1000 Hz (1 sample/ms), providing high temporal resolution in identifying the ends of saccades. Second, to avoid on-line image generation time, all 330 versions of each image were pre-stored in a ViewGraphics ViewStore 1 GB randomly accessible image memory and display controller. Third, display changes were initiated within 5 ms of the end of each saccade, based on the results of a separate study (McConkie & Loschky, 2002) showing that this deadline eliminated perception of image-update motion. Fourth, our Conrac Mars 9320 monitor had a 60 Hz refresh rate, and updates could be made at any point during the 17 ms refresh cycle. Thus, an area-of-interest update was completed within 22 ms of the start of a fixation.

These methods allow greater confidence that our observed effects are the result of the spatial characteristics of the images rather than the result of inaccurate area-of-interest placement or perceived image motion.

Procedures

The procedures were designed to measure participants' detection of peripheral filtering during dynamic image viewing. On most fixations, the normal, constant high-resolution image was displayed. The gaze-contingent multiresolutional display image was presented only on occasional, unpredictable, single fixations. Participants' two tasks were to (1) prepare for a long-term memory change detection test, and (2) do a concurrent peripheral blur detection task. The memory task was to motivate participants to make many eye movements. Participants were told they would be tested after seeing all 18 images; however, no final test was actually given.

Figure 1 is a schematic of the occasional gaze-contingent multiresolutional display paradigm. Each trial was a single picture presented for 30 s, beginning with a fixation point at the centre of the screen. When ready, the participant pushed a button to begin the trial. After 750 ms, a constant high-resolution picture appeared. After a randomly varying interval of 9–11 saccades, at the start of the next fixation, a gaze-contingent multiresolutional version of the image was presented with its point of highest resolution at the centre of gaze. The original constant high-resolution image then returned during the next saccade. Thus, the blurred image was present for a single *critical fixation*. The next critical fixation occurred after another 9–11 saccades, and this pattern continued until the end of the trial. Participants had to press a button as quickly as possible whenever they noticed image blur.

Each set of six images constituted a block, and the eyetracker was also recalibrated before each block. Participants were given detailed instructions and practice with the tasks prior to the experiment.

In the analyses of the data, a probability level of .05 was used for all inferential statistics.

RESULTS

Precursors

In this experiment, we tested the limits of visual resolution during free viewing of natural scenes, the effects of above-threshold filtering on eye movements, and the time course of those effects. We first cleaned the data by making several exclusions. We excluded all data from one of the 18 images (5.25% of the total) because of an artifact (a slight vertical shift between image versions). We excluded blinks, and cases with fixation durations in the 99th percentile and saccade lengths in the 1st percentile (3164 ms, and 0.16°, respectively). We kept short fixation durations to study their effect on detection, and excluded the very shortest saccades to avoid possible display change errors. We also analysed area-of-interest positioning error and excluded all cases with errors $\geq 2°$ (3.6% of the total). After all exclusions (12% of the original total), 925 critical fixations remained, with 94% having an area-of-interest positioning error $<1°$ ($M = 0.50°$, median = 0.45°, mode = 0.32°).

Perceptibility of image degradation as a function of filtering level

A key question is whether the Yang et al. (2001) model of visual sensitivity, based on studies using simple grating and Gabor patch stimuli, predicts the perceptibility of image filtering during dynamic free viewing of natural scene images. To answer this question, we measured viewers' explicit detection task responses, and their eye fixation durations. The detection task is a direct measure of perceptibility, but viewers' fixation durations might reveal an implicit sensitivity to filtered spatial frequencies absent from their explicit detection responses (cf. Hayhoe, Bensinger, & Ballard, 1998; Hollingworth, Williams, & Henderson, 2001). This combination of explicit and implicit measures should give a more detailed picture of viewers' sensitivity across the field of view in natural scene viewing. As we will show below, the results are largely consistent with the Yang et al. (2001) model.

Detection as a function of filtering level. The Yang et al. (2001) model predicted that filtering level 1 would be undetectably degraded, that level 2 would be at detection threshold, and that levels 3–5 would be increasingly detectable. We analysed detection rates as a function of filtering level using a multivariate ANOVA (five levels of filtering) on participants' arcsine transformed proportion detection data (Table 1). Raw proportions are shown in Figure 4, indicating a robust and significant main effect for filtering level on detection. Linear and cubic trends for level of filtering on detection were significant, linear trend, $F(1, 11) = 619.58$, $p < .001$, effect size = .98; cubic trend, $F(1, 11) = 83.11$, $p < .001$, effect size = .88, indicating that as filtering E_2 decreased, detection increased, but with both low and high detection asymptotes. The error bars in this figure represent 95% confidence intervals for the difference of each point against all others. Though the Yang et al. model predicted that filtering level 2 would be at detection threshold, there was no significant difference between filtering levels 1 and 2 (E_2 = 6.22 vs. 3.11°), neither of which was statistically different from zero detection. However,

TABLE 1
Multivariate ANOVA on mean proportion detection as a function of filtering level (E_2 = 0.39–6.22)

Effect	*Pillai's trace*	*F*	*Hypothesis df*	*Error df*	*Sig.*	*Partial eta squared*
Filtering level	.986	139.515	4.000	8.000	.000	.986

consistent with the model, filtering level 1 (E_2 = 6.22) was essentially never detected—the five detections across all participants were essentially identical to the six false alarms in the no-area-of-interest condition. There was no significant difference in detection between filtering levels 4 and 5 (E_2 = 0.78 vs. 0.39°) because both were at ceiling, which is consistent with the model's predictions of high detection levels for both. Consistent with the model's predictions, filtering levels 2–4 showed increasing detectability with significant differences between the three middle levels (E_2 = 3.11° vs. 1.55° and E_2 = 1.55° vs. 0.78°).

Fixation durations as a function of filtering level. Although the Yang et al. (2001) model makes no predictions about eye movements, assuming fixation durations implicitly measure perception, it would predict the same or stronger effects of filtering on fixation durations as for explicit detection. We analysed mean fixation durations on the critical fixation as a function of filtering level

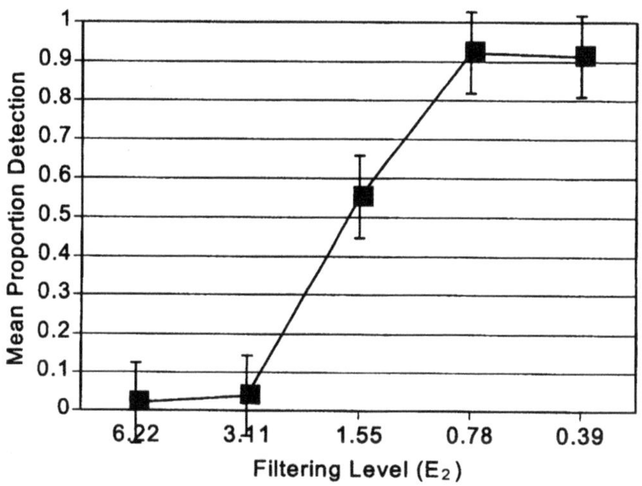

Figure 4. Proportion detection of occasionally presented gaze-contingent multiresolutional images as a function of peripheral filtering level (E_2 = 0.39–6.22°). Error bars represent 95% confidence intervals for significance of all comparisons.

TABLE 2
Multivariate ANOVA on mean fixation duration as a function of filtering level (E_2 = 0.39–6.22) with the no-area-of-interest condition

Effect	*Pillai's trace*	*F*	*Hypothesis df*	*Error df*	*Sig.*	*Partial eta squared*
Filtering level	.871	13.524	4.000	8.000	.001	.871

(five levels plus no-area-of-interest condition) with a multivariate ANOVA (Table 2). The results showed a significant main effect of image filtering on fixation durations. Follow-up, planned comparisons of each filtering level versus the no-area-of-interest condition are shown in Figure 5, with error bars indicating 95% confidence intervals of the difference for each condition versus the no-area-of-interest condition. The pattern of results looks very similar to that for detection, with two differences. First, strangely, fixation durations in the no-area-of-interest (all high-resolution) condition were longer than those in the least-filtered condition (level 1, E_2 = 6.22°), possibly reflecting different processing on these trials, compared with those experimental trials in which the area-of-interest actually appeared occasionally. To avoid this possible processing difference in the no-area-of-interest trials, we created a control condition

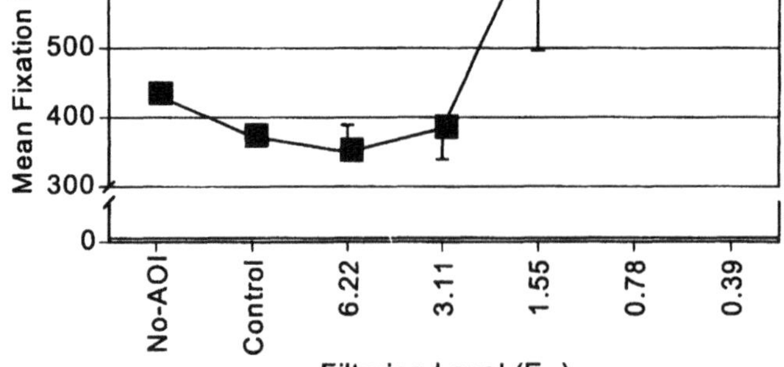

Figure 5. Mean fixation duration on occasionally presented gaze-contingent multiresolutional images as a function of peripheral filtering level (E_2 = 0.39–6.22°), the no-area-of-interest condition, and the control condition (based on the sixth and seventh fixations following each occasional gaze-contingent multiresolutional image across all filtering conditions). no-AOI = no-area-of-interest.

TABLE 3
Multivariate ANOVA on mean fixation duration as a function of filtering level (E_2 = 0.39–6.22) with the control condition

Effect	*Pillai's trace*	*F*	*Hypothesis df*	*Error df*	*Sig.*	*Partial eta squared*
Filtering level	.952	27.966	5.000	7.000	.000	.952

from the experimental trials by taking the sixth and seventh fixations following each critical fixation (i.e., the fixation on which the gaze-contingent image filtering was present) from each filtering condition (E_2 = 6.22–0.39°). These selected fixations had no filtering or manual responses associated with them, came several fixations after both the critical fixation and any manual responses that may have been made, and were several fixations prior to the next critical fixation. Using this control condition (see Figure 5), we reanalysed participants' mean fixation durations with a multivariate ANOVA (Table 3). The results were the same as before, except that, consistent with the detection results and our predictions, there was no significant difference between the least filtered condition (E_2 = 6.22) and the control. Also consistent with the detection results, there was a significant difference between filtering levels 2 and 3 (E_2 = 3.11° vs. 1.55°). However, contrary to the detection results and our predictions, there was no significant difference between filtering levels 3 and 4 (E_2 = 1.55° vs. 0.78°).

The time course of processes in detecting image filtering

The effect of fixation durations on filtering detection rates. Our first question was whether the duration of a critical fixation affects the frequency of detecting image filtering. In particular, is there some minimal fixation duration (or stimulus presentation duration) below which no detections occur, and some fixation duration above which asymptotic detection performance is reached? The answer to both questions is "Yes." As shown below, in the most detectably filtered conditions, detections did not occur for fixations much below 100 ms, and reached asymptote by roughly 200 ms. These values were greater in the threshold filtering condition.

Three filtering level conditions were produced by combining data for conditions whose detection levels were very similar, filtering levels 1 and 2 (E_2 = 6.22° and 3.11°), and filtering levels 4 and 5 (E_2 = 0.78° and 0.39°): Filtering level 3 (E_2 = 1.55°) constituted a third condition. The first condition, containing filtering levels 1 and 2, was virtually never detected at any fixation duration and is therefore uninformative for this analysis. Figure 6 shows, for the second and third groups, the frequencies of detections and nondetections as a function of fixation duration in 10 ms bins for the first 500 ms of viewing.

The top panel of Figure 6 shows that for the detection threshold filtering level (E_2 = 1.55°), the first detection occurred at roughly 200 ms, and detections increased thereafter. The bottom panel of Figure 6 shows that for the two highest filtering levels (E_2 = 0.78 and 0.39°) detections first occurred for fixations slightly less than 100 ms, and rapidly increased for fixations between 140 and 200 ms long. Figure 7 shows a more global representation of these trends, by combining the fixations into bins of 40 ms and showing the proportion detection for each bin.

The effect of filtering detectability on fixation durations. Our next question was, how does perceptible filtering affect the time course of fixation durations? As detailed below, perceptible image degradation had no effect on fixation durations before 160 ms, after which there was a gradual dampening of saccadic activity until roughly 400 ms, followed by a long lull in saccadic activity lasting up to 1000–2000 ms. Notably, differences between the threshold and highly detectable filtering conditions were much less than between those conditions and the undetectable filtering conditions.

Our approach was similar to research on the time course of perceptual and cognitive processes in reading, which showed how text manipulations affected fixation durations, using hazard curve analyses (Yang, 2002; Yang & McConkie, 2001, 2004). Hazard values give the probability of finality as a function of time, and can be compared across conditions. Example domains include demography (e.g., probability of death as a function of age, compared across gender, nationality, etc.) and engineering (e.g., probability of light bulb failure as a function of hours used, compared across filament type, brand, etc.). Here we analysed the probability of ending an eye fixation (i.e., making a saccade) as a function of its duration in milliseconds, compared across image filtering levels. Fixations were put into 40 ms intervals, with the interval labelled by the maximum value. Hazard values were calculated as the interval frequency divided by the number of remaining cases (total frequency minus cumulative frequency). Because successive values are based on fewer and fewer cases, the analysis was terminated at the 95th percentile. Data were smoothed using the average over a three-interval window, with the current interval doubly weighted. For comparability with our preceding analysis of the effects of fixation durations on detection, we used the same three filtering level groups: The undetectable filtering levels (E_2 = 6.22–3.11°), the detection threshold filtering level (E_2 = 1.55°), and the well-above-detection-threshold filtering levels (E_2 = 0.78–0.39°), plus the control condition.

The overall results of this analysis are shown in Figure 8, with the top panel showing the entire time course and the bottom panel showing greater detail during the first 600 ms. The hazard curves for all conditions start very low and

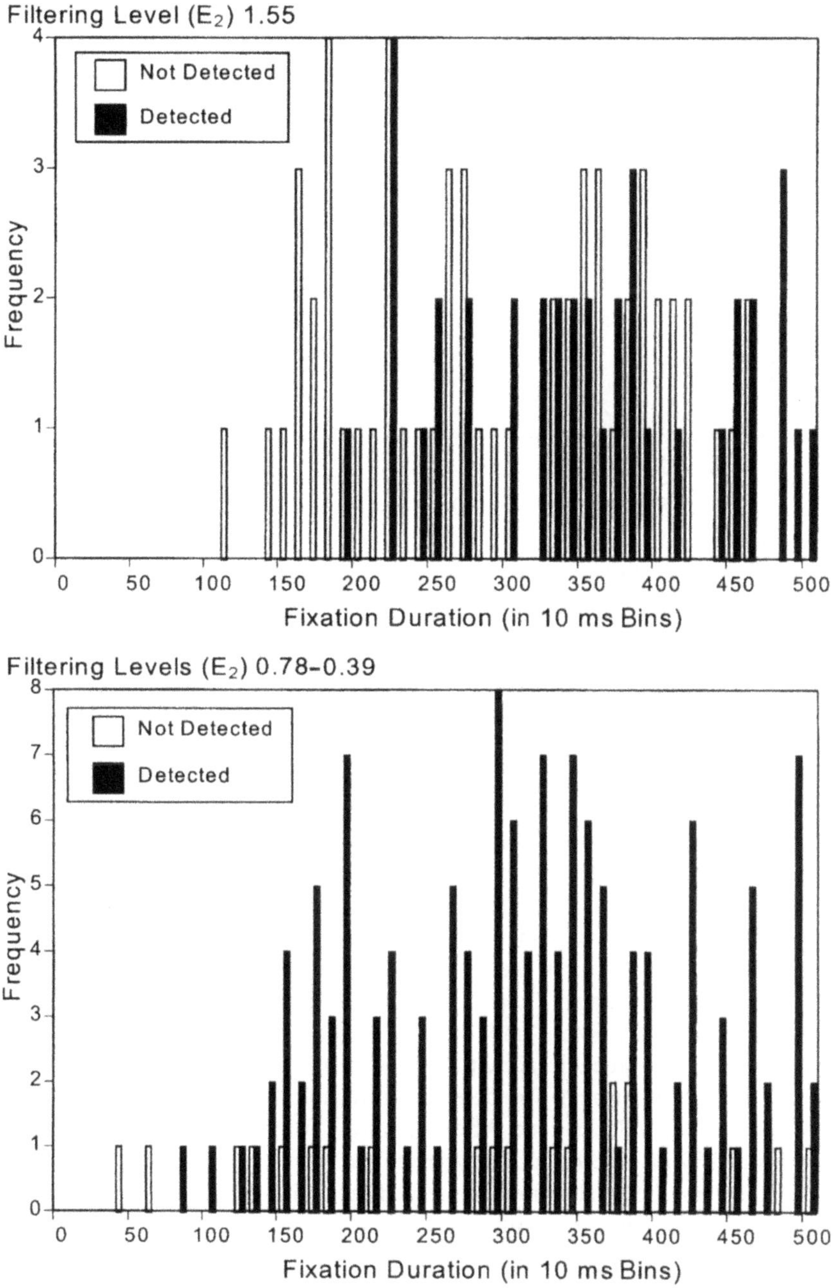

Figure 6. Frequencies of detected and undetected occasionally presented gaze-contingent multi-resolutional images as a function of grouped filtering level ($E_2 = 1.55°$, and 0.78–0.39°) and fixation duration (in 10 ms bins).

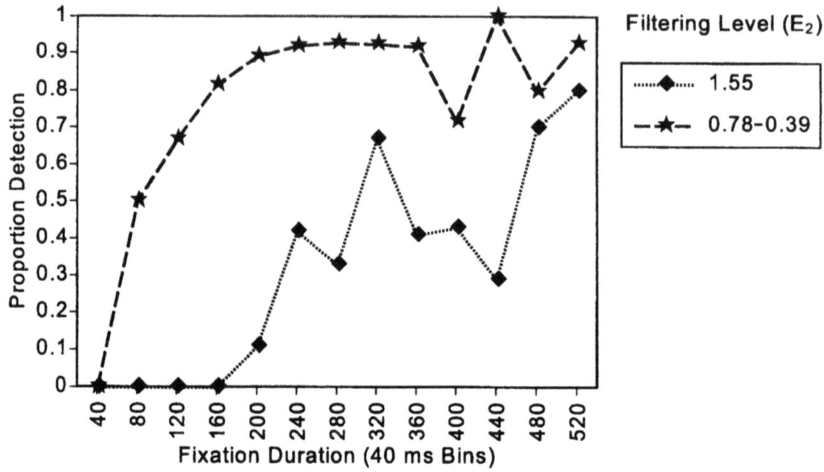

Figure 7. Proportion detection of occasionally presented gaze-contingent multiresolutional images as a function of grouped filtering level (E_2 = 1.55°, and 0.78–0.39°) and fixation duration (in 40 ms bins).

then begin rising from around 120 ms (i.e., few saccades are made before then). In the undetectable filtering levels (E_2 = 6.22–3.11°), the curve rises rapidly until about 320 ms, showing increasing saccadic activity. After that, the hazard curve is nearly asymptotic, suggesting a saturation of saccadic activity. Note that the rise in the curve around 520 ms occurs when 85% of the saccades have already been made and the data are less stable. In the highly detectable filtering levels (E_2 = 0.78–0.39°), from 160 ms the hazard curve begins rising more slowly than in the control and undetectable conditions, indicating relatively lower saccadic activity than in the undetectably filtered conditions (i.e., fixation durations start lengthening). Then from 360 ms, the curve initially drops followed by a bumpy asymptote until 95% of fixations have ended by 2280 ms, indicating a drop in saccadic activity followed by an extended lull in activity (i.e., long fixation durations). The detection-threshold filtering level (E_2 = 1.55°) shows a generally similar pattern. After separating from the undetectably filtered conditions at 160 ms, the hazard curve rises more slowly than those conditions until 400 ms, then drops, showing suppressed saccadic activity until roughly 920 ms. After that, the curve rises again and then reaches an unstable asymptote, suggesting a long period of fluctuating but relatively low saccadic activity.

A key question is whether or not the time courses of information accrual and eye movement control are the same. That is, how does the minimum effective fixation duration for detection compare with the time at which filtering first affects fixation durations? The bottom panel of Figure 8 shows that the first separation between the hazard curves for the detectably filtered conditions (E_2 =

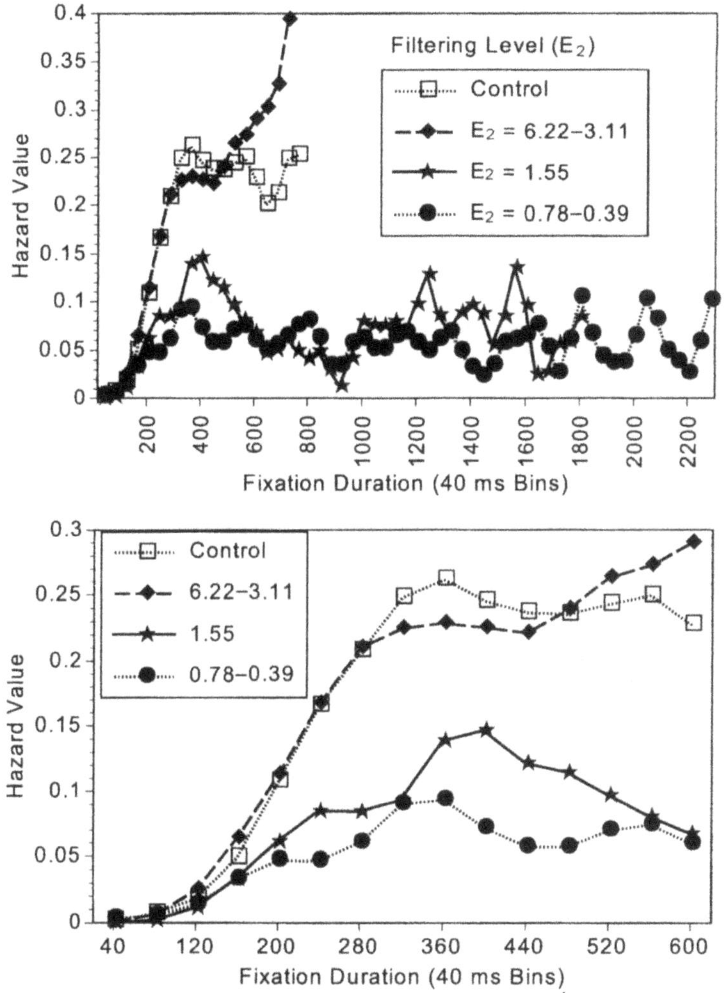

Figure 8. Top panel: Fixation duration hazard curves as a function of grouped filtering level (E_2 = 6.22–3.11°, 1.55°, and 0.78–0.39°) with curves carried out to the 95th percentile for each filtering level group. Fixation duration intervals labelled by the maximum value of bins 40 ms in width, with the data smoothed (average over three-bin window with the centre bin weighted double). Control = control condition based on the sixth and seventh fixations following each occasional gaze-contingent multiresolutional image across all filtering conditions. Bottom panel: Same as top panel, but focused on the first 600 ms.

1.55–0.39°) and the control and undetectably filtered conditions (E_2 = 6.22–3.11°) is at roughly 160–200 ms. Thus, in the highly detectable filtering conditions (E_2 = 0.78–0.39°), although an 80–120 ms fixation duration might provide enough temporal integration of sensory information to detect image filtering (Figure 7), it would be too short to show an effect of filtering on eye movements (bottom panel of Figure 8). In the case of the threshold filtering condition (E_2 = 1.55°), the timings are much closer together. The minimum effective fixation duration for detecting filtering is about 200 ms (Figure 7), which is about when fixation durations in this condition begin differing from the control and undetectable filtering conditions (bottom panel of Figure 8).

The effect of a manual response on fixation durations. Our next question was whether making a manual response had any effect on fixation durations (Fischer, 1999), and if so, at what point in time? Regarding the time course of detection, our most fundamental prediction would be that any response-related effects on fixation durations should occur after perceptual processing effects, since perception of filtering must logically precede any explicit detection response to it. However, we would also predict that response-related effects on fixation durations might start prior to actually executing the manual response, since both the detection decision process and motor programming of the manual response would occur prior to the actual button press, and either might affect fixation durations. In fact, as we will detail below, making a manual response did appear to lengthen fixation durations. This lengthening occurred in close temporal proximity to the manual response, either during a long fixation that preceded and continued beyond the manual response, or during the fixation immediately following the manual response.

To test the predictions listed above, we can look at Figure 9. This adds to the top panel of Figure 8 the manual response time (RT) hazard curves for the highly detectable (E_2 = 0.78–0.39°) and threshold detection filtering levels (E_2 = 1.55°) on the same time scale. Manual RTs show a flat hazard curve before 520 (E_2 = 0.78–0.39°) to 600 ms (E2 = 1.55°), indicating no responses before then, with rapid increases in manual response activity thereafter. In contrast to these results, recall that the separation between fixation duration hazard curves in the detectable and undetectable conditions occurred at roughly 160–200 ms. It seems unlikely that this initial drop in saccadic activity at 160–200 ms is caused by an imminent manual response 360–400 ms later. Rather, the drop in the fixation duration hazard curves for the detectable filtering conditions after 400–440 ms seems more plausibly related to the increasing manual response hazard curves at 520–600 ms. We see the drop in the fixation duration hazard curve for the highly detectable filtering conditions (E_2 = 0.78–0.39°) at 400 ms occurs 120 ms before the corresponding manual RT hazard curve first rises at 520 ms. Likewise, the drop in the fixation duration hazard curve for the detection-threshold filtering condition (E_2 = 1.55°) at 440 ms occurs 160 ms before

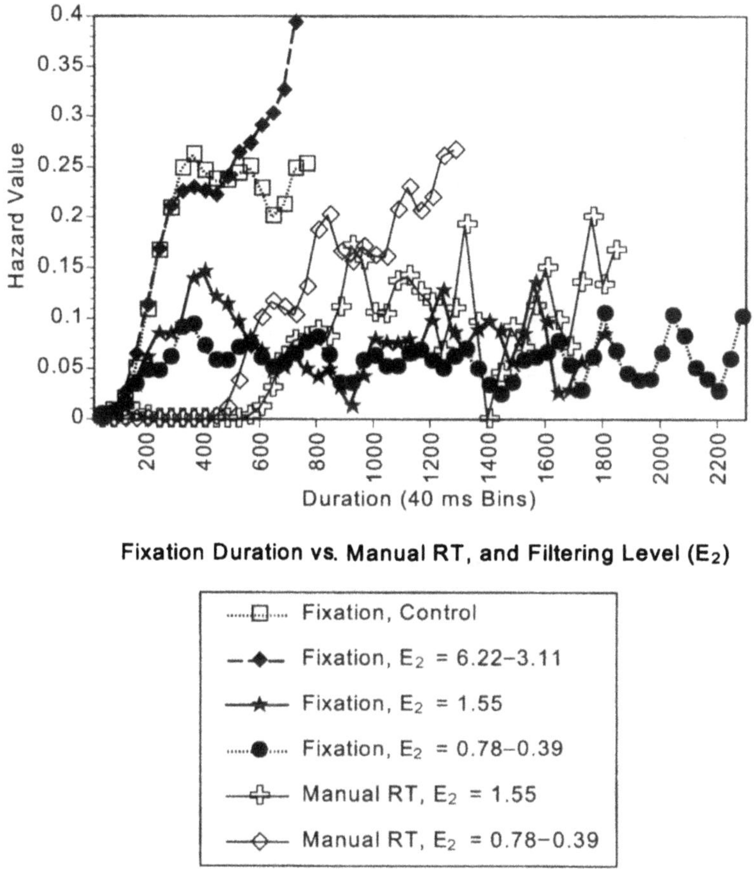

Figure 9. Fixation duration and manual response time hazard curves as a function of grouped filtering level (E_2 = 6.22–3.11°, 1.55°, and 0.78–0.39°) with curves carried out to the 95th percentile for each filtering level group. Manual response time hazard curves given for those conditions having many detection responses (E_2 = 1.55°, and 0.78–0.39°). Fixation duration intervals labelled by the maximum value of bins 40 ms in width, with the data smoothed (average over three-bin window with the centre bin weighted double). Fixation = fixation duration data. Manual RT = manual reaction time data. Control = control condition based on the sixth and seventh fixations following each occasional gaze-contingent multiresolutional image across all filtering conditions.

the rise in the corresponding manual RT hazard curve at 600 ms. These results are consistent with the above predictions regarding manual response-related effects. Specifically, the drop in saccadic activity at 400–440 ms is consistent with a manual response effect since it occurs after perceptual processing effects (at asymptote by 200 ms), but before the actual button press (starting at 520–600 ms).

If making a manual response interferes with making a saccade, then does this interference disappear as soon as the manual response is made? Interestingly, it does not. Figure 9 shows that in the detection threshold filtering condition (E_2 = 1.55°) the longest latencies for both manual responses and eye movements are very similar (both around 1800 ms), whereas in the highly detectable filtering conditions (E_2 = 0.78–0.39°), the longest eye movement latencies are about 1000 ms longer than the longest manual response latencies (roughly 2200 ms vs. 1200 ms, respectively). These findings suggest that, at least in the highly detectable filtering conditions, something besides the manual response is lengthening fixation durations, probably the image filtering itself.

To distinguish between the effect of the image filtering and the effect of the manual response on fixation durations, one can examine fixations immediately following the critical fixation. We will call these critical fixation +1, or n+1 fixations. Importantly, there was no image filtering present on those fixations. Our first interest is in that subset of n+1 fixations in which the previously viewed filtering was detected, but the manual response was *after* the critical fixation, n. Manual responses to detected filtering occurred *after* the critical fixation, n in 59% of cases in the highly detectable filtering conditions (E_2 = 0.78–0.39°). Figure 10 compares the hazard curves in the highly detectable filtering conditions (E_2 = 0.78–0.39°) for fixations n and n+1. Note that that in fixation n+1 a manual response was imminent (and occurred in 56% of these n+1 fixations), but there was no filtering present. As shown in Figure 10, even when there was no filtering present, the fact that a manual response occurred, or was imminent, was enough to reduce the fixation duration hazard curve relative to the control condition, suggesting interference with saccadic activity. On the other hand, this reduction in the fixation duration hazard value was only about half that shown for the critical fixation, suggesting that the manual response alone cannot account for all the reduction in saccadic activity. The remainder of this effect was assumedly due to the image filtering itself.

We can further explore the impact of the manual responses on fixation n+1 by comparing the fixation duration hazard curve for the highly detectable filtering conditions (E_2 = 0.78–0.39°) with the corresponding manual RT hazard curve on the same fixation (Figure 11). Here, manual RTs are given relative to the start of fixation n+1. The earliest manual RTs are in the –40 ms bin, meaning they occurred during the saccade preceding fixation n+1. This figure shows that the drop in saccadic activity early in the fixation corresponds with rising manual response activity.

Finally, we compare this same fixation n+1 condition with the fixation n+1 condition in which the manual response *was* made during the critical fixation n (Figure 12). In this latter condition, there is no image filtering present, nor is a manual response made or even imminent on fixation n+1. Thus, we might expect normal saccadic activity in this condition. Surprisingly, *both* types of n+1 fixations show reduced saccadic activity for the first 360 ms. Thus, even in the

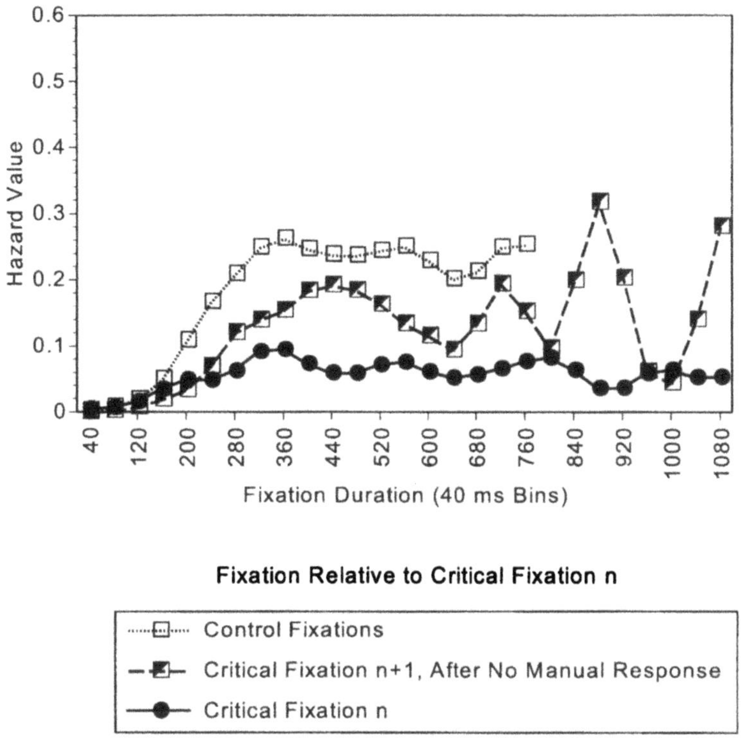

Figure 10. Fixation duration hazard curves as a function of fixation relative to the critical fixation n (for filtering levels $E_2 = 0.78–0.39°$). Fixation duration intervals labelled by the maximum value of bins 40 ms in width, with the data smoothed (average over three-bin window with the centre bin weighted double). Control fixations = control condition based on the sixth and seventh fixations following each occasional gaze-contingent multiresolutional image across all filtering conditions. Critical fixation n+1, after no manual response = fixation immediately following a critical fixation on which no manual response was made, but a manual response was imminent. Critical fixation n = fixation on which filtering occurred.

absence of image filtering, and with the manual response having been made during the previous fixation, there is still reduced saccadic activity for 360 ms—an apparent spillover effect from the critical fixation n onto fixation n+1. However, after 360 ms, saccadic activity becomes normal in this condition, whereas it is still reduced for another 360 ms when a manual response is made or is imminent.

In sum, even in the absence of filtering, if a manual response occurs or is imminent, saccadic activity is reduced on fixation n+1. Furthermore, the suppression of saccadic activity found on fixation n+1 corresponds with the timeline for late manual responses made on that fixation. Finally, there appears to be a

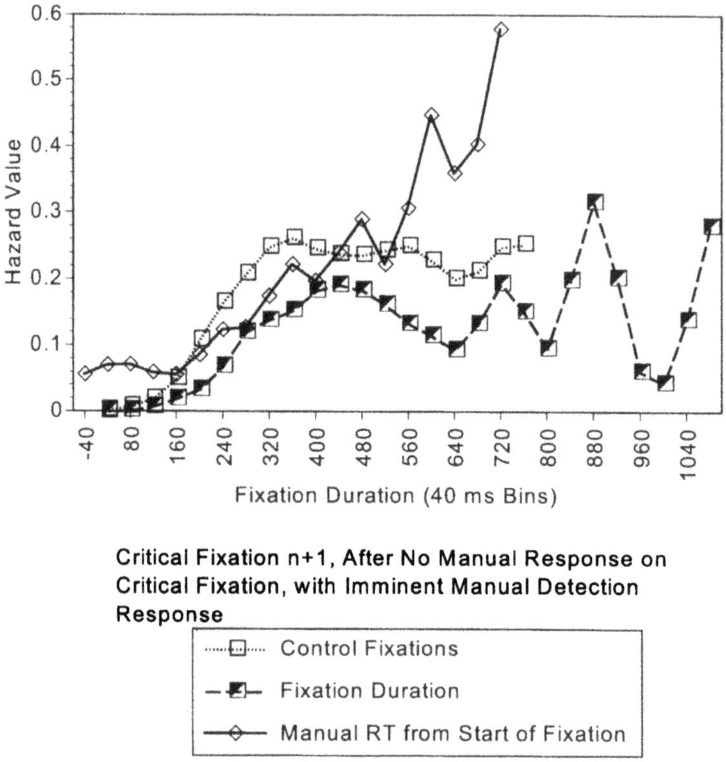

Figure 11. Fixation duration and manual response time hazard curves as a function of fixation relative to the critical fixation (for filtering levels E_2 = 0.78–0.39°). Fixation duration intervals labelled by the maximum value of bins 40 ms in width, with the data smoothed (average over three-bin window with the centre bin weighted double). Control fixations = control condition based on the sixth and seventh fixations following each occasional gaze-contingent multiresolutional image across all filtering conditions. Fixation duration = fixation immediately following a critical fixation on which no manual response was made, but a manual response was imminent. Manual RT from start of fixation = manual response time relative to the beginning of the current fixation (in ms) for responses on the current fixation.

spillover effect reducing saccadic activity on fixation n+1 when both image degradation and a manual response occurred immediately prior to it.

DISCUSSION

The limits of resolution in dynamic natural scene viewing

A central goal of the current study was to determine the limits of visual resolution in free viewing of natural scenes, and in so doing to test the Yang et al. (2001) model of eccentricity-dependent contrast sensitivity. The detection and

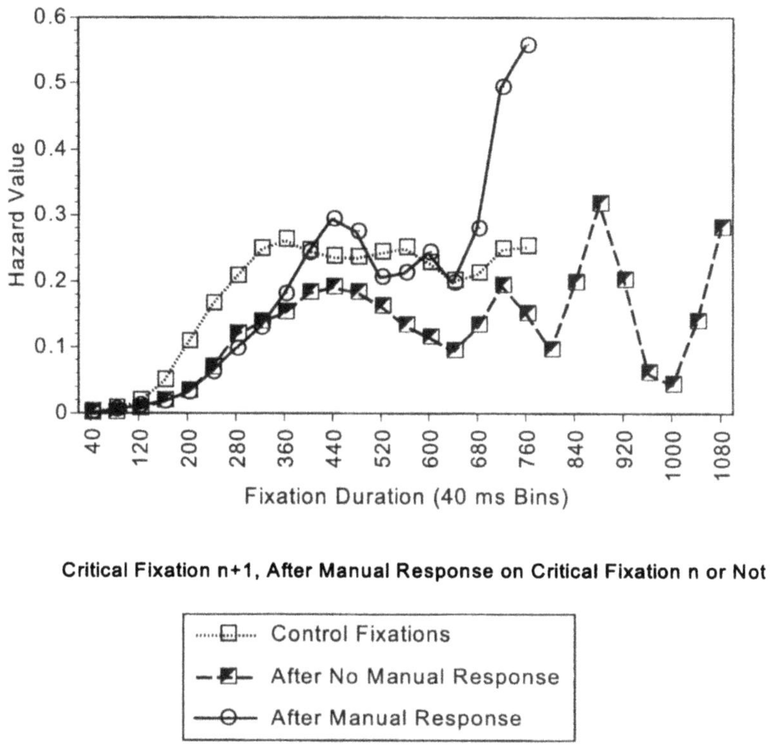

Figure 12. Fixation duration hazard curves as a function of whether a response was made on the preceding fixation (for filtering levels E_2 = 0.78–0.39°). Fixation duration intervals labelled by the maximum value of bins 40 ms in width, with the data smoothed (average over three-bin window with the centre bin weighted double). Control fixations = control condition based on the sixth and seventh fixations following each occasional gaze-contingent multiresolutional image across all filtering conditions. After no manual response = fixation immediately following a critical fixation on which no manual response was made, but a manual response was imminent. After manual response = fixation immediately following a critical fixation on which a manual response was made.

fixation duration results were generally consistent with the predictions of the model, which was based on earlier studies that measured viewers' eccentricity-dependent CSFs using grating or Gabor patch stimuli (Anderson et al., 1991; Banks et al., 1991; Cannon, 1985; Peli et al., 1991; Pointer & Hess, 1989; Robson & Graham, 1981). Filtering level 1 (E_2 = 6.22°), which removed only higher spatial frequencies than the predicted contrast threshold at each retinal eccentricity, was almost never detected, and fixation durations showed a similar pattern. Conversely, filtering levels 4–5 (E_2 = 0.78–0.39°), which removed spatial frequencies predicted to be 2–3 octaves lower than the contrast threshold, were at asymptote for both detection (above 90%) and fixation durations.

The current results are also consistent with earlier studies using briefly flashed natural images with peripheral filtering based on eccentricity-dependent

CSFs (Peli & Geri, 2001; Sere et al., 2000; Yang et al., 2001). Those studies found that viewers were somewhat less sensitive to peripheral image filtering than predicted based on studies using isolated grating or Gabor patch stimuli. Consistent with those studies, our *predicted* threshold filtering level—level 2 (E_2 = 3.11°)—was *almost never* detected, and the fixation duration results showed the same pattern (i.e., no difference from the control). Rather, the first filtering level to remove lower spatial frequencies than the predicted threshold (level 3—E_2 = 1.55°)—was detected 55% of the time, and appears to be the actual threshold filtering level. Lateral masking from neighbouring natural image regions may account for this lower-than-predicted sensitivity (Peli & Geri, 2001; Yang et al., 2001). A related possibility involving attention is the well-known narrowing of the functional field of view by image clutter (Ball, Beard, Roenker, Miller, & Griggs, 1988; Bertera & Rayner, 2000; Mackworth, 1965; Sekuler, Bennett, & Mamelak, 2000).

The similarity of these results under dynamic viewing conditions to those found with statically presented multiresolutional images (Peli & Geri, 2001; Sere et al., 2000; Yang et al., 2001) is somewhat surprising, since, for the technical reasons discussed in our Apparatus section, one might predict higher detection rates under dynamic viewing conditions than under static ones. Our replication of lower-than-predicted sensitivity to image filtering, found in previous studies using static presentation of multiresolutional images, may be because we managed to eliminate low-level artifacts from our gaze-contingent multiresolutional display. As noted earlier, we updated our images as quickly as possible using an update initiation deadline of 5 ms after the end of a saccade (McConkie & Loschky, 2002), thus likely eliminating detections based on motion transients. We also used an eyetracking system (Crane & Steele, 1978) with high spatial resolution and a relatively fine grid of spatial locations for the centre of high-resolution, thus achieving a relatively small degree of error in placing the centre of high-resolution at the centre of gaze.

In sum, the Yang et al. (2001) model is largely supported by the data, though based on the current results we may need to adjust the parameters (e.g., k, representing the rate of visual resolution drop-off). We estimate that the resolution limit at each eccentricity in natural scene viewing is approximated by Equation 3 (restated below) with an E_2 value of 1.55, since that was the filtering threshold E_2 value for both detection and fixation durations:

$$f_c = 43.1 * 1.55/(1.55 + E)$$

Again, f_c is the frequency cutoff, or resolution limit, and E is retinal eccentricity. Further research may more precisely pinpoint the perceptual threshold using more filtering levels or perhaps different underlying hypothesized CSFs, but the current estimate provides a good first approximation.

The limits of visual resolution across the visual field and scene perception. The above estimate of the limits of visual resolution in dynamic scene perception provides an important limit on the perceptible information assumed to be available in extrafoveal scene regions. This limit provides important constraints for theories of scene perception dealing with scene gist and layout acquisition, extrafoveal object recognition, and eye movement guidance in extended scene viewing. These limits are relevant to theories of scene processing whether or not they consider extrafoveal object recognition to be important.

Consider theories in which extrafoveal recognition of diagnostic objects plays an important role in scene gist acquisition (e.g., Biederman, Rabinowitz, Glass, & Stacy, 1974; Friedman, 1979). Studies have suggested an ideal spatial frequency band for object recognition (Braje et al., 1995; Gold et al., 1999; Olds & Engel, 1998). (While such studies tend to express these bandwidths in cycles/object, rather than cycles/degree, as in the current study, it is possible to translate between these scales, with the simplest case being an object 1° wide.) If object recognition is also isolated from scene semantic constraints (Hollingworth & Henderson, 1999), it would suggest that when a scene-diagnostic object is at an eccentricity where the central frequency of the critical bandwidth is below threshold, object recognition should suffer, and therefore scene gist acquisition should suffer as well.

Next, consider theories in which parafoveal or peripheral object information is used to guide the eyes in extended scene viewing, such as in a search task (Rao, Zelinsky, Hayhoe, & Ballard, 2002; Wolfe, 1994). If the eyes are targeted on the basis of discriminating target and nontarget objects, knowing the limits of resolution in scenes should be important for such a theory (e.g., Rao et al., 2002). Discrimination of a peripheral distractor from the target should be influenced by (1) the spatial frequency bandwidth necessary for discrimination, (2) the retinal eccentricity and contrast of the target and distractor, and (3) the limits of resolution in scenes. Thus, if diagnostic object recognition, or object feature discrimination, is assumed to play an important role in acquiring scene gist, or in guiding the eyes during extended scene viewing, knowing the limits of visual resolution across the visual field in scene viewing is critical.

However, even in theories of scene gist acquisition or eye guidance positing no role for object recognition or object feature discrimination, knowing the limits of visual resolution should be important. For example, Schyns and Oliva (1994) have argued that most frequently, low spatial frequency information is used early in determining scene gist. However, in roughly a third of their trials, high spatial frequency information was used for that purpose. A possible explanation for this asymmetry in information use is in terms of the visual resolution drop-off across the visual field. That is, high spatial frequency information is available in only a small part of the visual field, near the centre of vision, whereas low spatial frequency information is available across the entire visual field. Thus, if high spatial frequency information useful for acquiring scene gist is only resolvable, because near the fovea, about one third of the time, the visual system would need to rely on the resolvable lower spatial frequency information the remaining two thirds of the time. In this way, the limits of visual resolution in scene viewing, as estimated in the current study, may explain a reliance on low spatial frequency information in acquiring gist (Schyns & Oliva, 1994).

Similarly, an interesting question for theories of gist acquisition that posit an important role for global texture discrimination (e.g.,Walker & Malik, 2002) is whether the texture spatial frequency profile for a scene type varies with the point of fixation, or whether any initial fixation point would provide a match. In either case, an accurate description of the texture spatial frequency profiles used to identify scenes of various categories will need to incorporate the limits of visual resolution. A similar argument would apply to the spatial envelope model of gist acquisition (Oliva & Torralba, 2001).

In the case of eye movement guidance in scene viewing, an interesting possibility is that the limits of visual resolution, or more generally CSFs at different eccentricities, play a role in determining the salience, or attractiveness, of peripheral targets for eye movements with or without recognizable peripheral target objects (see Deubel, 1984; Loschky, 2003; Loschky & McConkie, 2002). An interesting question for further research is whether the filtering threshold for detecting image blur in the current study is the same as the filtering threshold for affecting saccade targeting in other studies (e.g., Loschky & McConkie, 2002). Such a determination would allow one to estimate the useful information for selecting saccade targets in natural scenes.

The limits of visual resolution and eye movements

The current study helps create a bridge between the literatures on spatial vision and eye movement control in the context of scene perception. A number of earlier studies have shown that image degradation of one kind or another lengthens fixation durations (Loftus, 1985; Loftus et al., 1992; Loschky & McConkie, 2002; Mannan et al., 1995; Parkhurst et al., 2000). However, the current study adds to an understanding of what determines such durations by relating these oculomotor effects to a model of visual sensitivity across the field of view. The results suggest a close correspondence between the conscious perceptibility of image degradation, when filtering goes beyond the limits of visual resolution, and the duration of fixations.

Comparison of detection rates and fixation durations. To provide a robust test of the limits of visual resolution in dynamic scene viewing, we used two very different measures, one explicit, blur detection, and the other implicit, fixation durations, and we found a similar pattern of results for both. One interpretation of this similarity is that the networks of brain areas involved in both (1) making a detection response in our task, and (2) deciding when to move the eyes, may rely on the same set of low-level processes sensitive to the spatial frequency content of images, perhaps in the primary visual cortex (Itti, Koch, & Braun, 2000).

An alternative explanation for the similar pattern of responses for detection and fixation duration measures is that the one has influenced the other. Specifically, it is possible that making a manual response may have interfered with making saccades, thus increasing fixation durations whenever the filtering was detectable. This effect might help explain the chief difference between the results for the two measures, namely a significant difference in detection between the threshold filtering level ($E_2 = 1.55°$) and the next higher level ($E_2 = 0.78°$), but no such difference for fixation durations. If the manual detection responses lengthened fixation durations, it would reduce the variability of the fixation durations in all conditions having detectable filtering. Since our results are consistent with the idea that manual responses lengthened fixation durations (discussed below), we cannot exclude this explanation for the similar detection and fixation duration results. Note, however, that the two above explanations are not necessarily mutually exclusive. As suggested by our analyses of the time course of processes in detecting image filtering, there appear to be separate effects of filtering and manual response on fixation durations.

The time course of processes in detecting image filtering

The current study helps understand the time course of perceptual and oculomotor processes invoked by removing information beyond the limits of resolution, and making a manual response to it.

The effect of fixation durations on filtering detection rates. The current results suggest that fixation durations do affect perception of filtering, and in a way that is compatible with research on stimulus duration effects. Beyond a minimum effective fixation duration, we see a "liftoff effect" on detection rates similar to that found in experiments on the effects of stimulus duration (Bundesen & Harms, 1999; Loftus & Ruthruff, 1994). Furthermore, as the degree of image filtering is reduced to the detection threshold, the minimum effective fixation duration increases and the slope of the detection x fixation duration function becomes shallower, again consistent with earlier research showing tradeoffs between stimulus contrast, duration, and visual performance (Loftus & Ruthruff, 1994). Nevertheless, our minimum effective fixation durations may be longer than the minimum effective stimulus durations in a static presentation, due to postsaccadic suppression, which can last for 20–50 ms from the start of a fixation (Diamond, Ross, & Morrone, 2000; McConkie &

Loschky, 2002; Shioiri, 1993). Interestingly, for the most detectable filtering levels (E_2 = 0.78–0.39°), the minimum effective fixation duration is close to what many researchers consider the minimum fixation duration (~100 ms), and asymptotic near-perfect detection rates are achieved at normal-to-short fixation durations (~200 ms). Thus, the typical distribution of fixation durations in scene viewing seems well suited to perceiving the above-threshold spatial frequency profile of images. If other basic perceptual processes have similar minimum effective fixation durations of roughly 100–200 ms, then it might explain why fixation durations have sometimes been argued to have little impact on perception and memory (Loftus, 1972), since most variation in fixation durations will be for greater durations. However, due to the small number of observations in our data for these very short fixation durations, these proposals require further verification.

The effect of filtering detectability on fixation durations. The analysis of fixation duration hazard curves showed that filtering began lengthening fixations from roughly 160–200 ms onward for all detectably degraded conditions. For the most detectable filtering levels (E_2 = 0.78–0.39°), this oculomotor response was somewhat slower than the minimum effective fixation duration of 80–120 ms, suggesting a delay between the information integration and eye movement response processes. However, in the filtering detection threshold condition (E_2 = 1.55°), this delay was not found, most likely because the minimum effective fixation duration for detection was longer, about 200 ms, allowing the oculomotor system more time to respond to the filtering. This result suggests that there are rather fixed delays in registering peripheral degradation in the oculomotor system, which are somewhat independent of information thresholds for consciously registering image degradation.

The effect of a manual response on fixation durations. The results of the current study suggest that, if a fixation is long enough to be close in time to a manual response, it will interfere with executing a saccade, thus lengthening the fixation. Thus, motor programming for a manual response may interfere with that for a saccade. Furthermore, this interference can be long lasting and spill over to the next fixation. Thus, whatever interference is caused by both the manual response and the image degradation is not immediately erased by making an eye movement, but instead can persist for several hundred milliseconds. Such motor interference between hands and eyes could occur in the presupplementary motor area and/or the supplementary eye field, which have been shown in monkeys to be involved in both eye and hand movements (Fujii, Mushiake, & Tanji, 2002).

Further research is needed to investigate the role of response interference from hands to eyes. Indeed we have conducted a second experiment with the same participants using the same levels of image filtering but without a detection task or manual response (Loschky, 2003, Exp. 3). That experiment showed a similar pattern of increased fixation durations as a function of filtering detectability, but the effects, while statistically significant, were roughly a level of magnitude smaller than those in the current experiment, consistent with the claim that the manual response in the current experiment caused longer fixations. However, the second experiment presented a multiresolutional image on every fixation, instead of only on occasional fixations, as in the current study. Thus, a future study might compare the effects of an occasional gaze-contingent multiresolutional display with and without a detection task and manual response.

Theoretical accounts of effects on fixation durations

There are two general approaches to accounting for the onset times of saccades. The first, often referred to as a cognitive account, assumes that the programming of a saccade is initiated when some critical cognitive event occurs (Henderson, 1993, 1996; Henderson & Ferreira, 1993; Rayner, 1998). In reading, this event may be the completion of word identification (Morrison, 1984) or the assessment of the familiarity of a word (Reichle, Pollatsek, Fisher, & Rayner, 1998; Reichle, Rayner, & Pollatsek, 1999, 2003); in pictures, it may be the identification of an object (Loftus, 1981).

The second, sometimes referred to as an oculomotor inhibition account, assumes that saccades are generated at random intervals, with the average times controlled by parameters, and that they can be inhibited if processing difficulty is encountered (McConkie & Dyre, 2000; Yang, 2002; Yang & McConkie, 2001, 2004). The current results suggest that the cognitive account needs to be modified in some way to allow extrafoveal information on the current fixation to influence the saccade onset time (e.g., Kennedy, 1998, 2000); either by allowing characteristics of more peripheral information to influence the rate of foveal processing in order to delay the saccade-triggering event, or by allowing extrafoveal information to affect saccade onset time in some more direct fashion. The same is true for the effect of motor response interference such as pressing a button; this action must either compete for processing resources and thus delay the critical saccade-triggering processing event, or it must delay saccades in some other more direct fashion (Fischer, 1999).

Based on the current results, the oculomotor inhibition account would need a minor modification such that either the lack of normally available spatial frequencies is directly detected, or their lack creates difficulty in perception, either of which causes inhibition of saccadic activity. A more interesting modification of the account would be to allow for oculomotor inhibition caused by motor activity in a different effector modality (e.g., the hands), and for the inhibition produced in association with making a motor response to take some time after the response (several hundred ms or more) to dissipate. In both cases, the degree and duration of the inhibition would be revealed by fixation duration hazard curves. The hazard curves in the current study show similarities to hazard curves from abnormal stimulus patterns encountered during reading (Yang, 2002; Yang & McConkie, 2001; Yang & McConkie, 2004), with the effect of the stimulus abnormality (e.g., a drop of hazard level below that of the control condition) only appearing after 160–200 ms after the onset of the fixation in both data sets. An interesting question for the current study is whether these effects reflect disruption of low-level perceptual processes, or whether the effect is due to a reduced ability to identify objects. Thus, the current study provides information useful for further developing current theories of eye movement control.

REFERENCES

Anderson, S. J., Mullen, K. T., & Hess, R. F. (1991). Human peripheral spatial resolution for achromatic and chromatic stimuli: Limits imposed by optical and retinal factors. *Journal of Physiology (London)*, *442*, 47–64.

Antes, J. R. (1974). The time course of picture viewing. *Journal of Experimental Psychology*, *103*(1), 62–70.

Ball, K. K., Beard, B. L., Roenker, D. L., Miller, R. L., & Griggs, D. S. (1988). Age and visual search: Expanding the useful field of view. *Journal of the Optical Society of America*, *5*(12), 2210–2219.

Banks, M. S., Sekuler, A. B., & Anderson, S. J. (1991). Peripheral spatial vision: Limits imposed by optics, photoreceptors, and receptor pooling. *Journal of the Optical Society of America*, *8*(11), 1775–1787.

Beard, B. L., Levi, D. M., & Klein, S. A. (1997). Vernier acuity with non-simultaneous targets: The cortical magnification factor estimated by psychophysics. *Vision Research*, *37*(3), 325–346.

Bertera, J. H., & Rayner, K. (2000). Eye movements and the span of the effective stimulus in visual search. *Perception and Psychophysics*, *62*(3), 576–585.

Biederman, I., Rabinowitz, J., Glass, A., & Stacy, E. (1974). On the information extracted from a glance at a scene. *Journal of Experimental Psychology*, *103*, 597–600.

Boyce, S., & Pollatsek, A. (1992). Identification of objects in scenes: The role of scene background in object naming. *Journal of Experimental Psychology: Learning, Memory, and Cognition*, *18*, 531–543.

Boyce, S., Pollatsek, A., & Rayner, K. (1989). Effect of background information on object identification. *Journal of Experimental Psychology: Human Perception and Performance*, *15*, 556–566.

Braje, W. L., Tjan, B. S., & Legge, G. E. (1995). Human efficiency for recognizing and detecting low-pass filtered objects. *Vision Research*, *35*(21), 2955–2966.

Bundesen, C., & Harms, L. (1999). Single-letter recognition as a function of exposure duration. *Psychological Research*, *62*(4), 275–279.

Campbell, F. W. (1983). Why do we measure contrast sensitivity? *Behavioural Brain Research*, *10*(1), 87–97.

Campbell, F. W., & Robson, J. G. (1968). Application of Fourier analysis to the visibility of gratings. *Journal of Physiology*, *197*, 551–566.

Cannon, M. W. (1985). Perceived contrast in the fovea and periphery. *Journal of the Optical Society of America*, *2*(10), 1760–1768.

Crane, H. D., & Steele, C. S. (1978). Accurate three-dimensional eyetracker. *Applied Optics*, *17*, 691–705.

Deubel, H. (1984). *Wechselwirkung von sensorik und motorik bei sakkadischen augenbewegungen: Eine kyberntische beshreibun des visuellen systems*. Unpublished doctoral dissertation, Technical University of Munich, Germany.

Deubel, H., & Schneider, W. X. (1996). Saccade target selection and object recognition: Evidence for a common attentional mechanism. *Vision Research, 36*(12), 1827–1837.

DeValois, R. L., & DeValois, K. K. (1988). *Spatial vision*. New York: Oxford University Press.

Diamond, M. R., Ross, J., & Morrone, M. C. (2000). Extraretinal control of saccadic suppression. *Journal of Neuroscience, 20*(9), 3449–3455.

Fischer, M. H. (1999). An investigation of attention allocation during sequential eye movement tasks. *Quarterly Journal of Experimental Psychology, 52*(A), 649–677.

Friedman, A. (1979). Framing pictures: The role of knowledge in automatized encoding and memory for gist. *Journal of Experimental Psychology: General, 108*(3), 316–355.

Fujii, N., Mushiake, H., & Tanji, J. (2002). Distribution of eye- and arm-movement-related neuronal activity in the SEF and in the SMA and pre-SMA of monkeys. *Journal of Neurophysiology, 87*(4), 2158–2166.

Geisler, W. S., & Perry, J. S. (1998). A real-time foveated multi-resolution system for low-bandwidth video communication. *Proceedings of the SPIE: The International Society for Optical Engineering, 3299*, 294–305.

Gold, J., Bennett, P. J., & Sekuler, A. B. (1999). Identification of band-pass filtered letters and faces by human and ideal observers. *Vision Research, 39*(21), 3537–3560.

Hayhoe, M. M., Bensinger, D. G., & Ballard, D. H. (1998). Task constraints in visual working memory. *Vision Research, 38*(1), 125–137.

Henderson, J., Pollatsek, A., & Rayner, K. (1987). The effects of foveal priming and extrafoveal preview on object identification. *Journal of Experimental Psychology: Human Perception and Performance, 13*, 449–463.

Henderson, J., Pollatsek, A., & Rayner, K. (1989). Covert visual attention and extrafoveal information use during object identification. *Perception and Psychophysics, 45*(3), 196–208.

Henderson, J. M. (1993). Visual attention and saccadic eye movements. In J. V. R. Gery d'Ydewalle (Ed.), *Studies in visual information processing: Vol. 4. Perception and cognition: Advances in eye movement research* (pp. 37–50). Amsterdam: North-Holland/Elsevier Science Publishers.

Henderson, J. M. (1996). Visual attention and the attention-action interface. In K. A. Akins (Ed.), *Perception* (Vol. 5, pp. 290–316). New York: Oxford University Press.

Henderson, J. M., & Ferreira, F. (1993). Eye movement control during reading: Fixation measures reflect foveal but not parafoveal processing difficulty. *Canadian Journal of Experimental Psychology, 47*(2), 201–221.

Henderson, J. M., & Hollingworth, A. (1998). Eye movements during scene viewing: An overview. In G. Underwood (Ed.), *Eye guidance in reading and scene perception* (Vol. xi, pp. 269–293). Oxford, UK: Elsevier Science Ltd.

Henderson, J. M., McClure, K. K., Pierce, S., & Schrock, G. (1997). Object identification without foveal vision: Evidence from an artificial scotoma paradigm. *Perception and Psychophysics, 59*(3), 323–346.

Henderson, J. M., Williams, C. C., Castelhano, M. S., & Falk, R. J. (2003). Eye movements and picture processing during recognition. *Perception and Psychophysics, 65*(5), 725–734.

Hoffman, J. E., & Subramaniam, B. (1995). The role of visual attention in saccadic eye movements. *Perception and Psychophysics, 57*(6), 787–795.

Hollingworth, A., & Henderson, J. M. (1999). Object identification is isolated from scene semantic constraint: Evidence from object type and token discrimination. *Acta Psychologica, 102*(2–3), 319–343.

Hollingworth, A., Williams, C. C., & Henderson, J. M. (2001). To see and remember: Visually specific information is retained in memory from previously attended objects in natural scenes. *Psychonomic Bulletin and Review, 8*(4), 761–768.

Irwin, D. E., & Gordon, R. D. (1998). Eye movements, attention and trans-saccadic memory. *Visual Cognition, 5*(1–2), 127–155.

Itti, L., Koch, C., & Braun, J. (2000). Revisiting spatial vision: Toward a unifying model. *Journal of the Optical Society of America A—Optics and Image Science, 17*(11), 1899–1917.

Kennedy, A. (1998). The influence of parafoveal words on foveal inspection time: Evidence for a processing trade-off. In G. Underwood (Ed.), *Eye guidance in reading and scene perception* (pp. 149–179). Amsterdam: Elsevier Science.

Kennedy, A. (2000). Parafoveal processing in word recognition. *Quarterly Journal of Experimental Psychology A, 53A*(2), 429–455.

Kowler, E., Anderson, E., Dosher, B., & Blaser, E. (1995). The role of attention in the programming of saccades. *Vision Research, 35*(13), 1897–1916.

Levi, D. M. (1999). Progress and paradigm shifts in spatial vision over the 20 years of ECVP. *Perception, 28*(12), 1443–1459.

Levin, D. T., Takarae, Y., Miner, A. G., & Keil, F. (2001). Efficient visual search by category: Specifying the features that mark the difference between artifacts and animals in preattentive vision. *Perception and Psychophysics, 63*(4), 676–697.

Li, F. F., VanRullen, R., Koch, C., & Perona, P. (2002). Rapid natural scene categorization in the near absence of attention. *Proceedings of the National Academy of Sciences of the United States of America, 99*(14), 9596–9601.

Loftus, G., Kaufman, L., Nishimoto, T., & Ruthruff, E. (1992). Effects of visual degradation on eye-fixation durations, perceptual processing, and long-term visual memory. In K. Rayner (Ed.), *Eye movements and visual cognition: Scene perception and reading* (pp. 203–226). New York: Springer-Verlag.

Loftus, G. R. (1972). Eye fixations and recognition memory for pictures. *Cognitive Psychology, 3*(4), 525–551.

Loftus, G. R. (1981). Tachistoscopic simulations of eye fixations on pictures. *Journal of Experimental Psychology: Human Learning and Memory, 7*(5), 369–376.

Loftus, G. R. (1985). Picture perception: Effects of luminance on available information and information-extraction rate. *Journal of Experimental Psychology: General, 114*(3), 342–356.

Loftus, G. R., & Ruthruff, E. (1994). A theory of visual information acquisition and visual memory with special application to intensity/duration trade-offs. *Journal of Experimental Psychology: Human Perception and Performance, 20*(1), 33–49.

Loschky, L. C. (2003). *Investigating perception and eye movement control in natural scenes using gaze-contingent multi-resolutional displays*. Unpublished doctoral dissertation, University of Illinois, Urbana-Champaign.

Loschky, L. C., & McConkie, G. W. (2002). Investigating spatial vision and dynamic attentional selection using a gaze-contingent multi-resolutional display. *Journal of Experimental Psychology: Applied, 8*(2), 99–117.

Mackworth, N. H. (1965). Visual noise causes tunnel vision. *Psychonomic Science, 3*, 67–68.

Maekelae, P., Naesaenen, R., Rovamo, J., & Melmoth, D. (2001). Identification of facial images in peripheral vision. *Vision Research, 41*(5), 599–610.

Mannan, S., Ruddock, K. H., & Wooding, D. S. (1995). Automatic control of saccadic eye movements made in visual inspection of briefly presented 2-D images. *Spatial Vision, 9*(3), 363–386.

McConkie, G. W., & Dyre, B. P. (2000). Eye fixation durations in reading: Models of frequency distributions. In A. Kennedy & R. Radach (Eds.), *Reading as a perceptual process* (pp. 683–700). Amsterdam: Elsevier Science Ltd.

McConkie, G. W., & Loschky, L. C. (2002). Perception onset time during fixations in free viewing. *Behavioural Research Methods, Instruments, and Computers, 34*(4), 481–490.

McConkie, G. W., & Rayner, K. (1975). The span of the effective stimulus during a fixation in reading. *Perception and Psychophysics, 17*(6), 578–586.

Morrison, R. E. (1984). Manipulation of stimulus onset delay in reading: Evidence for parallel programming of saccades. *Journal of Experimental Psychology: Human Perception and Performance, 10*(5), 667–682.

Nelson, W. W., & Loftus, G. R. (1980). The functional visual field during picture viewing. *Journal of Experimental Psychology: Human Learning and Memory*, *6*(4), 391–399.

Olds, E. S., & Engel, S. A. (1998). Linearity across spatial frequency in object recognition. *Vision Research*, *38*(14), 2109–2118.

Oliva, A., & Schyns, P. G. (2000). Diagnostic colors mediate scene recognition. *Cognitive Psychology*, *41*(2), 176–210.

Oliva, A., & Torralba, A. (2001). Modeling the shape of the scene: A holistic representation of the spatial envelope. *International Journal of Computer Vision*, *42*(3), 145–175.

Parkhurst, D., Culurciello, E., & Niebur, E. (2000). Evaluating variable resolution displays with visual search: Task performance and eye movements. In A. T. Duchowski (Ed.), *Proceedings of the Eye Tracking Research and Applications symposium* (pp. 105–109). New York: Association for Computing Machinery.

Parkhurst, D. J., & Niebur, E. (2002). Variable-resolution displays: A theoretical, practical, and behavioural evaluation. *Human Factors*, *44*(4), 611–629.

Peli, E., & Geri, G. A. (2001). Discrimination of wide-field images as a test of a peripheral-vision model. *Journal of the Optical Society of America A–Optics and Image Science*, *18*(2), 294–301.

Peli, E., Yang, J., & Goldstein, R. B. (1991). Image invariance with changes in size: The role of peripheral contrast thresholds. *Journal of the Optical Society of America*, *8*(11), 1762–1774.

Pointer, J. S., & Hess, R. F. (1989). The contrast sensitivity gradient across the human visual field: With emphasis on the low spatial frequency range. *Vision Research*, *29*(9), 1133–1151.

Rao, R. P. N., Zelinsky, G. J., Hayhoe, M. M., & Ballard, D. H. (2002). Eye movements in iconic visual search. *Vision Research*, *42*(11), 1447–1463.

Rayner, K. (1975). The perceptual span and peripheral cues in reading. *Cognitive Psychology*, *7*(1), 65–81.

Rayner, K. (1998). Eye movements in reading and information processing: 20 years of research. *Psychological Bulletin*, *124*, 372–422.

Reichle, E. D., Pollatsek, A., Fisher, D. L., & Rayner, K. (1998). Toward a model of eye movement control in reading. *Psychological Review*, *105*(1), 125–157.

Reichle, E. D., Rayner, K., & Pollatsek, A. (1999). Eye movement control in reading: Accounting for initial fixation locations and refixations within the E-Z Reader model. *Vision Research*, *39*(26), 4403–4411.

Reichle, E. D., Rayner, K., & Pollatsek, A. (2003). The E-Z Reader model of eye-movement control in reading: Comparisons to other models. *Behavioural and Brain Sciences*, *26*(4), 445–526.

Reingold, E. M., Loschky, L. C., McConkie, G. W., & Stampe, D. M. (2003). Gaze-contingent multi-resolutional displays: An integrative review. *Human Factors*, *45*(2), 307–328.

Robson, J. G., & Graham, N. (1981). Probability summation and regional variation in contrast sensitivity across the visual field. *Vision Research*, *21*(3), 409–418.

Sanocki, T. (2003). Representation and perception of spatial layout. *Cognitive Psychology*, *47*(1), 43–86.

Sanocki, T., & Epstein, W. (1997). Priming spatial layout of scenes. *Psychological Science*, *8*(5), 374–378.

Schyns, P., & Oliva, A. (1994). From blobs to boundary edges: Evidence for time- and spatial-scale-dependent scene recognition. *Psychological Science*, *5*, 195–200.

Sekuler, A. B., Bennett, P. J., & Mamelak, M. (2000). Effects of aging on the useful field of view. *Experimental Aging Research*, *26*(2), 103–120.

Sere, B., Marendaz, C., & Herault, J. (2000). Nonhomogeneous resolution of images of natural scenes. *Perception*, *29*(12), 1403–1412.

Shioiri, S. (1993). Postsaccadic processing of the retinal image during picture scanning. *Perception and Psychophysics*, *53*(3), 305–314.

Shioiri, S., & Ikeda, M. (1989). Useful resolution for picture perception as a function of eccentricity. *Perception*, *18*, 347–361.

Thorpe, S. J., Gegenfurtner, K. R., Fabre-Thorpe, M., & Buelthoff, H. H. (2001). Detection of animals in natural images using far peripheral vision. *European Journal of Neuroscience, 14*(5), 869–876.

Treisman, A. M., & Gelade, G. (1980). A feature-integration theory of attention. *Cognitive Psychology, 12*(1), 97–136.

Van Diepen, P. M. J., & Wampers, M. (1998). Scene exploration with Fourier-filtered peripheral information. *Perception, 27*(10), 1141–1151.

Walker, L. L., & Malik, J. (2002). When is scene recognition just texture recognition? *Journal of Vision, 2*(7). Abstract no. 255a. Retrieved from http://journalofvision.org/2/7/255/

Wilson, H. R., Levi, D., Maffei, L., Rovamo, J., & DeValois, R. (1990). The perception of form: Retina to striate cortex. In L. Spillmann, J. S. Werner et al. (Eds.), *Visual perception: The neurophysiological foundations* (pp. 231–272). San Diego, CA: Academic Press, Inc.

Wolfe, J. M. (1994). Guided Search 2.0: A revised model of visual search. *Psychonomic Bulletin and Review, 1*(2), 202–238.

Yang, J., Coia, T., & Miller, M. (2001). Subjective evaluation of retinal-dependent image degradations. In *Proceedings of PICS 2001: Image Processing, Image Quality, Image Capture Systems conference* (pp. 142–147). Springfield, VA: Society for Imaging Science and Technology.

Yang, J., Qi, X., & Makous, W. (1995). Zero frequency masking and a model of contrast sensitivity. *Vision Research, 35*(14), 1965–1978.

Yang, J., & Stevenson, S. B. (1998). Effect of background components on spatial-frequency masking. *Journal of the Optical Society of America, 15*(5), 1027–1035.

Yang, S.-N. (2002). *Inhibitory control of saccadic eye movements in reading: A neurophysiologically based interaction-competition theory of saccade programming*. Unpublished doctoral dissertation, University of Illinois, Urbana-Champaign.

Yang, S. N., & McConkie, G. W. (2001). Eye movements during reading: A theory of saccade initiation times. *Vision Research, 41*(25–26), 3567–3585.

Yang, S. N., & McConkie, G. W. (2004). Saccade generation during reading: Are words necessary? *European Journal of Cognitive Psychology, 16*(1–2), 226–261.

Yeshurun, Y., & Carrasco, M. (1999). Spatial attention improves performance in spatial resolution tasks. *Vision Research, 39*(2), 293–306.

VISUAL COGNITION, 2005, 12 (6), 1093–1123

Quantifying the contribution of low-level saliency to human eye movements in dynamic scenes

Laurent Itti

Departments of Computer Science, Pyschology, and Neuroscience Graduate Program, University of Southern California, Los Angeles, USA

We investigated the contribution of low-level saliency to human eye movements in complex dynamic scenes. Eye movements were recorded while naive observers viewed a heterogeneous collection of 50 video clips (46,489 frames; 4–6 subjects per clip), yielding 11,916 saccades of amplitude $\geq 2°$. A model of bottom-up visual attention computed instantaneous saliency at the instant each saccade started and at its future endpoint location. Median model-predicted saliency was 45% the maximum saliency, a significant factor 2.03 greater than expected by chance. Motion and temporal change were stronger predictors of human saccades than colour, intensity, or orientation features, with the best predictor being the sum of all features. There was no significant correlation between model-predicted saliency and duration of fixation. A majority of saccades were directed to a minority of locations reliably marked as salient by the model, suggesting that bottom-up saliency may provide a set of candidate saccade target locations, with the final choice of which location of fixate more strongly determined top-down.

Humans and other mammals make extensive use of rapid eye movements to direct the highest-resolution region of their foveated eyes towards locations and objects of current behavioural interest. A productive approach for gaining a better understanding of foveated vision at the system level has been to use eyetracking devices, to evaluate image statistics at the locations visited by the eyes (Motter & Belky, 1998; Shen, Reingold, & Pomplun, 2000). In the context of natural scenes, most such research has thus far largely focused on characterizing local image properties at fixated locations. For instance, Zetzsche et al. (Barth, Zetzsche, & Rentschler, 1998; Zetzsche, Schill, Deubel, Krieger, Umkehrer, & Beinlich, 1998) inferred from human eyetracking recordings that the eyes preferentially fixated regions with multiple superimposed orientations,

Please address all correspondence to: L. Itti, Departments of Computer Science, Psychology, and Neuroscience Graduate Program, University of Southern California, Los Angeles, CA 90089-2520, USA. Email: itti@usc.edu

Supported by NSF, NEI, NIMA, the Zumberge Fund, and the Charles Lee Powell Foundation.

DOI:10.1080/13506280444000661

like corners. These authors then proceeded to derive nonlinear local operators to detect these regions. In a related study, Reinagel and Zador (1999) analysed the local grey-level distributions along eye scanpaths generated by humans while free-viewing greyscale images. They found the spatial contrast at fixated locations to be significantly higher than, on average, at random locations, and the pairwise pixel correlations (image uniformity) to be significantly lower. A complementary approach was proposed by Privitera and Stark (2000), who computed the linear combination of a collection of image processing operators (e.g., local cross detector, Laplacian of Gaussian, local entropy measure, etc.) that maximized overlap between regions of high algorithmic responses and regions fixated by human observers. Their study provided a new tool for creating feature detectors that would capture some of the local image properties that attracted the observer's attention and eye movements.

One extension of these approaches, further investigated here, consists of explicitly taking into account the long-range interactions across possibly distant locations in the visual field, as extensively documented in monkey electrophysiology and human psychophysics (Cannon & Fullenkamp, 1991; Gallant, Connor, & van Essen, 1998; Gilbert & Wiesel, 1989; Li & Gilbert, 2002; Sillito, Grieve, Jones, Cudeiro, & Davis, 1995; Polat & Sagi, 1993). That is, instead of attempting to characterize fixated locations solely in terms of their local image properties, computational models have been applied to correlate human fixations to measures of context-dependent saliency (conspicuity) of objects embedded within background clutter. A recent example is the study of Parkhurst, Law, and Niebur (2002), which compared eye movement scanpaths recorded from human observers presented with static colour scenes to a saliency map derived from a computational model of bottom-up attention (Itti, Koch, & Niebur, 1998). An important aspect of the model used in this study is that it explicitly incorporates a spatial competition for saliency, by which foci of high neural activity at possibly distant locations in the visual field compete for representation in the saliency map (Itti & Koch, 2001b). Parkhurst et al. found a significantly elevated model-predicted saliency at human fixations compared to random fixations (also see Peters, Itti, & Koch, 2002, for related results), an effect which was stronger for the first few fixations on a given image than for subsequent fixations.

Bottom-up saliency is only one of the many factors that contribute to the spatiotemporal deployment of attention and eye movements over complex dynamic visual scenes (Henderson & Hollingworth, 1999; Itti & Koch, 2001a; Rensink, 2000). The rapid understanding of the scene's gist and rough layout is also thought to provide priors on where objects of current behavioural interest may be located and to facilitate their recognition; for example, a vase is likely to be located on a tabletop (Friedman, 1979; Hollingworth & Henderson, 1998; Oliva & Schyns, 1997; Potter & Levy, 1969; Torralba, 2003), but see (Biederman, Teitelbaum, & Mezzanotte, 1983). Further, as specific objects are searched for, low-level visual processing may be biased for the features of these objects (Ito & Gilbert, 1999; Moran & Desimone, 1985; Motter, 1994; Reynolds, Pasternak, & Desimone, 2000; Treue & Martinez Trujillo, 1999; Treue & Maunsell, 1996; Yeshurun & Carrasco, 1998). This top-down modulation of bottom-up processing results in an ability to guide search towards targets of interest (Wolfe, 1994, 1998; Wolfe, Cave, & Franzel, 1989). In more general terms, task has been widely shown to affect eye movements (Peebles & Cheng, 2003; Tanenhaus, Spivey-Knowlton, Eberhard, & Sedivy, 1995; Yarbus, 1967), and training and general expertise as well (Moreno, Reina, Luis, & Sabido, 2002; Nodine & Krupinski, 1998; Savelsbergh, Williams, van der Kamp, & Ward, 2002). In addition, eye movements may often be executed in a purely volitional manner; for example, recognizing one object and knowing its typical spatial relationship to another target object may allow an observer to orient towards the target. Reading is an important example where eye movements are strongly influenced by the known spatial arrangement of words on a page and by sentence parsing and comprehension processes (Legge, Hooren, Klitz, Mansfield, & Tjan, 2002; Reichle, Rayner, & Pollatsek, 2003). Similarly, memorizing the locations of objects found along an initial exploratory scanpath on a new scene may allow observers to later orient back to some of these objects in an efficient manner (Henderson & Hollingworth, 2003; Hollingworth, Williams, & Henderson, 2001). Finally, eye movements from different observers exhibit different idiosyncrasies, which may result from possibly different internal world representations (Noton & Stark, 1971), different search strategies, and other factors (Andrews & Coppola, 1999).

Given all the factors that contribute to eye movements, it is hence expected and generally widely accepted that top-down influences may much more strongly determine gaze allocation than bottom-up influences. Our focus in this study is an attempt to quantify the contribution of low-level saliency to human scanpaths over complex dynamic scenes, and to analyse which low-level visual features contributing to saliency (such as colour, intensity, motion) are more strongly correlated with human eye movements. Our main contributions are two-fold: First, we study dynamic scenes (video clips); previous work has focused on static scenes (still images). Adding a temporal dimension brings a number of complications in trying to realistically predict saliency with a computational model. Also, when viewing natural video scenes in a realistic setting, top-down influences certainly are expected to be the largely predominant factor in determining the next fixation location, such that it is unclear whether bottom-up saliency would play a significant role at all. Second, we push the analysis beyond the observation that, overall, model-predicted saliency is elevated at the locations of human eye fixations, in particular by analysing the distribution of saliency at eye fixations and by attempting to tease out which low-level visual features were strongest at locations fixated by humans.

METHODS

We recorded eye movements from human observers watching a heterogeneous collection of 50 video clips, and segmented the recordings into saccade, blink and fixation/smooth-pursuit periods. For each human recording, we then employed a neurobiological model of bottom-up attention to explicitly predict bottom-up saliency at every scene location. At the beginning of each human saccade, we sampled the model-predicted saliency around the location of the future endpoint of that saccade. Our analysis focuses on comparing these saliency samples at saccade targets to maximum and randomly sampled saliency measures. These measures were analysed to determine the extent to which humans tended to saccade towards locations in their visual field that the model predicted were salient.

Our purpose in this study is not to attempt to maximize agreement between humans and model, but to quantitatively estimate the contribution of bottom-up processing to everyday eye movements. We hypothesized that if saliency was higher at saccade targets compared to other locations in the visual field, it would suggest that saliency may have contributed to choosing the saccade targets. Because of the many other factors mentioned above that influence eye movements, most of which are generally considered to be top-down, our starting hypothesis was that the fraction of eye movements directed to the most salient location in the scene would probably be very small—that is, top-down factors would dominate in determining the spatiotemporal allocation of gaze. One of our goals was to estimate that fraction in a quantitative manner. Here we assumed that our available computational model was a reasonable approximation of bottom-up processing, but we did not attempt to tune it to the specific data used in this study.

Human eye movement recording

Human subjects were naïve to the purpose of the experiment and were USC students and staff (three females, five males, mixed ethnicities, ages 23–32, normal corrected or uncorrected vision). They were instructed to watch the video clips, and to attempt to follow the main actors and actions presented in the clips, as they would later be asked some general questions about what they had watched. It was emphasized that the question would not pertain to small details (e.g., specific small objects, colours of clothing, or text messages), but would instead help us evaluate their general understanding of the contents of the clips. These instructions had two main purposes. First, we avoided instructionless free viewing, since in preliminary testing this had often yielded largely idiosyncratic patterns of eye movements, as subjects would lose interest and disengage from the experiment over time. Second, as mentioned in our introduction, our purpose in this study was not to attempt to maximize agreement between human eye movements and model predictions, but rather to quantify the extent to which low-level bottom-up saliency computed from responses of very simple feature detectors may contribute to the guidance of eye movements in normal viewing conditions. This is why we asked subjects to ''follow the main actors and actions'' rather than to ''attend to the most salient image locations'', and why we did not explain to subjects our computational notion of bottom-up saliency. While the notion of main actor is essentially a top-down one, requiring an evaluation of all actors and a decision of which one is the most interesting, our data hence quantifies the extent to which low-level saliency may bias this decision. While the bulk of our results should be interpreted carefully given our task definition, we also tested one control subject to whom no other instructions than to watch the video clips and enjoy were given (subject PA, not included in any of the group analyses). The procedure was approved by USC's Internal Review Board, and informed consent was obtained from all subjects. A set of calibration points and clips not part of the experiment were shown to familiarize the subjects with the displays.

Stimuli were presented on a 22-inch computer monitor (LaCie Corp; 640 × 480, 60.27 Hz double-scan, mean screen luminance 30 cd/m^2, room 4 cd/m^2). Subjects were seated on an adjustable chair at a viewing distance of 80 cm (28° × 21° usable field-of-view) and rested on a chinrest. A nine-point eye-tracker calibration was performed every five clips. Each calibration point consisted of fixating first a central cross, then a blinking dot at a random point on a 3 × 3 matrix. The experiment was self-paced and subjects could stretch before any nine-point calibration. For every video clip, subjects fixated a central cross, pressed a key to start, at which point the eyetracker was triggered, the cross blinked for 1206 ms, and the clip started. Stimuli were presented on a Linux computer, under SCHED_FIFO scheduling (process would keep 100% of the CPU for as long as needed; Finney, 2001). Each clip (MPEG-1 encoded) was entirely preloaded into memory. Frame displays were hardware-locked to the vertical retrace of the monitor (one movie frame was shown for two screen retraces, yielding a playback rate of 30.13 fps). Microsecond-accurate (Finney, 2001) timestamps were stored in memory as each frame was presented, and later saved to disk to check for dropped frames. No frame drop ever occurred and all timestamps were spaced by 33,185 $\pm$ 2 μs.

Eye position was tracked using a 240 Hz infrared-video-based eyetracker (ISCAN, Inc. model RK-464). Point of regard (POR) was estimated from comparative tracking of both the centre of the pupil and the specular reflection of the infrared light source on the cornea. This technique renders POR measurements immune to small head translations (tested up to $\pm$10 mm in our laboratory). Thus, no stricter restraint than a chinrest was necessary, which is important as head restraint has been shown to alter eye movements (Collewijn,

Steinman, Erkelens, Pizlo, & van der Steen, 1992). All analysis was performed offline. Linearity of the POR-to-stimulus coordinate mapping was excellent, as previously tested using a 7 × 5 calibration matrix, justifying the use of a 4 × 3 matrix here. The eyetracker calibration traces were filtered for blinks, then automatically segmented into two fixation periods (the central cross, then the flashing point), or discarded if that segmentation failed a number of quality control criteria. From the nine (or fewer) resulting calibration points, an affine POR-to-stimulus transform was computed in the least-square sense, outlier calibration points were eliminated, and the affine transform was recomputed. If fewer than six points remained after outlier elimination, recordings were discarded until the next calibration. A thin-plate-spline nonlinear warping algorithm was then computed to account for any small residual nonlinearity (Bookstein, 1989). Only 25% of the thin-plate-spline deformation was applied to the data (0% would induce no warping, 100% would ensure a zero residual error on all calibration points), as observers often slightly jittered around a given calibration point. Thus, we assumed almost perfect linearity of the tracker, but possible noise in individual calibration traces. Data was discarded until the next calibration if residual errors greater than 20 pixels on any calibration point or 10 pixels overall remained. Eye traces for the five video clips following a calibration were remapped to screen coordinates, or discarded if they failed some quality control criteria (excessive eye-blinks, loss of tracking due to head motion or excessive wetting of the eye, loss of corneal reflection due to excessive squinting). Calibrated eye traces were visually inspected when superimposed with the corresponding video clips, but none was discarded based upon that subjective inspection.

Fifty video clips were selected from a database of eighty-five, with the only selection criterion of maximizing diversity. All clips had been digitized from analogue interlaced NTSC video sources using a consumer-grade framegrabber (WinTV Go, Hauppage, Inc.) and no attempt was made at deinterlacing or colour-correcting them. The clips included:

beverly: Daytime outdoors scenes filmed at a park in Beverly Hills
gamecube: Various video games (first-person, racing, etc.)
monica: Outdoors day/night scenes at the Santa Monica Promenade
saccadetest: A synthetic disk drifting on a textured background
standard: Daylight scenes filmed at a crowded openair rooftop bar
tv-action: An action scene from a television movie
tv-ads: Television advertisements
tv-announce: A television programmes announcement
tv-music: A music video interleaved with some football scenes
tv-news: Various television newscasts
tv-sports: Televised basketball and football games
tv-talk: Television talk-shows and interviews.

Each clip comprised between 164 and 2814 frames (5.5–93.9 s), for a total of 46,489 distinct frames (25 min and 42.7 s). Selection of clips was made prior to the experiments and was not altered after recording and data analysis. Subjects viewed each clip at most once, to ensure that they were naïve to its contents. Five subjects viewed all clips and three only viewed a few; after our quality-control criteria were applied, calibrated eye movement data was available for four to six subjects on each clip, yielding a total of 235 calibrated eye movement traces.

The calibrated eye movement data was further segmented into saccades, eye blinks, and fixation/smooth pursuit periods (Sparks, 2002). In a preprocessing step, eye blinks were eliminated from the recordings. Blinks were characterized in our recordings by a measured pupil diameter falling to zero. These periods were marked in the datasets, so that visual input to the model would be turned off during blinks. Our dataset contained occurrences of blinked saccades, that is, blinks (as defined by the measured pupil diameter of zero) with distant eye position at the end of the blink period compared to its beginning. This slightly complicated the extraction of saccades from the dataset, as, typically, high-velocity saccadic eye motion preceded and followed these particular blinks (that is, these blinks occurred within a fraction of the temporal extent of a saccade). We note that, behaviourally, blinking during saccades seems an efficient strategy in terms of minimizing the total amount of time during which visual information will not be reliably exploitable by the central nervous system. For this reason, in our preprocessing, we linearly interpolated eye position during these blinked saccades, so that they could be detected as legitimate saccades in the next processing step. Similarly, we also linearly interpolated eye position during normal blinks, so that fixation periods could be reliably detected. Thus, the output of this preprocessing step consisted of fully continuous eye position traces, with blink periods marked so that visual input would be turned off in the model.

The segmentation of saccades was based on instantaneous eye velocity computed from a low-pass-filtered version of the blink-interpolated eye movement traces. Although more sophisticated methods have been proposed (Salvucci & Goldberg, 2000), this simple approach worked reliably with our high sampling rate of 240 Hz. Periods of the recordings where smoothed eye velocity exceeded $20^\circ/s$ were marked as candidate saccade periods. The eye position at the end of each saccade was considered the saccade target location in our analysis. Candidate saccades of amplitude smaller than 2° (a threshold chosen rather arbitrarily) were not considered as saccades in our analysis, but were instead reverted to the pool of fixation/smooth pursuit periods. The rationale for this decision to not consider very small saccades was that they were unlikely to indicate a shift of foveation from one object to another (possibly salient) object, but rather, we hypothesized, would mainly reflect small adjustments of eye position onto various parts of a same object (no matter its saliency). Thus, very small saccades would be of little interest in determining whether humans tended to orient towards salient objects or not.

In summary, our data recording and segmentation yielded a total of 11,916 saccades, for each of which the starting eye position, starting time, and target location were known.

Bottom-up attention model

The model computes a topographic saliency map (Figure 1), which quantitatively predicts how conspicuous every location in the input scenery is. Its implementation has been previously described in details (Koch & Ullman, 1985; Itti & Koch, 2000, 2001a; Itti et al., 1998). Retinal input is processed in parallel by 12 multiscale low-level feature maps, which detect local spatial discontinuities using simulated centre-surround neurons (Hubel & Wiesel, 1962; Kuffler). The 12 neuronal features implemented are sensitive to colour contrast (red/green and blue/yellow, separately), temporal flicker (onset and offset of light intensity, combined), intensity contrast (light-on-dark and dark-on-light, combined), four orientations (0°, 45°, 90°, 135°), and four oriented motion energies (up, down, left, right) (Itti & Koch, 2001a; Itti et al., 1998). Centre and surround scales are obtained using dyadic pyramids with nine levels (from level 0, the original image, to level 8, reduced by a factor 256 horizontally and vertically). Centre-surround differences are then computed as pointwise differences across pyramid levels, for combinations of three centre scales ($c = \{2, 3, 4\}$) and two centre-surround scale differences ($\delta = \{3, 4\}$); thus, six feature maps are computed for each of the 12 features, yielding a total of 72 feature maps. After long-range competition for salience, all feature maps are summed (Itti & Koch, 2001b) into the unique scalar saliency map that guides attention. The saliency map is at scale 4 (40×30 pixels given our 640×480 video frames).

Each feature map is endowed with internal dynamics that operate a strong spatial within-feature and within-scale competition for activity, followed by within-feature, across-scale competition (Itti & Koch, 2001b). These long-range within-feature competitive interactions are a crucial component of our computational framework, as they are mainly responsible in the model for "pop-out" effects observed behaviourally in visual search psychophysics (Treisman & Gelade, 1980; Wolfe, 1998). In a first approximation, the computational effect of these interactions is similar to applying a soft winner-take-all to each feature map before it is combined into the saliency map (Itti & Koch, 2001b). Consider for instance a simplistic scene with a single red disc and a single blue disc on a black background. Our model would predict that both discs are equally salient, if the red and blue colours are matched for luminance and their chrominances are optimally tuned to the model's red/green and blue/yellow feature detectors, respectively. The red disc would be mainly represented in the red/green feature maps with approximately matching centre-surround receptive field sizes, while

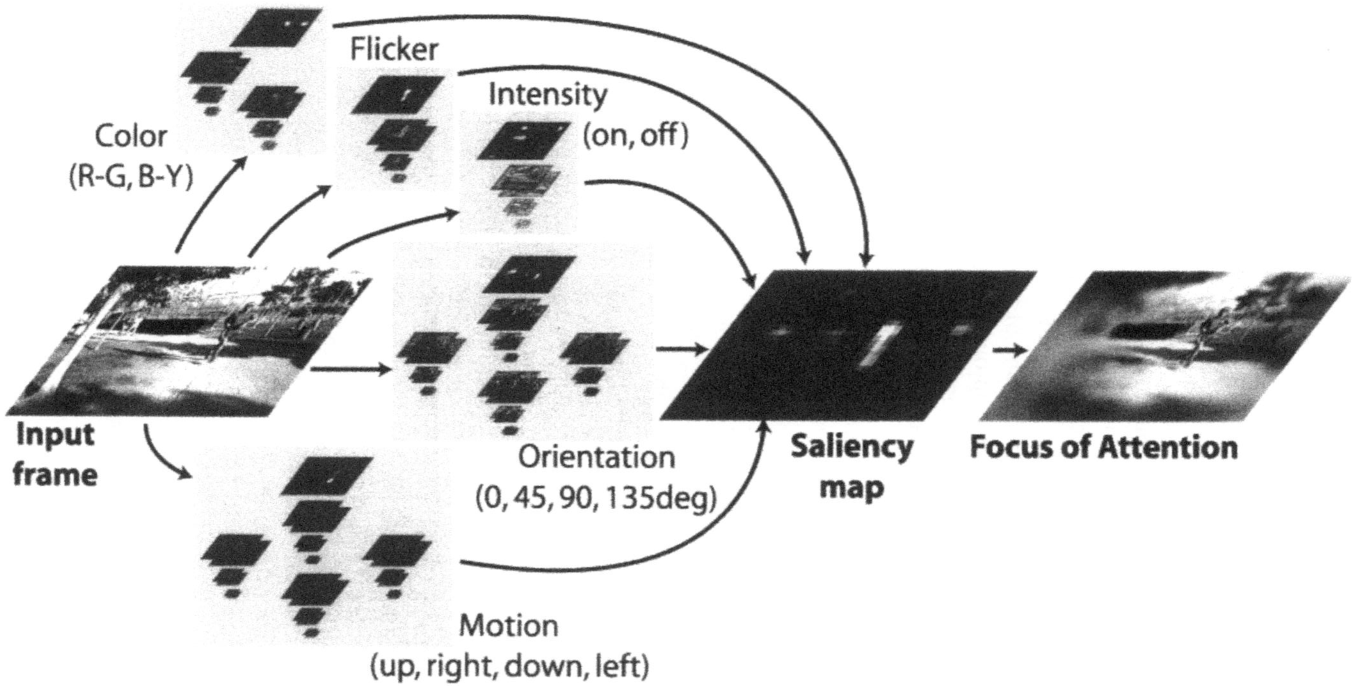

Figure 1. Overview of the model. Low-level visual features are computed from incoming input frames in a set of multiscale feature maps tuned to colour opponencies, intensity contrast, orientation contrast, temporal change (flicker) and oriented motion energy. After strong nonlinear within-feature competition for saliency, all feature maps provide input to the unique saliency map.

the blue disc would mainly excite blue/yellow feature maps with matching receptive field sizes. Starting from this baseline image, now consider the addition of three additional red discs identical to the first one and placed at distant locations from the initial two discs. The model's long-range interactions in the red/green feature maps would yield competition among the four red discs present in this new display, resulting in reduced saliency for each of the red discs. In contrast, no such competition would be triggered in the model's blue/yellow feature maps, and the blue disc would remain as salient as it initially was. Resultingly, after the various feature maps are summed into the saliency map, the blue disc would be more salient than any of the red discs, and would be the most salient location in the display (pop-out effect). Thus, although our low-level feature detectors do respond to local image properties, the addition of nonlinear competition for saliency in our model renders the predicted saliency of a given object dependent on every other object and on background clutter present in the input images (Itti & Koch, 2001b).

We have previously proposed several implementations for these long-range interactions (Itti & Koch, 2000, 2001b; Itti et al., 1998), yielding saliency maps with different overall sparseness (Itti, 2004). While Parkhurst et al. (2002) used fairly weak interactions, yielding relatively dense saliency maps where many salient locations are highlighted (Itti et al., 1998), here we use one iteration of the iterative scheme proposed by Itti and Koch (2001b), which yields much sparser saliency maps that typically highlight only 3–5 locations in the entire image.

The saliency map itself is modelled as a two-dimensional layer of leaky integrator neurons (Itti & Koch, 2000). It receives as input the sum of all feature maps, after that sum is subjected one more time to nonlinear competition for saliency. Its main function is to provide temporal smoothing of the saliency values, and no further interactions across distant locations are implemented at this stage. Consequently, in this study we directly evaluate the saliency values at current human eye position, even though some reaction delay exists between appearance of a stimulus element and the possible decision to orient the eyes towards that location. Thus, we here do not investigate in detail the effects of stimulus-to-response latency, but instead temporally low-pass filter the saliency map. The integrators are simulated using difference equations at a temporal resolution of 0.1 ms of simulated time (Itti & Koch, 2000).

Comparing model predictions to human eye movements

The computational model was used in the following manner to evaluate bottom-up saliency at the target locations of human saccades. Each of the 235 human eye movement recordings was considered in turn. For a given recording, the corresponding video clip was presented to the model, at the same resolution and framerate as it had been presented to human subjects.

It is important to note that in this study we do not attempt to replicate many of the details of human foveated active vision. In particular, the video images are processed in their native frame of reference (as opposed to a retinal frame of reference that would be centred at current human eye position), and previously explored model components (Itti, Dhavale, & Pighin, 2003) for foveation, overt inhibition-of-return (IOR), and saccadic suppression were not used here. The rationale for this decision was that some of these factors may have top-down components (e.g., IOR) which would make the interpretation of our results in terms of bottom-up processing more difficult. Rather than employing a hybrid framework (e.g., including foveation but not IOR nor coordinate transforms) like, for example, Parkhurst et al. (2002) did, in the present study we attempt to compute a pure form of stimulus saliency, which is an intrinsic property of the video stimuli and is independent of the observer. It is understood that the actual bottom-up computations performed by humans observers are modulated by foveation, coordinate transforms, short-term visual memory, and other factors, so that the internal representation of saliency may substantially differ from the intrinsic image saliency computed here. Hence the saliency computed here does not so much attempt to replicate the internal representation of saliency in the human brain than the external intrinsic stimulus saliency.

RESULTS

At the beginning of each human saccade, several measurements were made on the model's current internal state, to quantify the extent to which that human saccade was aimed towards a location of high model-predicted saliency. The following samples were taken:

- S_h: Saliency at human eye position, computed as the maximum over a circular aperture of diameter 5.6° (nine pixels in the saliency map) of the model's dynamical saliency map sampled at the moment a saccade starts and around the location of the future endpoint of that saccade. Thus, a high value of S_h indicates that, when it starts, a saccade is targeted towards a location that is highly salient at that moment.
- S_r: Saliency at a random location, computed in exactly the same manner as S_h except that a random endpoint within the image (with uniform probability) is considered rather than a given human saccade endpoint. Thus, a high value of S_r indicates that, at the moment a saccade starts, saliency is high around a randomly chosen location within the image. This provides a control for the evaluation of human saccades, by estimating the overall spatial density of active locations in a saliency map, i.e., the degree of sparseness or selectivity of the map.

- S_{max}: Maximum saliency over entire frame, computed as the maximum over the spatial extent of the entire dynamical saliency map, at the same instant as the other measurements were taken. This provides a very strict test on the saliency of a saccade target, in that $S_h = S_{max}$ if and only if the saccade is directed towards the currently most salient location in the entire visual field. It also allows us to compare saliency values across frames and video clips, by normalizing S_h, and S_r relative to S_{max}.

Analysis by saccade amplitude

A first observation concerns the distribution of saccade amplitudes, which shows a strong bias for shorter saccades (Figure 2). Given this bias, we next asked whether saccade length would be a factor in our measures of saliency at the endpoint of saccades. Figure 3a shows a histogram of the ratios of saliency sampled, at the time each human saccade began, around the location of the future saccade endpoint (S_h), compared to the maximum saliency over the entire saliency map at the same moment (S_{max}). For comparison, the histogram of the same ratios but using the saliency at random saccade endpoint locations (S_r) is

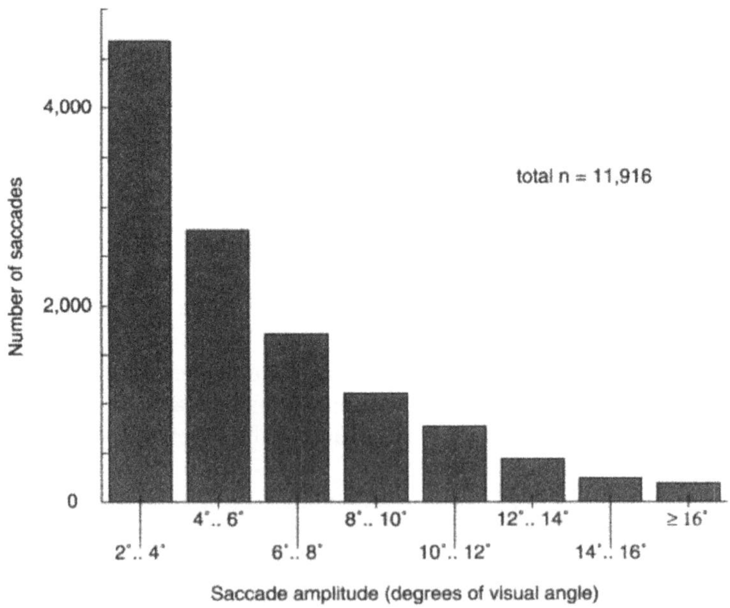

Figure 2. The distribution of observed saccade amplitudes during our experiments shows a strong bias for shorter saccades. Saccades shorter than 2° were treated as fixation/smooth-pursuit in our analysis. The count for each bar is inclusive of the lower bound but exclusive of the higher bound (e.g., the bar labelled 8°.. 10° shows the number of saccades where 8° ≤ amplitude < 10°).

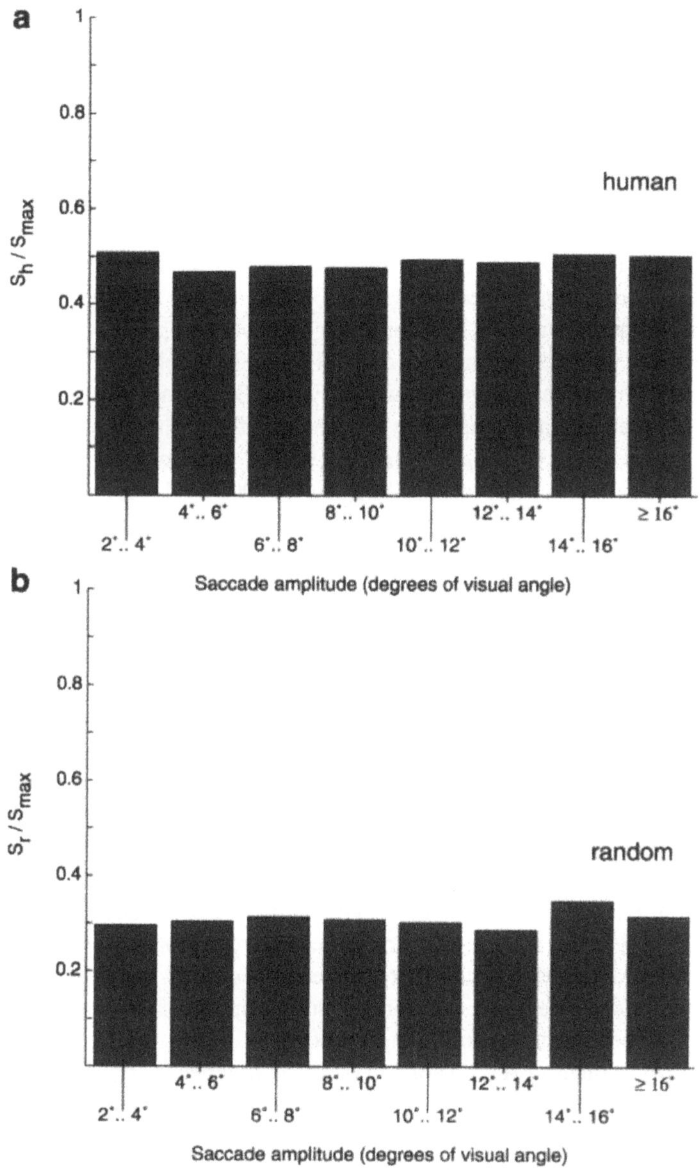

Figure 3. Distribution, by saccade amplitude, of the ratios of saliency at human saccade target locations to the maximum saliency (S_h/S_{max}; panel a), and of the ratios of saliency at random locations to the maximum saliency (S_r/S_{max}; panel b). For each saccade, all measurements were taken at the moment the saccade began. Overall, S_h/S_{max} was higher than S_r/S_{max}, in a highly significant manner (see text). Shorter and longer human saccades show slightly higher S_h/S_{max} (significant effect in a one-way ANOVA; see text), but random saccades do not. Note that the histogram in panel b is indexed by human saccade amplitude like that of panel a (so that a given amplitude range includes the same number of saccades in both panels), not by random saccade amplitude.

shown in Figure 3b. Looking at the first order statistics of the plotted distributions, saliency around saccade endpoints was higher than around random locations, with a median S_h/S_{max} of 0.45 over all 11,916 saccades, compared to a median S_r/S_{max} of 0.22 (Table 1). A paired-sample nonparametric sign test rejected the null hypothesis that the samples S_h/S_{max} and the samples S_r/S_{max} could have been drawn from distributions with same medians, in a highly significant manner ($p < 10^{-232}$). This difference was also highly significant for all saccade amplitude ranges shown in Figure 3. That is, the sign test suggested a significant difference with $p < 10^{-10}$ or better also when applied only to saccades with amplitudes between 2° and 4°, as well as when applied to saccades with amplitudes in any of the ranges shown in Figure 3. A one-way ANOVA suggested that human saccade amplitude (binned into the eight ranges of Figure 3) had a significant effect onto S_h/S_{max}, $F(7, 11908) = 5.22$, $p < 10^{-5}$, eight groups and 11,916 saccades, but not onto S_r/S_{max}, $F(7, 11908) = 1.86$, n.s. Indeed, very short and very long saccades tended to exhibit slightly higher S_h/S_{max}, but that trend was not present in the S_r/S_{max} data.

A breakdown of the data into various subsamples is shown in Table 1. Overall, for all subsamples except one (the clip tv-announce), median S_h/S_{max} was still significantly elevated compared to median S_r/S_{max}. This breakdown as well as the previous breakdown by saccade amplitude suggest that our main finding of an overall elevated S_h/S_{max} compared to S_r/S_{max} was not primarily driven by a subset of the recorded saccades; instead, significant differences between humans and random were found for all saccade amplitudes, all observers, and all but one classes of video clips. This suggests a strong contribution of intrinsic image saliency, as defined by our computational model, to human eye movements in a wide range of situations. In particular, the first, first five, and first ten saccades onto a given video clip corresponded to higher ratios of medians, indicating a stronger correlation between bottom-up saliency and human saccade targets during the first few seconds of viewing a new video clip. Further comparisons between the various subgroups of saccades presented in Table 1 are presented in the following sections.

Our control subject (PA in Table 1), studied independently and not part of any of the group analyses, was given no instruction to follow the main actors and actions in the video clips, yet fell within the range of variability of other subjects. However, this subject showed slightly higher overall saccade frequency (approximately 2.4 saccades/s on average while the average of all other subjects was approximately 1.7 saccades/s). This seems reasonable as the instruction of "following" the main actors and actions may have incited the other subjects to lock onto these scene elements for extended periods of time.

This first set of results confirms with dynamic scenes previous studies which suggested that, overall, humans look are image regions of higher saliency than would be expected by chance (Parkhurst et al., 2002). To summarize, median S_h/S_{max} was about twice the median S_r/S_{max}, a factor significantly greater than unity.

TABLE 1

Average and median S_h/S_{max} and S_r/S_{max} for various subsamples of our overall pool of saccades

Samples	*n*	*Median* S_h/S_{max}	*Median* S_r/S_{max}	*Sign test*	*Ratio of medians*	*Kullback-Leibler distance*
All	11,916	0.45	0.22	$p < 10^{-232}$	2.03	0.195 ± 0.006
First	235	0.57	0.24	$p < 4 \times 10^{-16}$	2.32	0.414 ± 0.062
First 5	1,173	0.48	0.21	$p < 3 \times 10^{-52}$	2.31	0.279 ± 0.021
First 10	2,322	0.49	0.22	$p < 2 \times 10^{-90}$	2.25	0.278 ± 0.015
CZ	2,369	0.46	0.22	$p < 5 \times 10^{-66}$	2.12	0.219 ± 0.016
JC	1,527	0.41	0.23	$p < 8 \times 10^{-34}$	1.76	0.147 ± 0.011
JZ	1,965	0.44	0.21	$p < 9 \times 10^{-61}$	2.10	0.186 ± 0.015
NM	578	0.47	0.22	$p < 3 \times 10^{-24}$	2.19	0.267 ± 0.031
ND	1,602	0.45	0.23	$p < 2 \times 10^{-42}$	2.00	0.197 ± 0.014
RC	3,006	0.47	0.22	$p < 7 \times 10^{-95}$	2.12	0.218 ± 0.013
VN	731	0.48	0.23	$p < 2 \times 10^{-29}$	2.07	0.237 ± 0.027
VC	138	0.28	0.20	$p < 7 \times 10^{-4}$	1.42	0.101 ± 0.041
beverly	593	0.55	0.15	$p < 6 \times 10^{-39}$	3.71	0.522 ± 0.045
gamecube	3,442	0.50	0.17	$p < 2 \times 10^{-130}$	2.87	0.281 ± 0.014
monica	770	0.52	0.23	$p < 2 \times 10^{39}$	2.28	0.303 ± 0.029
saccadetest	72	1.00	0.00	$p < 2 \times 10^{-8}$	∞	1.284 ± 0.338
standard	876	0.43	0.27	$p < 3 \times 10^{-15}$	1.58	0.116 ± 0.014
tv-action	105	0.53	0.22	$p < 6 \times 10^{-6}$	2.43	0.408 ± 0.101
tv-ads	844	0.45	0.26	$p < 4 \times 10^{-18}$	1.73	0.161 ± 0.017
tv-announce	153	0.38	0.35	n.s.	1.08	0.083 ± 0.028
tv-music	244	0.41	0.34	$p < .02$	1.23	0.146 ± 0.030
tv-news	2,556	0.37	0.23	$p < 8 \times 10^{-48}$	1.60	0.114 ± 0.009
tv-sports	1,292	0.46	0.26	$p < 2 \times 10^{-39}$	1.76	0.145 ± 0.015
tv-talk	989	0.46	0.23	$p < 10^{-34}$	1.98	0.233 ± 0.020
PA	3,395	0.43	0.22	$p < 6 \times 10^{-89}$	1.99	0.172 ± 0.011

Sign test: Significance level with which the null hypothesis was rejected that the samples S_h/S_{max} and S_r/S_{max} could have been drawn from distributions with equal median, as tested with a nonparametric paired-sample sign test. Ratio of medians: Ratio of the median S_h/S_{max} to the median S_r/S_{max}. Kullback-Leibler distance: KL distance between the histogrammed distributions of S_h/S_{max} and of S_r/S_{max} (higher distances indicate more different distributions). All: All 11,916 saccades recorded in our experiments. First: all the first saccades of our 235 individual recordings (50 clips and 4–6 subjects per clip). First 5, First 10: All the first five and first ten saccades of our 235 recordings (some recordings had fewer than five or ten saccades). CZ to VC: Breakdown by human subject. beverly to tv-talk: Breakdown by category of video clips. Median S_h/S_{max} is significantly higher than median S_r/S_{max} for all but one clip (tv-announce). PA: Data from a control subject (not included in any other analysis) who was given no instructions before watching the video clips.

Analysis by ratios to maximum saliency

The observation of elevated median saliency at human saccade targets compared to random targets is a fairly crude measure that does not take into account possible differences in the shapes of the human and random distributions that would not affect the median.

Hence, we further pushed our analysis to determine the fraction of saccades that had been made to locations of various saliency relative to the maximum saliency. In Figure 4, saccades were binned by their saliency ratio to the maximum saliency, both for human and corresponding random saccades. One striking feature in this figure is that the number of saccades made to absolutely not salient locations was much lower for human than random saccades; conversely, the number of saccades made to the single most salient location in the display was much higher for human than random saccades.

To quantify these differences in shape of distributions, we computed the Kullback-Liebler distance between the histogrammed distribution of $S_h/S_{\max}$ values and that of $S_r/S_{\max}$ values. The Kullback-Leibler (KL) distance is a general metric by which shape similarity between two arbitrary distributions may be measured (Kullback, 1959). For two discrete distributions X and Y with probability density functions x_k and y_k, the KL distance between X and Y is nothing else than the relative entropy of X with respect to Y:

$$KL(X, Y) = \sum_k x_k \log\left(\frac{x_k}{y_k}\right) \tag{1}$$

Note that this actually is not a distance, as it is not symmetric. Hence, in what follows, the shorthand denomination of "KL distance" always refers to the KL distance from the human $S_h/S_{\max}$ distribution (X) to the random $S_r/S_{\max}$ distribution (Y).

Table 1 (rightmost column) reports the KL distance for all saccades as well as the various subgroups of saccades studied. A larger KL distance indicates that the distribution of relative saliency at human saccade targets (as shown, for example, in Figure 4a for all saccades) was more different from the distribution of relative saliency at random saccade targets (as shown in Figure 4b for all saccades). A KL distance of zero indicates no difference between the two distributions under consideration. To obtain an estimate of the relevant accuracy at which the KL distances should be interpreted, we repeated the random saccade generation process 100 times, and each time measured the corresponding KL distances; the standard deviation computed over the 100 repeated distance measurements for each subgroup of saccades then quantifies the dependence of the KL metric onto the random saccade generation process.

The KL distance basically followed the same trend as the ratio of medians for the saccade subgroups shown in Table 1 (although it will not always be the

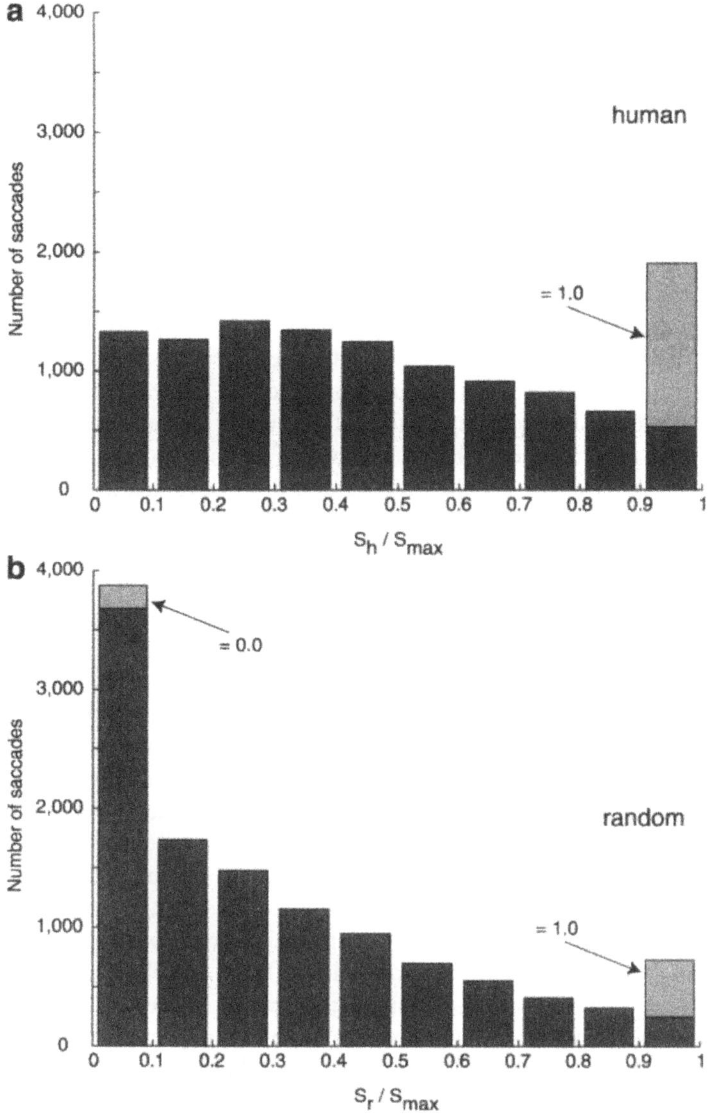

Figure 4. Distributions, by ratios S_h/S_{max} (panel a) and S_r/S_{max} (panel b) of human and random saccades. Each bar includes the lower bound and excludes the higher bound (e.g., the bar between 0.4 and 0.5 is for $0.4 \leq$ ratio < 0.5). In lighter shades are those saccades with ratio < 0.1 where the absolute saliency was zero (noted as "=0.0"; saccades to an absolutely not salient location), and those saccades with ratio ≥ 0.9 where the absolute saliency was S_{max} (noted as "=1.0"; saccades to the single most salient location). The histogram for human saccades show a largely increased number of saccades to the most salient location and a largely decreased number of saccades to not salient locations, compared to the random saccades, as further analysed in Table 2.

case in following sections). All KL distances in the table were significantly higher than zero (t-test, $p < .01$ or better). The first few saccades showed higher KL distance in addition to the higher ratios of medians noted above. Distributions of saliency at saccade targets for every observer taken individually were also significantly different from the random distributions. The same result was observed with subgroups of video clips. The *saccadetest* clip consisted of a single coloured disc jumping to various places on a uniformly textured blue background, and provided a simple control as there always was only salient location to look at. For this clip, median S_h/S_{max} was 1.0, indicating that more than half of the human saccades had been directed to the single most salient location in the display (the jumping disc); in contrast, median S_r/S_{max} was 0.0, indicating that more than half of the random saccades had been made to locations of zero model-predicted saliency. The corresponding KL distance was 1.284 ± 0.338, which gives an estimate of the values that may be obtained in ideal conditions. Values obtained for other subgroups were generally lower, indicating again that, although strong, bottom-up influences were clearly not the only driver of eye movements, and top-down influences also played a strong role as most other clips contained many potentially interesting objects and actors.

A breakdown by ratios to maximum saliency is presented in Table 2. For a given threshold value (0.25, 0.75, or 1.00), this table shows the number of human and of random saccades targeted to regions with relative saliency above threshold. A first interesting point from this analysis is that the frequency of saccades directed to the single most salient location in the saliency map was small, only 11.5% of all saccades. This suggests that a simple strategy consisting of directing the eyes to the most active location in our saliency map would only account for a small fraction of the observed eye movement traces, though in a very reliable manner (only 4.0% of the random saccades were targeted to the most salient location, a factor of almost three times smaller than the human saccades). Given the sparseness of our saliency maps, as exemplified in Figures 5 and 6, it is reasonable to consider that a relative saliency value of 0.25 or more indicates a location that was reliably detected by the model as salient (regions with relative saliency below 0.25 essentially appear as black in the figures). Considering all saccades, 72.3% of the human saccades were targeted towards regions above that threshold, compared to 46.6% of random saccades. While it probably is a wrong conclusion that all of these human saccades were primarily driven bottom-up, certainly it seems that by and large human subjects were more likely to target regions marked by the model as salient than would be expected by chance. Further considerations of bottom-up compared to top-down influences are presented in the Discussion.

A second message of our study, thus, is that the proportion of saccades to the most salient location in the display was small (11.5%) but overall much greater than what would have been expected by chance alone. The proportion of human

TABLE 2
Breakdown by ratios S_h/S_{max} and S_r/S_{max}

		S_h/S_{max}			S_r/S_{max}		
Samples	*n*	≥ 0.25	≥ 0.75	$= 1.00$	≥ 0.25	≥ 0.75	$= 1.00$
All	11,916	72.3%	24.5%	11.5%	46.6%	10.6%	4.0%
First	235	85.1%	33.2%	14.9%	49.4%	12.3%	5.1%
First 5	1,173	77.7%	26.6%	13.2%	44.8%	10.6%	4.3%
First 10	2,322	77.6%	26.9%	13.2%	46.3%	10.7%	4.3%
CZ	2,369	72.8%	26.9%	12.7%	46.2%	11.7%	4.7%
JC	1,527	69.9%	21.0%	8.8%	47.9%	10.3%	3.5%
JZ	1,965	71.5%	24.2%	11.9%	45.5%	10.0%	3.0%
NM	578	74.7%	28.2%	13.0%	45.8%	10.0%	4.2%
ND	1,602	72.5%	23.5%	12.1%	47.3%	11.1%	4.2%
RC	3,006	73.1%	25.2%	11.6%	46.7%	10.0%	4.1%
VN	731	76.3%	23.9%	11.4%	47.2%	10.7%	4.5%
VC	138	54.3%	7.2%	2.2%	42.0%	10.1%	5.1%
beverly	593	79.1%	35.4%	21.1%	39.3%	9.9%	4.4%
gamecube	3,422	70.5%	29.5%	16.0%	41.2%	10.1%	4.0%
monica	770	81.0%	28.3%	10.1%	47.0%	11.2%	4.2%
saccadetest	72	68.1%	65.3%	54.2%	12.5%	8.3%	6.9%
standard	876	72.8%	17.8%	5.0%	53.0%	11.6%	4.0%
tv-action	105	86.7%	32.4%	8.6%	46.7%	11.4%	3.8%
tv-ads	844	74.2%	22.7%	8.8%	51.4%	10.7%	3.2%
tv-announce	153	69.3%	9.2%	2.6%	61.4%	14.4%	5.2%
tv-music	244	77.0%	17.2%	5.3%	61.1%	11.9%	3.3%
tv-news	2,556	67.9%	18.0%	8.1%	47.2%	10.7%	4.4%
tv-sports	1,292	74.1%	21.8%	8.6%	50.9%	11.1%	3.6%
tv-talk	989	72.9%	25.2%	12.0%	48.4%	9.5%	3.8%
PA	3,395	69.6%	23.8%	10.5%	45.5%	10.0%	3.7%

The row labels are the same as in Table 1.

saccades directed to overall salient locations was large (72.3%) and also much larger than expected by chance (a factor of 1.55).

Contributions of individual low-level features

To study whether particular features in the model would be better predictors of human saccadic targeting than others, we repeated the analysis, but using variants of the model which only included one feature channel. The feature channels thus considered separately were: color contrast (red/green and blue/yellow double opponencies, combined), intensity contrast (light on dark and dark on light, combined), orientation 0°, 45°, 90°, and 135°, combined), flicker

Figure 5. Example frames from our video clips. From left to right: Input to the model, input plus explanatory markers generated by the model, and instantaneous saliency map. From top to bottom: beverly01 (frame 31/490), gamecube04 (frame 914/2083), monica03 (frame 483/1526), saccadetest (frame 63/516), standard03 (frame 112/515), and tv-action01 (frame 142/567). During each frame shown here, a human saccade was initiated, which would in subsequent frames follow the arrow shown on each frame. Yellow circle at arrowhead (and also white circle in saliency map) shows the circular aperture within which saliency at the future location of saccade target was sampled.

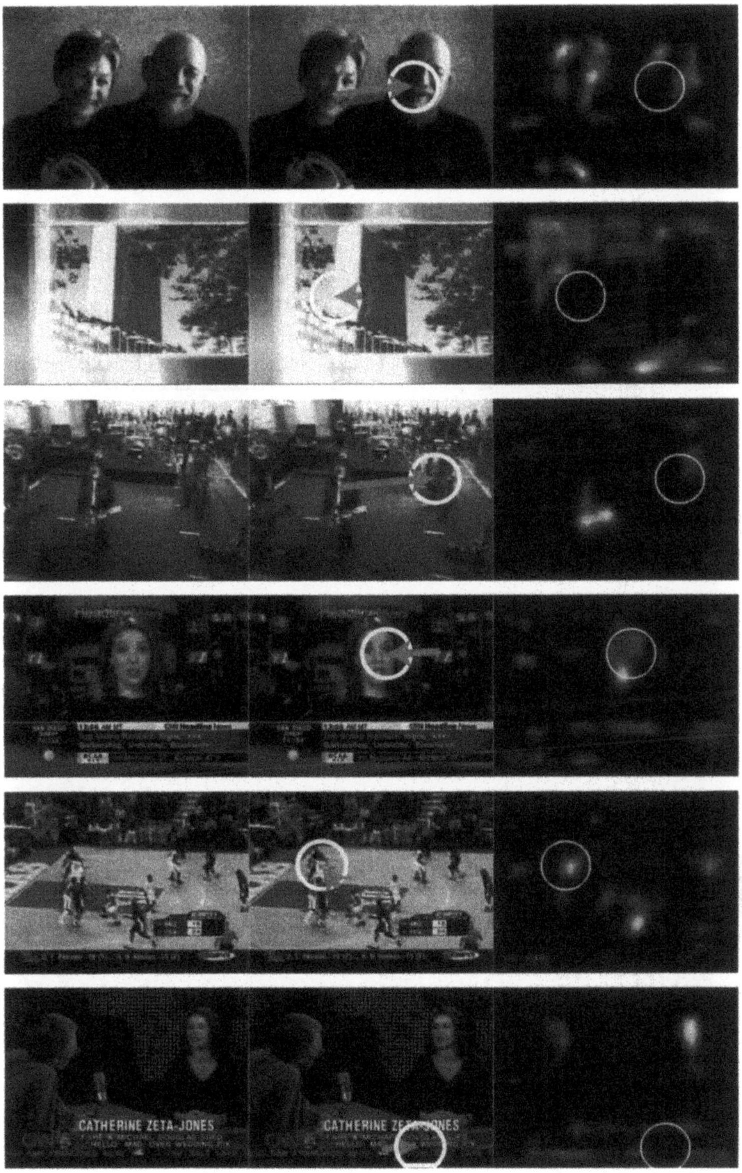

Figure 6. More example frames from our video clips. Format is identical to Figure 5. From top to bottom: tv-ads02 (frame 209/387), tv-announced01 (frame 18/434), tv-music01 (frame 909/1022), tv-news03 (frame 70/1444), tv-sports01 (frame 379/579), and tv-talk01 (frame 121/1651).

TABLE 3
Comparison among models

Model	*Median* S_h/S_{max}	*Median* S_r/S_{max}	*Sign test*	*Ratio of medians*	*Kullback-Leibler distance*
CIOFM	0.45	0.22	$p < 10^{-232}$	2.03	0.195 ± 0.006
C	0.29	0.13	$p < 5 \times 10^{-146}$	2.20	0.075 ± 0.003
I	0.33	0.15	$p < 3 \times 10^{-167}$	2.14	0.089 ± 0.003
O	0.32	0.16	$p < 2 \times 10^{-162}$	1.98	0.084 ± 0.004
F	0.33	0.09	$p < 10^{-232}$	3.71	0.178 ± 0.005
M	0.34	0.09	$p < 10^{-232}$	3.70	0.176 ± 0.005

CIOFM: Model with all feature channels. C: Colour only; I: Intensity only; O: Orientation only; F: Flicker only; M: Motion only. Columns are a subset of those in Table 1.

(absolute difference between successive frames), and motion energy (up, down, left, and right, combined).

Results are presented in Table 3. Flicker and motion each taken separately yielded higher ratios of medians and KL distances than intensity, colour, and orientation taken separately. In fact, the ratios of medians for flicker and motion were even higher than for the full model which includes all features. To a large extent, this was due to the increased sparseness of the motion and flicker saliency maps compared to the full-model saliency maps. Indeed, the number of human saccades targeted to the most salient location was similar (12.1% for the motion-only model and 12.1% for the flicker-only, compared to 11.5% for the full model reported in Table 2, with random saccades all around 4.0%). In contrast, there were many more random saccades targeted to nonsalient locations, hence driving the median saliency at random saccade endpoints down, and the ratio of medians up. The KL distance is useful here to provide a finer analysis of which version yielded the largest differences in saliency distribution between human and random scanpaths. Motion and flicker, each individually, yielded KL distances quite close to that obtained with the full model, although significantly lower, as shown in Table 4. These results indicate that motion and image transients are much more reliable predictors of human saccadic targeting than colour, intensity, or orientation contrasts, with the best predictor being the sum of all these features.

Duration of subsequent fixations

A last analysis was to determine whether the duration of a fixation following a saccade would correlate with the saliency at that saccade's endpoint. With our video clips, average duration of fixation was short, approximately 135 ms (although with a long tail towards longer durations). Figure 7 shows the average duration of fixation following saccades to locations with given model-predicted

TABLE 4
Contributions of features

	CIOFM	*C*	*I*	*O*	*F*	*M*
CIOFM	—					
C	$p < 4 \times 10^{-23}$	—				
I	$p < 4 \times 10^{-23}$	$p < 3 \times 10^{-22}$	—			
O	$p < 4 \times 10^{-23}$	$p < 3 \times 10^{-18}$	$p < 4 \times 10^{-8}$	—		
F	$p < 1 \times 10^{-20}$	$p < 4 \times 10^{-23}$	$p < 4 \times 10^{-23}$	$p < 4 \times 10^{-23}$	—	
M	$p < 3 \times 10^{-22}$	$p < 4 \times 10^{-23}$	$p < 4 \times 10^{-23}$	$p < 4 \times 10^{-23}$	n.s.	

Reported are the p values with which the null hypothesis was rejected for a t-test that any two model variants had yielded the same KL distance. All model variants were significantly different from each other, except for M (motion) compared to F (flicker). Notations are as in Table 3.

saliency, grouped in 10 bins by saliency. A one-way ANOVA suggested no significant effect of saliency onto duration, both for human saccades, $F(9, 11906) = 0.75$, n.s., 11,916 saccades and 10 bins, and for random saccades, $F(9, 11906) = 1.64$, n.s. Only considering the 2306 saccades with duration 200 ms or more did not alter this conclusion: Human, $F(9, 2296) = 0.76$, n.s.; random: $F(9, 2296) = 1.68$, n.s. Hence, for our video clips, subjects, and viewing instructions, we found no correlation between the model-predicted saliency of a saccade target location and the duration of the subsequent fixation onto that location.

DISCUSSION

Our study confirms previous findings that humans overall tend to look at salient objects in their visual environment. In particular, our study confirms and reinforces, under conditions of sustained viewing and dynamic colour scenes, the findings of Parkhurst et al. (2002) using an earlier version of our model and static scenes. In addition, we here have pushed the analysis further, by specifically counting those saccades that were made to the single most salient location in the scene, and using the KL distance to quantify how the distributions of saliency at human and random saccade endpoints differed. With the simulation framework used here, we found that occurrences of saccades to the most salient location in the scene were rare, although much more frequent than expected by chance. Motion and flicker were found to be better correlated with human saccades than colour, intensity, and orientation, but not as good as all features combined in terms of KL distance. No correlation was found between the model-predicted saliency of a location and the amount of time during which that location was subsequently fixated. Subjective visual analysis of the eye movement traces yielded a number of additional observations, illustrated in Figures 5

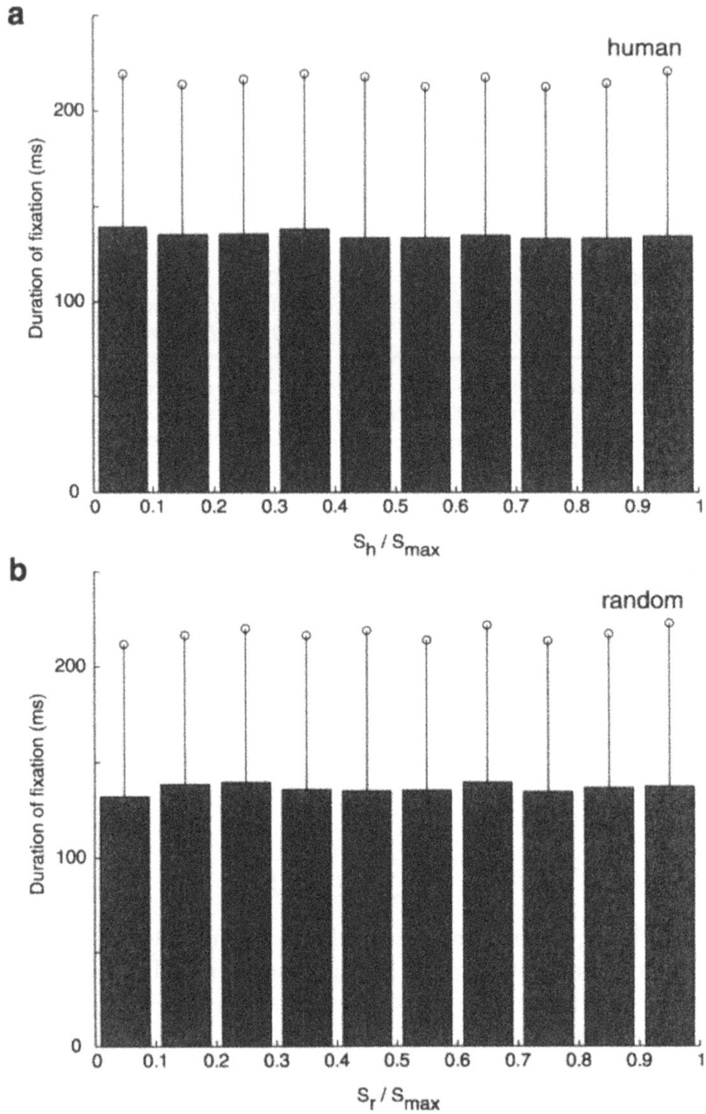

Figure 7. Distribution of the durations of fixations (mean + standard deviation) following saccades targeted towards locations of a given saliency. We found no significant effect of target saliency onto subsequent fixation duration (one-way ANOVA; see text).

and 6 and discussed here, which may serve as guidelines for extensions of the present study.

We used a fairly small saliency sampling window (5.6° diameter), and in some instances it was clear that the observer had saccaded towards a salient object, but had targeted such part of the object that the saliency sample at the saccade endpoint had partially missed the representation of the object in our low-resolution saliency map (e.g., second and fifth rows of Figure 5). Using a larger sampling window would certaintly address this problem, but at the cost of possibly artificially increasing the saliency reading when observers had saccaded to a nonsalient object (e.g., a larger radius of the circle in the first or second rows of Figure 6 would increase the sampled saliency value, while in these examples clearly the observer had saccaded to a location that the model considered not salient). An additional complication in this respect was that humans sometimes were fairly inaccurate in targeting their saccades, especially for long ones (e.g., fourth row of Figure 5; this was not an inaccuracy of the eyetracking, as the saccade shown here was followed by another short saccade that better centred the eye onto the object). One limitation of the current study which will need to be further explored is the lack of a systematic investigation of how sampling window size would affect the results reported here. Our comparison between human and random saccades, however, aims at minimizing the effects of window size onto our conclusions (as different window sizes would affect both the human and random histograms).

We measured a minority but not negligible number of human saccades targeted to locations deemed not salient by the model (27.7% of all saccades human saccades were to locations with $S_h/S_{\max} < 0.25$). In some cases, these saccades were predictive of moving object trajectories; that is, instead of focusing onto a moving object, observers looked at an empty location slightly ahead of that object. This top-down predictive capability is not captured by our computational model. Building neural models that can embody it remains an open issue (although many mathematical models, such as Kalman filters, may be used to solve this problem).

Our measure of saliency was very strict, in that a unique sample was taken at the precise moment when each saccade began. An interesting extension of our analysis will be to investigate a more relaxed question, such as whether a given saccade target location has been among the five most salient locations at any given time during the second preceding a given saccade. Intuitively, we often are confronted with situations where several objects are of interest at a given moment, but we employ a serial strategy to focus onto each in turn. Adding some account for such memory process in our measure of saliency may reveal a larger fraction of saccades made to objects that were among the few most salient at some point in the recent past. One difficulty in developing such more relaxed measure lies in the necessity to be able to reliably segment the few most salient objects in the saliency map, on a continual basis.

In the present study we have decided to remain agnostic to putative saccade generation models, which has limited how far the analysis could be pushed. Our results suggest, however, that a winner-take-all strategy often used in covert attention models (Itti & Koch, 2001a), directing gaze towards the most salient location in the display, would only account for a small fraction of human saccades (around 12%). To estimate the fraction of human and random saccades directed to salient locations, we have explored three arbitrary S_h/S_{max} thresholds of 0.25, 0.75 and 1.00 in Table 2. It is important to realize that these indeed are arbitrary and are here only intended to summarize, for many subgroups of saccades, the more detailed picture provided by the distributions in Figure 4 for all saccades. Furthermore, applying a nonlinearity to the saliency values would in many cases change the figures in Table 2. Hence, while it is certainly tempting to attribute some of the saccades to bottom-up influences, and the rest to top-down influences, such attribution may not be very meaningful. Indeed, changing the cut-off threshold, or applying a nonlinearity to the saliency values, could yield a wide range of bottom-up/top-down ratios. For example, if one considers that only saccades to the maximum saliency can be attributed to bottom-up, then the ratio is 11.5% bottom-up for 88.5% top-down, or approximately 1:8. In contrast, if all saccades to regions reliably marked by the model as salient ($S_h/S_{max} \geq 0.25$) are attributed to bottom-up, then the ratio becomes 72.3% to 27.7% or approximately 2.5:1. The value of 0.25 seems to be an appropriate threshold, since, given the sparseness of our saliency maps exemplified in Figures 5 and 6, any location with saliency above that threshold is likely to be a nonaccidental detection. More quantitatively, it is interesting to note from Table 2 that a majority of random saccades were to locations with $S_r/S_{max} < 0.25$ (for all saccades, all first, first 5, and first 10 saccades, and all observers taken individually, although not for all classes of video clips taken individually), confirming that overall the number of pixels (or surface area) in the saliency map above that threshold were in minority (especially given our aperture-based sampling scheme).

In reality, it is likely that bottom-up and top-down influences simultaneously contribute to directing every saccade. Intuitively, it is clear that top-down influences can be very strong—for instance, we can command our gaze to any location we like, we can make eye movements based purely on cognitive reasoning (e.g., from knowing the typical spatial relationship between the object of current fixation and a desired object), and there are many other top-down factors that have been clearly shown to influence eye movements, as reviewed in introduction. Yet, our results suggest that, if one is to accept the threshold of 0.25 just discussed, a comfortable majority of human saccades (72.3% overall) was directed to a minority of locations that had been highlighted in our bottom-up saliency map (and to which a minority of 46.6% of random saccades went). This relative rarity of occurrence of saccades to locations that were not bottom-up salient suggests that the bottom-up saliency map may be used as a mask, highlighting a set of potentially interesting locations in the scene, with top-down influences mainly responsible for deciding upon one specific location among these candidates. This hypothesis seems well in line with our finding that the most salient location in the display was not attracting gaze very often; hence it may only have been one among several bottom-up candidate locations, with the choice of which candidate would eventually be fixated primarily mediated top-down rather than by bottom-up saliency.

Our discussion thus far is compatible with a number of saccade generation models. For example, the area activation model of Pomplun, Shen, and Reingold (2003), proposed in the context of visual search, suggests that scanpaths visit peaks of an activation map that are above a given threshold, with the selection of the next peak based on proximity to current fixation rather than absolute peak activity. The activation map in that model is not purely bottom-up, but substantially modulated by the nature of the search task and of target and distractor items; thus, it may be considered a combined bottom-up and top-down map. The greedy choice of the next saccade target based on proximity brings the idea that there are also other factors to consider, as this mechanism may not be considered purely top-down nor bottom-up—rather, it is an additional bias that may influence saccade target selection in cases where one is faced with a number of equally bottom-up salient and top-down interesting candidate targets (such as the distractor items in cases of difficult visual search). Similarly, Rao, Zelinsky, Hayhoe, and Ballard (2002) select saccade targets by subjecting an activation map to a softmax operator (which further emphasizes strong activation peaks and weakens regions of lower activation) and then aiming for the centre of gravity of the transformed map. This model, in particular by reproducing centre-of-gravity effects observed experimentally when several items simultaneously attract gaze, suggests that in general the entire spatial distribution of values in the saliency (or activation) map is likely to contribute to saccadic orienting, rather than only a selected subset of locations or the single location of maximum saliency. In previous work, we have proposed that attention and eye movements may be guided by an attention guidance map (AGM) defined as the pointwise product between a bottom-up saliency map (SM) and a top-down task-relevance map (TRM; Navalpakkam & Itti, 2002). Hence, a location would have to be both somewhat salient and somewhat relevant to become a candidate saccade target. The SM might be modulated top-down using feature weighing similar to Guided Search (Wolfe et al., 1989), and the TRM could be populated based on information from the gist and rough layout of the scene (Hollingworth & Henderson, 1998) as well as top-down priors on the likely locations of desired objects (Torralba, 2003). Our present observations are compatible with this hypothesis (also see Rensink, 2000), although many of the details still need to be elucidated.

These considerations reinforce the idea that a metric like the KL distance, which considers dissimilarity of shape of the full distributions of saliency at human compared to random saccade targets, is a more appropriate metric for comparing models than, for example, the simpler ratio of median saliencies. Indeed, even if the distribution of model-predicted activation at human saccade targets were to differ from random by exhibiting a larger number of human saccades towards low model-predicted activation values, that information could still be exploited by a saccade generation mechanism. The mechanism would just have to reflect the fact that low-activation regions actually should be strong attractors of gaze. Thus, in principle at least, any shape difference between the human and random distributions obtained for a given model in a manner similar to that of Figure 4 could be exploited to develop a gaze generation model, not just a shift towards higher values. Under these conditions, we found that models based purely on motion energy or on temporal change (flicker) performed much better than models relying solely on colour, intensity, or orientation information. While it has often been posited that motion was deemed to be one of the strongest attractors of attention, our study allows us to quantify this in terms of KL distance: The distance between human and random obtained for the motion and flicker models was about twice that obtained for the colour, intensity, or orientation models. Yet, in terms of KL distance, the best model remained that which incorporated all features. This is not surprising, as strong motion cues were not always present in our stimuli. During periods of rather still video content, colour, intensity, and orientation certainly were better predictors of human saccade targets than flicker and motion which essentially yielded no output during these periods.

As reviewed in our introduction, bottom-up saliency is only one of the many factors that may influence eye movements. The main contribution of our study is to attempt to quantify the extent to which it may continually contribute to human eye movements. Our general conclusion in this respect is that a comfortable majority of human saccades were made to a minority of fairly salient locations in the saliency map, suggesting that bottom-up saliency plays a non-negligible role in determining saccade targets. In addition, we found that few saccades were directed to the most salient location in the visual input, hence suggesting that absolute saliency may play only a minor role in picking one target among a set of fairly salient candidates, a process likely to be primarily driven top-down. We found that motion energy and temporal change were strong attractors of attention, though the model that also included colour, intensity, and orientation features performed overall best. Finally, we did not find any evidence of correlation between saliency of a saccade target and duration of subsequent fixation of that target, suggesting that bottom-up saliency does not contribute to duration of fixation. All these observations are compatible with the idea that a combination of bottom-up and top-down influences determines every saccade target location, rather than some saccades being primarily directed bottom-up while others are primarily targeted top-down.

REFERENCES

Andrews, T. J., & Coppola, D. M. (1999). Idiosyncratic characteristics of saccadic eye movements when viewing different visual environments. *Vision Research, 39*(17), 2947–2953.

Barth, E., Zetzsche, C., & Rentschler, I. (1998). Intrinsic two-dimensional features as textons. *Journal of the Optical Society of America: Part A. Optics Image Science and Vision, 15*(7), 1723–1732.

Biederman, I., Teitelbaum, R. C., & Mezzanotte, R. J. (1983). Scene perception: A failure to find a benefit from prior expectancy or familiarity. *Journal of Experimental Psychology: Learning of Memory Cognition, 9*(3), 411–429.

Bookstein, F. L. (1989). Principal warps: Thin-plate splines and the decomposition of deformations. *IEEE Translations on Pattern Analysis and Machine Intelligence, 11*, 567–585.

Cannon, M. W., & Fullenkamp, S. C. (1991). A transducer model for contrast perception. *Vision Research, 31*(6), 983–998.

Collewijn, H., Steinman, R. M., Erkelens, C. J., Pizlo, Z., & van der Steen, J. (1992). Effect of freeing the head on eye movement characteristics during three-dimensional shifts of gaze and tracking. In A. Berthoz, W. Graf, & P. P. Vidal (Eds.), *The head-neck sensory motor system*. Oxford, UK: Oxford University Press.

Finney, S. A. (2001). Real-time data collection in Linux: A case study. *Behavioral Research Methods, Instruments and Computers, 33*, 167–173.

Friedman, A. (1979). Framing pictures: The role of knowledge in automatized encoding and memory for gist. *Journal of Experimental Psychology: General, 108*(3), 316–355.

Gallant, J. L., Connor, C. E., & van Essen, D. C. (1998). Neural activity areas V1, V2 and V4 during free viewing of natural scenes compared to controlled viewing. *Neuroreport, 9*(9), 2153–2158. (Reprinted with corrections from *Neuroreport*, 1998, *9*(1), 85–90)

Gilbert, C. D., & Wiesel, T. N. (1989). Columnar specificity of intrinsic horizontal and cortico-cortical connections in cat visual cortex. *Journal of Neuroscience, 9*(7), 2432–2442.

Henderson, J. M., & Hollingworth, A. (1999). High-level scene perception. *Annual Review of Psychology, 50*, 243–271.

Henderson, J. M., & Hollingworth, A. (2003). Global transsaccadic change blindness during scene perception. *Psychological Science, 14*(5), 493–497.

Hollingworth, A., & Henderson, J. M. (1998). Does consistent scene context facilitate object perception? *Journal of Experimental Psychology: General, 127*(4), 398–415.

Hollingworth, A., Williams, C. C., & Henderson, J. M. (2001). To see and remember: Visually specific information is retained in memory from previously attended objects in natural scenes. *Psychonomic Bulletin and Review, 8*(4), 761–768.

Hubel, D. H., & Wiesel, T. N. (1962). Receptive fields, binocular interaction and functional architecture in the cat's visual cortex. *Journal of Physiology, 160*, 106–154.

Ito, M., & Gilbert, C. D. (1999). Attention modulates contextual influences in the primary visual cortex of alert monkeys. *Neuron, 22*(3), 593–604.

Itti, L. (2004). Automatic foveation for video compression using a neurobiological model of visual attention. *IEEE Transactions on Image Processing, 13*, 1304–1318.

Itti, L., & Koch, C. (2000). A saliency-based search mechanism for overt and covert shifts of visual attention. *Vision Research, 40*(10–12), 1489–1506.

Itti, L., & Koch, C. (2001a). Computational modeling of visual attention. *Nature Review Neuroscience, 2*(3), 194–203.

Itti, L., & Koch, C. (2001b). Feature combination strategies for saliency-based visual attention systems. *Journal of Electronic Imaging, 10*(1), 161–169.

Itti, L., Koch, C., & Niebur, E. (1998). A model of saliency-based visual attention for rapid scene analysis. *IEEE Transactions on Pattern Analysis and Machine Intelligence, 20*(11), 1254–1259.

Itti, L., Dhavale, N., & Pighin, F. (2003, Aug.). Realistic avatar eye and head animation using a neurobiological model of visual attention. In *Proceedings of the SPIE 48th annual international symposium on Optical Science and Technology* (pp. 64–78). San Diego, CA: SPIE.

Koch, C., & Ullman, S. (1985). Shifts in selective visual attention: Towards the underlying neural circuitry. *Human neurobiology, 4*(4), 219–227.

Kuffler, S. W. (1953). Discharge patterns and functional organization of mammalian retina. *Journal of Neurophysiology, 16*, 37–68.

Kullback, S. (1959). *Information theory and statistics*. New York: Wiley.

Legge, G. E., Hooven, T. A., Klitz, T. S., Mansfield, J. S., & Tjan, B. S. (2002). Mr. Chips 2002: New insights from an ideal-observer model of reading. *Vision Research, 42*(18), 2219–2234.

Li, W., & Gilbert, C. D. (2002). Global contour saliency and local colinear interactions. *Journal of Neurophysiology, 88*(5), 2846–2856.

Moran, J., & Desimone, R. (1985). Selective attention gates visual processing in the extrastriate cortex. *Science, 229*(4715), 782–784.

Moreno, F. J., Reina, R., Luis, V., & Sabido, R. (2002). Visual search strategies in experienced and inexperienced gymnastic coaches. *Perceptual and Motor skills, 95*(3, Pt. 1), 901–902.

Motter, B. C. (1994). Neural correlates of attentive selection for color or luminance in extrastriate area V4. *Journal of Neuroscience, 14*(4), 2178–2189.

Motter, B. C., & Belky, E. J. (1998). The guidance of eye movements during active visual search. *Vision Research, 38*(12), 1805–1815.

Navalpakkam, V., & Itti, L. (2002). A goal oriented attention guidance model. *Lecture Notes in Computer Science, 2525*(Nov.), 453–461.

Nodine, C. F., & Krupinski, E. A. (1998). Perceptual skill, radiology expertise, and visual test performance with NINA and WALDO. *Academic Radiology, 5*, 603–612.

Noton, D., & Stark, L. (1971). Scanpaths in eye movements during pattern perception. *Science, 171*(968), 308–311.

Olivia, A., & Schyns, P. G. (1997). Coarse blobs or fine edges? Evidence that information diagnosticity changes the perception of complex visual stimuli. *Cognitive Psychology, 34*(1), 72–107.

Parkhurst, D., Law, K., & Niebur, E. (2002). Modeling the role of salience in the allocation of overt visual attention. *Vision Research, 42*(1), 107–123.

Peebles, D., & Cheng, P. C. (2003). Modeling the effect of task and graphical representation on response latency in a graph reading task. *Human Factors, 45*(1), 28–46.

Peters, R. J., Itti, L., & Koch, C. (2002, Nov.). Eye movements are influenced by short-range interactions among orientation channels. In *Proceedings of the Society for Neuroscience annual meeting (SFN'02)* (p. 715.12). Orlando, FL: Society for Neuroscience.

Polat, U., & Sagi, D. (1993). Lateral interactions between spatial channels: Suppression and facilitation revealed by lateral masking experiments. *Vision Research, 33*(7), 993–999.

Pomplun, M., Shen, J., & Reingold, E. M. (2003). Area activation: A computational model of saccadic selectivity in visual search. *Cognitive Science, 27*, 299–312.

Potter, M. C., & Levy, E. I. (1969). Recognition memory for a rapid sequence of pictures. *Journal of Experimental Psychology, 81*(1), 10–15.

Privitera, C. M., & Stark, L. W. (2000). Algorithms for defining visual regions-of-interest: Comparison with eye fixations. *IEEE Transactions on Pattern Analysis and Machine Intelligence, 22*(9), 970–982.

Rao, R. P., Zelinsky, G., Hayhoe, M., & Ballard, D. H. (2002). Eye movements in iconic visual search. *Vision Research, 42*(11), 1447–1463.

Reichle, E. D., Rayner, K., & Pollatsek, A. (2003). The E-Z reader model of eye-movement control in reading: Comparisons to other models. *Behavioural Brain Sciences, 26*(4), 445–476; discussion 477–526.

Reinagel, P., & Zador, A. M. (1999). Natural scene statistics at the centre of gaze. *Network: Computation in Neural Systems, 10*, 341–350.

Rensink, R. A. (2000). The dynamic representation of scenes. *Visual Cognition*, *7*(1/2/3), 17–42.

Reynolds, J. H., Pasternak, T., & Desimone, R. (2000). Attention increases sensitivity of V4 neurons. *Neuron*, *26*(3), 703–714. Comment in: *Neuron*, 2000, *26*(3), 548–550.

Salvucci, D. D., & Goldberg, J. H. (2000). Identifying fixations and saccades in eye-tracking protocols. In *Proceedings of the Eye Tracking Research and Applications symposium* (pp. 71–78). New York: ACM Press.

Savelsbergh, G. J., Williams, A. M., van der Kamp, J., & Ward, P. (2002). Visual search, anticipation and expertise in soccer goalkeepers. *Journal of Sports Sciences*, *20*(3), 279–287.

Shen, J., Reingold, E. M., & Pomplun, M. (2000). Distractor ratio influences patterns of eye movements during visual search. *Perception*, *29*(2), 241–250.

Sillito, A. M., Grieve, K. L., Jones, H. E., Gudeiro, J., & Davis, J. (1995). Visual cortical mechanisms detecting focal orientation discontinuities. *Nature*, *378*(6556), 492–496.

Sparks, D. L. (2002). The brainstem control of saccadic eye movements. *Nature Reviews Neuroscience*, *3*(12), 952–964.

Tanenhaus, M. K., Spivey-Knowlton, M. J., Eberhard, K. M., & Sedivy, J. C. (1995). Integration of visual and linguistic information in spoken language comprehension. *Science*, *268*(5217), 1632–1634.

Torralba, A. (2003). Modeling global scene factors in attention. *Journal of the Optical Society of America: Part A Optics, Image Science and Vision*, *20*(7), 1407–1418.

Treisman, A. M., & Gelade, G. (1980). A feature-integration theory of attention. *Cognitive Psychology*, *12*(1), 97–136.

Treue, S., & Martinez Trujillo, J. C. (1999). Feature-based attention influences motion processing gain in macaque visual cortex. *Nature*, *399*(6736), 575–579.

Treue, S., & Maunsell, J. H. (1996). Attentional modulation of visual motion processing in cortical areas MT and MST. *Nature*, *382*(6591), 539–541.

Wolfe, J. M. (1994). Visual search in continuous, naturalistic stimuli. *Vision Research*, *34*(9), 1187–1195.

Wolfe, J. M. (1998). Visual memory: What do you know about what you saw? *Current Biol*, *8*(9), R303–R304.

Wolfe, J. M., Cave, K. R., & Franzel, S. L. (1989). Guided Search: An alternative to the feature integration model for visual search. *Journal of Experimental Psychological Human Perception Performance*, *15*(3), 419–433.

Yarbus, A. (1967). *Eye movements and vision*. New York: Plenum Press.

Yeshurun, Y., & Carrasco, M. (1998). Attention improves or impairs visual performance by enhancing spatial resolution. *Nature*, *396*(6706), 72–75.

Zetzsche, C., Schill, K., Deubel, H., Krieger, G., Umkehrer, E., & Beinlich, S. (1998). investigation of a sensorimotor system for saccadic scene analysis: An integrated approach. In *Proceedings of the fifth international conference on Simulation of Adaptive Behavior: Vol. 5 From animals to animats* (pp. 120–126). Cambridge, MA: MIT Press.

VISUAL COGNITION, 2005, 12 (6), 1124–1142

Semantic guidance of attention within natural scenes

Elizabeth Walter and Paul Dassonville

Department of Psychology and Institute of Neuroscience, University of Oregon, Eugene, OR, USA

When viewing real-world scenes composed of a myriad of objects, detecting changes can be quite difficult, especially when transients are masked. In general, changes are noticed more quickly and accurately if they occur at the currently (or a recently) attended location. Here, we examine the effects of explicit and implicit semantic cues on the guidance of attention in a change detection task. Participants first attempted to read aloud a briefly presented prime word, then looked for a difference between two alternating versions of a real-world scene. Helpful primes named the object that changed, while misdirecting primes named another (unchanging) object in the picture. Robust effects were found for both explicit and implicit priming conditions, with helpful primes yielding faster change detection times than misdirecting or neutral primes. This demonstrates that observers are able to use higher order semantic information as a cue to guide attention within a natural scene, even when the semantic information is presented outside of explicit awareness.

Our perceived visual world is a compromise between external input and internal expectations. It is constructed bottom-up by neuronal algorithms that capitalize on the many regularities in our visual environment in order to simplify the overwhelming amount of incoming perceptual information. In addition, top-down influences allow us to delineate objects and make sense of this incoming stream. Glance up for a moment and survey your surroundings. What do you see? Rather than noticing free-floating colours, or masses of lines, you probably noticed a chair or table, and maybe a tree or two outside the window. The sight of a particular arrangement of brown lines and green colour patches triggers not only the visual image of a weeping willow, but also the semantic content of

Please address all correspondence to: Elizabeth Walter, Department of Psychology, 1227 University of Oregon, Eugene, OR 97403, USA. Email: ewalter1@darkwing.uoregon.edu

We would like to thank Scott Huettel and colleagues for allowing us use of the stimuli from their change blindness study (Huettel, Güzeldere, & McCarthy, 2001). In addition, we would like to thank Edward Awh, Ulrich Mayr, Jadgeep K. Bala, and the reviewers, John Henderson and Andrew Hollingworth, for many helpful comments. This research was supported by NIH Neuroscience Training Grant 5-T32-GMO7257 (EW), and NSF Continuing Grant BCS-9996264 (PD).

DOI:10.1080/13506280444000670

"tree". Consciously, we perceive the world at the level of objects, rather than in terms of edges, colour swatches, and motion transients. Our overall perception is of a stable, meaningful environment, and in general we are quite confident in our ability to correctly perceive the world as it "truly" is (Levin, Momen, Drivdahl, & Simons, 2000).

Visual illusions exploit our brains' automatic processes and reveal occasions in which our perceptions do not match reality. Change blindness is one such phenomenon that has been recently investigated. In a typical change blindness "flicker paradigm" (Rensink, O'Regan, & Clark, 1997), two pictures, identical except for a difference in a single object (e.g., a large shift in the horizon or disappearing and reappearing towers), are repeatedly alternated in an ABAB or AABB fashion with a blank mask transient inserted between each picture. In general, participants will examine these picture pairs for long periods without detecting any anomaly.

Normally when something changes in the world, there is an accompanying motion signal or other transient, which automatically attracts attention (Klein, Kingstone, and Pontefract, 1992). Change blindness results when some interfering transient (e.g., a saccade, eyeblink, flicker, or mudsplat) swamps or overloads the normal transient produced by the change (Rensink, 2000). Many researchers have argued that we overlook changes in this paradigm because the visual system does not maintain a detailed representation of the world (Chun & Nakayama, 2000; Rensink, 2002; Rensink et al., 1997). Therefore, unless one is attending to the location of change at the time of the switch (or just prior to the switch), it is very difficult to notice the difference. Although our general feeling is of residing in a stable, intricate visual environment, this theory argues that we do not internalize as much of our surroundings as we might think we do.

Our conscious, subjective experience is that of seeing an intact environment populated by meaningful objects. At some level, semantic labels and categories for these "object files" must be activated. But *at what level* does this label activation occur? At what processing stage does the semantic content of an object in a scene become available for higher order control of attention? Some of the semantic meaning must be attached during the rapid "gist" labelling and overall spatial mapping of the scene during the first few hundred milliseconds of viewing (Potter, 1975, 1976). It seems then, that some semantic information is being linked to objects and organized extremely rapidly. Once the gist and layout of the scene has been ascertained, there are a number of objects that could potentially attract attention. How do we decide to which objects we should attend?

Previous studies of change blindness have shown that some types of changes tend to be noticed more quickly than others. Rensink and colleagues (O'Regan, Deubel, Clark, & Rensink, 2000) have proposed that changes to "centres of interest" in pictures tend to get noticed more quickly than changes of "marginal interest." Werner and Thies (2000) found differences in detection ability based on the participant's level of expertise within a domain (in this case, American football). Hollingworth, Williams, and Henderson (2001) found that changes to semantically inconsistent objects tended to be detected more often than changes to objects that were semantically consistent with the rest of the scene. In an attempt to understand what aspects of our environment are most naturally encoded, Aginsky and Tarr (2000) cued participants as to what *type* of change would occur (though not what *object* would change). They showed participants colour photographs of natural scenes, and explicitly precued the participant as to whether the change would be a colour change, object position change, or object presence/absence change. Only the times to detect colour changes were facilitated with this cueing paradigm. The authors interpreted their results as an indication that colour is encoded as a ''surface'' property, which is not as central to the overall layout and gist of the scene as are the ''configural'' properties (presence/absence and position change). It is thought that the overall layout of the scene is ascertained very rapidly, with finer details filled during later allocation of attention (Biederman, 1981). In this two-stage model, alterations of the layout of the scene would be automatically detected, and thus not greatly facilitated by a cue. On the other hand, colour is often not as important to the overall structure of the scene, and would not be automatically encoded during the gist labelling. Because of this, cueing participants to pay more attention to colour *would* be useful in a change detection task, and thus would decrease detection time.

Studies have shown that semantically related word primes do facilitate picture categorization and naming (cf. Carr, McCauley, Sperber, & Parmelee, 1982; Smith, Meiran, & Besner, 2000). In addition, as a control condition to ensure that changes were detectable, the first studies of the flicker paradigm explicitly informed participants where to look for the change (Rensink et al., 1997) and found that this cueing reduced detection times. Given that explicit semantic information can obviously affect attention, it is of particular interest whether implicit primes would have the same effect. In other words, does linking semantic information to a pictorial scene require conscious/explicit processes? Or, are words actually processed enough at an unconscious level to be useful when linked to picture representations?

EXPERIMENT 1

None of the change blindness studies to date have attempted to implicitly modulate semantic attention within a natural scene. The existence of implicit semantic priming would imply that the processing of word primes could somehow tap into the attentional-guidance abilities at work during a person's everyday life. Natural scenes are interesting experimental stimuli precisely because the objects and the relationships between objects carry abstract semantic information as well as perceptual information. In any real-world scene there will be many possible objects to attend to, and these must be segmented from the relatively chaotic perceptual input. Explicitly priming the participant with the name of the object that changes is expected to decrease detection time, by quickly limiting the set of attended objects. In an attempt to establish whether a similar guidance of attention can be controlled implicitly, we determined the effects of a briefly presented, masked prime on detection times in a subsequent change blindness flicker paradigm.

Method

Participants. Participants were University of Oregon students who took part in exchange for credit in an undergraduate psychology course. All 32 participants were naïve to the hypothesis, and reported normal or corrected-to-normal vision, and fluency in English. Experiment duration was approximately 1 hour.

Apparatus. Stimuli were presented at eye level to the participant sitting at a desk in a darkened room. Stimuli were back-projected onto a translucent screen by a Barco Cine 7 projector at a screen refresh rate of 60 Hz. Participants sat at a comfortable viewing distance (approximately 100 cm) away from the screen. When projected, the pictures each measured 33.7 cm high × 44.5 cm wide and therefore subtended visual angles of approximately 19° and 25°, respectively. Words were presented in black, lowercase 12-point Charcoal font, were 2.5 cm high, and between 5 cm to 16.5 cm in length (corresponding to a visual angle of 1.4° × 2.3–9.4°). Response times were recorded with a custom-built button box, with participants using the index finger on the dominant hand for responses. Stimuli were presented using the PsyScript programming language (Bates & D'Oliveiro, 2003), running from a Macintosh G4.

Picture stimuli. The change-blindness stimuli used in this experiment were photographs of real-world scenes. One hundred stimuli came from a set used by Huettel, Güzeldere, and McCarthy (2001). Thirty-two pictures from this set were modified (to make the changes more salient) and twenty additional stimuli were created by the experimenter. Pictures included scenes of artificial objects, people, cityscapes, landscapes, and other ordinary scenes. Picture pairs were nearly identical, differing only in the presence/absence of an object or object part, the location for a particular object, or the colour of an object. Some changes were more difficult to detect than others, but all were noticeable once pointed out.

Once a change has been detected in a task like this, it becomes quite obvious upon subsequent viewing. Because of this, each stimulus pair was presented only once to each participant. To counterbalance the stimuli, the set was broken into four groups. Two pilot studies were run (6 participants and 12 participants, respectively) to assess overall detectability of changes, and to ascertain the general reaction times to stimuli. After the first pilot study, it was found that some changes were not detected by any (or very few) of the participants. These stimuli were subsequently altered to make the change more salient, and the second pilot study served to norm the new versions into picture groups of approximately equal average reaction times. Each picture was assigned two primes, one *helpful* prime (naming the changing object) and the other *misdirecting* (naming another, unchanging object in the picture). Half of the participants saw a particular picture stimulus paired with the helpful prime, and the other half with the misdirecting prime.

Procedure. On each trial, participants encountered one of two prime types. Prime words were either related to the object that changed (helpful), or related to another object in the picture (misdirecting). Primes were either 33 ms in duration, immediately followed by a 250 ms mask of uppercase Xs, or were 200 ms in duration with an immediate 83 ms mask of Xs. Most participants were only able to read approximately one third of the 33 ms primes, but the 200 ms prime length gave all participants enough time to reliably read the word aloud, while still maintaining task difficulty. These parameters yielded four equally probable priming conditions (33 ms/helpful, 200 ms/helpful, 33 ms/misdirecting, and 200 ms/misdirecting). Four base lists of stimuli pairings were then created by matching each of the four picture groups with a particular priming condition. Presentation order of pictures within each base list was then randomized for each participant. In this experiment each picture pair was thus viewed by each participant once, but was randomized across participants to occur equally often after each of the four priming conditions.

To determine precisely which primes were consciously perceived (*explicit*) versus those which were *not* consciously perceived (*implicit*), the participant was required to say the prime word out loud (or to say "X" out loud if the word was unreadable) immediately after prime presentation, but before the change-blindness stimulus appeared. In this way, we precisely determined on a trial-by-trial basis which primes were truly implicit or explicit.

During each trial (Figure 1), a black fixation cross was presented for 500 ms on a mid-level grey screen, followed by a prime and a subsequent mask for a total prime/mask duration of approximately 283 ms. Next, a blank delay period of 500 ms was introduced while participants attempted to read the prime word aloud. After this delay, pictures from a particular picture pair began alternating as follows: 300 ms for picture 1, 100 ms grey screen, 300 ms for picture 2, 100 ms grey screen, etc., in an ABAB fashion. Participants were requested to press a button as soon as they noticed a change. This button press halted further flickering, and blanked the screen. At this point, participants were required to verbally describe the object or location that changed. Both verbal responses (to prime and location of change) were manually recorded by the experimenter. In

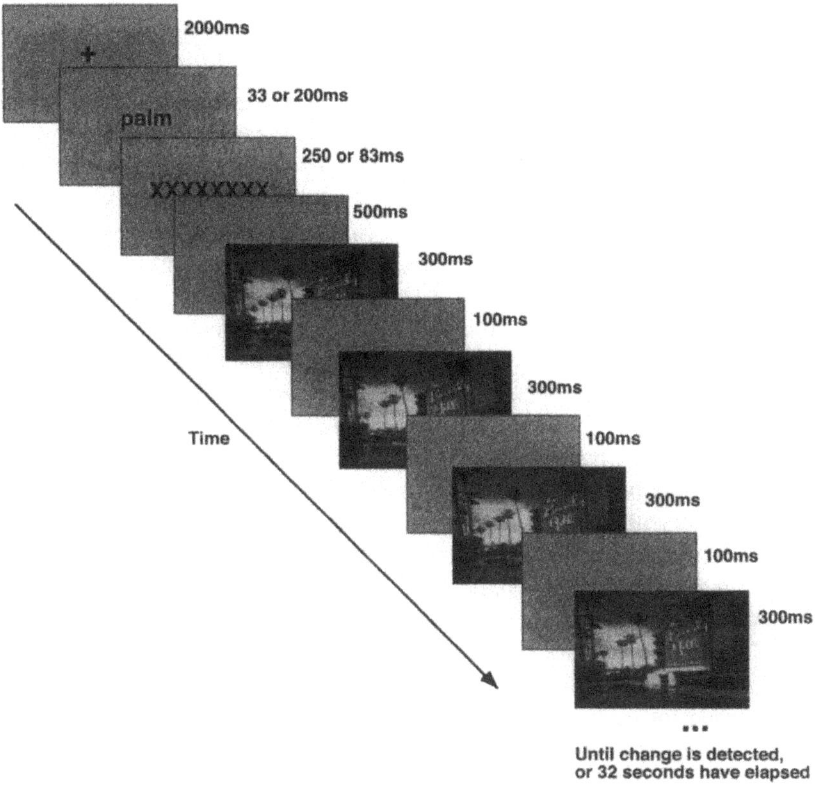

Figure 1. Experiment 1 trial layout. In this picture pair, the palm tree second from the left disappears and reappears in alternate pictures. In the pilot study with no priming, average reaction time for detecting this particular change was approximately 9 s.

four of the trials, the two pictures that flickered were identical. In these cases, as well as in cases in which the participant failed to notice the change, the flickering would halt after 32 s. At this point participants were requested to declare ''no change''. Each participant was given a demonstration consisting of 10 practice trials, followed by 124 experimental trials. At first, many participants found it difficult to report the prime, look for a change, press a button, and then describe the change. Those participants that needed extra practice in order to learn the proper response order were allowed two runs through demonstration pictures.

Results

The four trials in which no change occurred did not elicit any false alarms and were not used in any of the analyses. With one exception (which was excluded from subsequent data analyses), all long-duration (200 ms) primes were read by all participants, and were therefore coded as explicit during further analysis. Detection of the short-duration (33 ms) primes varied widely across participants (Figure 2). Most participants were able to read at least a few of the short-duration primes, and many reported that it became easier to read the briefly presented words as the experiment progressed. Some participants read as many as half of the short-duration primes, and a few participants read more than half. All short-duration primes that were read aloud were subsequently coded as explicit data points for further analysis. Incorrect reports of short-duration primes were also considered as explicit reports, as were occasions in which participants were capable of reading a fragment of the prime. Incorrectly read helpful primes were recoded as misdirecting, unless the response was a synonym for the helpful prime. Implicit data points refer only to those short-duration trials in which the participant said ''X'' instead of a correct or incorrect prime word.

In general, participants were able to detect the changes, seeing an average of 100 out of the 120 changes (83%) with detection accuracy ranging from 70% to 95% across participants. Across both explicit and implicit trials, participants detected significantly more changes that were preceded by a helpful prime (88%) versus those preceded by a misdirecting prime (77%), $t(31) = -5.633$, $p < .001$. Undetected changes were coded as having a reaction time of 32 s (cutoff of flicker). Because the reaction time distribution was quite skewed (Figure 3), the median reaction time for each condition for each participant was used as the dependent variable. Reaction times to short- and long-duration explicit primes were collapsed for each participant, since there were no significant differences in reaction time across these two prime durations.

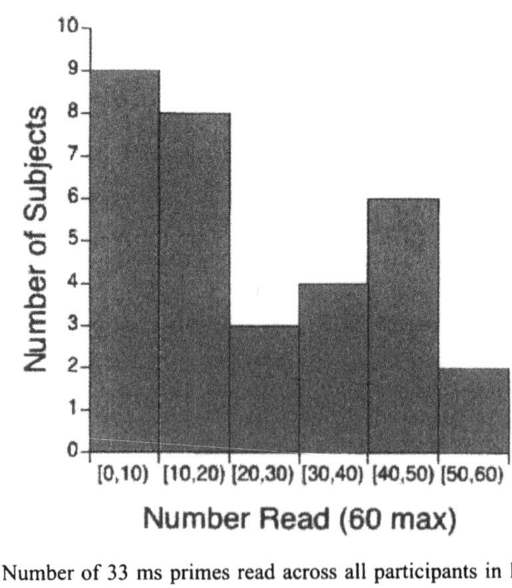

Figure 2. Number of 33 ms primes read across all participants in Experiment 1.

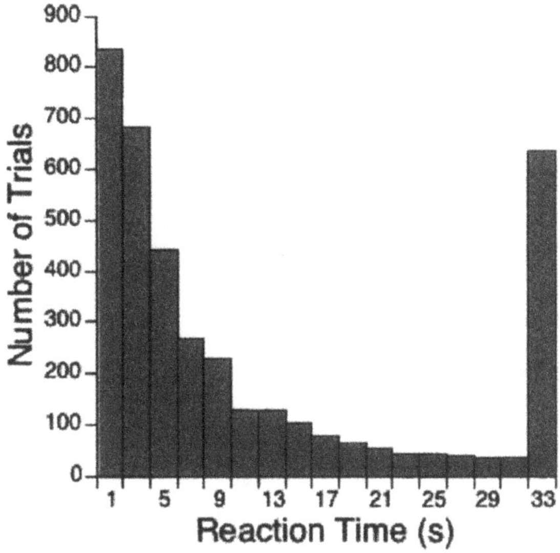

Figure 3. Distribution of reaction times across all participants and conditions in Experiment 1.

A two-factor repeated-measures ANOVA of detectability (explicit/implicit) by prime type (helpful/misdirecting) was performed on the data. Overall, there was a significant effect of detectability, $F(1,31) = 9.425$, $p < .01$, with implicit trials lasting approximately 1.9 s longer than explicit trials (Figure 4A). There was also a significant effect of prime type, $F(1,31) = 40.803$, $p < .001$, with helpful primes eliciting overall RTs approximately 4.8 s faster than the misdirecting primes. However, there was a significant interaction between detectability and prime type, $F(1,31) = 4.966$, $p < .05$, indicating that the difference between helpful and misdirecting primes was not equal across both explicit and implicit trials. When tested separately, though, the effect of prime type was significant in both the explicit, $F(1,31) = 111.225$, $p < .001$, as well as implicit, $F(1,31) = 5.173$, $p < .05$, trials.

As noted above, the 32 participants differed with respect to prime-detection ability (Figure 2); some participants were able to read many of the short-duration primes, while others were only able to read a few. The nature of implicit processing is thought to depend heavily on precise stimulus parameters and the distance of these parameters from an individual's explicit detection threshold. As a result, implicit primes might have different effect sizes due to individual differences in prime detection abilities. To investigate this possibility, participants were divided according to the number of short-duration primes that each was able to read. Those participants who read more than the median number were grouped as *near-threshold* participants, while those who read fewer than the median were grouped as *far-from-threshold* participants. Each group was

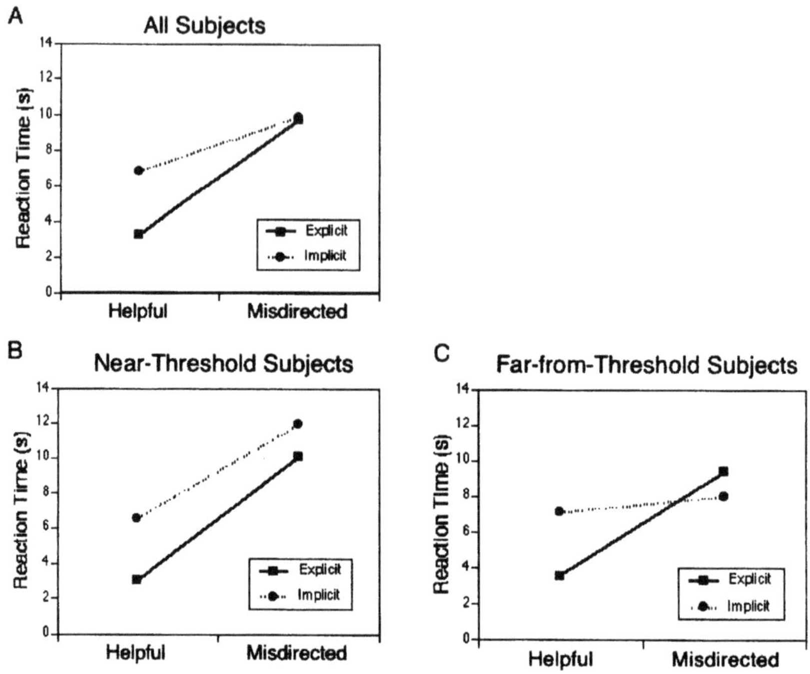

Figure 4. Experiment 1 results. (A) Across all participants, change detection times were shorter with helpful compared to misdirecting primes, in both the implicit and explicit conditions. (B) Those participants who could read more 33 ms primes (near-threshold) showed equal priming across both explicit and implicit trials. (C) Those participants who read fewer 33 ms primes (far-from-threshold) only had significant priming in the explicit trials.

then subjected to a separate two-factor, repeated-measures ANOVA of detectability (explicit/implicit) by prime type (helpful/misdirecting).

Near-threshold participants showed greater priming effects for implicit trials than did the far-from-threshold participants. Overall the near-threshold participants had a significant effect of prime type, $F(1, 15) = 20.732$, $p < .001$, with helpful primes eliciting faster reaction times than misdirecting (Figure 4B). In addition, there was a significant effect of detectability, $F(1, 15) = 6.221$, $p < .05$, with explicit trials eliciting faster reaction times than implicit trials. There was no interaction between detectability and prime type, suggesting that the explicit and implicit trials elicited semantic priming effects of the same magnitude for these near-threshold participants.

Far-from-threshold participants showed a very different pattern of results. Overall they did not have a significant effect of detectability, though they did have a significant effect of prime type, $F(1, 15) = 54.886$, $p < .001$, as well as a significant interaction, $F(1, 15) = 9.777$, $p < .01$. When the effect of prime type was tested separately, these participants showed a significant difference in the explicit trials, $F(1, 15) = 36.783$, $p < .001$, with helpful primes eliciting reaction times 5.9 s faster than misdirecting, but a nonsignificant 0.9 s difference in the implicit trials (Figure 4C).

Discussion

These findings replicate a previous study (Rensink et al., 1997) showing reduced change detection times after explicit cueing, and go further by demonstrating a similar reduction in detection times even in cases in which participants were not able to process a prime word enough to read it aloud (i.e., implicit trials). Importantly, this implicit priming effect seems to be driven by those participants who were able to read more of the short-duration (33 ms) primes. It is possible that for these participants, the short-duration primes were more fully processed by the visual system and thus were closer to their threshold for explicit detection. Even in those cases where these participants were not able to read the prime word aloud, they seem to have been able to process the prime to a greater degree than did those who needed more salient stimuli in order to be explicitly aware of them.

EXPERIMENT 2

When viewing the scenes in Experiment 1, participants reported that they tended to orient more quickly to an object related to the word they had previously read aloud, even though they had not been told that there would be any meaningful relationship between the prime and picture. Thus, when the immediately attended item was the object that changed, participants were naturally faster at detecting the difference. However, it is also possible that the difference in detection times between helpful and misdirecting primes could be at least partially explained by an additional effect. That is, the misdirecting primes slowed the rate of detection by misguiding participants and causing them to waste time with attention focused on an unchanging portion of the scene.

An earlier study may shed some light on why changes in pictures preceded by misdirecting primes took longer to detect than those preceded by helpful primes. Bruner and Potter (1964) presented participants with unfocused pictures, and then incrementally brought the pictures into focus. At each step they asked the participant to guess what the picture might be. They found that participants who started with the fuzziest representations needed the picture to be in greatest focus before realizing exactly what the scene depicted. Those participants who started with a medium amount of focus were able to more quickly discern the gist of the scene. Bruner and Potter took this as evidence that a participant's original, erroneous top-down ideas about a scene (formed when there wasn't enough information to make an accurate assessment) can override what would otherwise be quite obvious. In other words, participants seemed to have a difficult time rejecting their first incorrect hypotheses, even though they were based on very degraded cues. Perhaps this failure to reject incorrect ideas is analogous to how the explicit semantic priming worked in the current experiment. Reading a prime word out loud seemed to allow participants to quickly orient attention within the scene. In the helpful prime cases, this was useful and allowed faster detection time. In the misdirected cases, however, it might have caused a perseveration on the incorrectly cued item, thereby preventing attentional resources from being allocated to any other item, including the item that was actually changing.

In Experiment 1, helpful primes elicited faster detection times than did misdirecting primes. However, it is impossible to determine whether helpful primes sped up change detection times, misdirecting primes slowed change detection times, or both. Experiment 2 served as a replication of Experiment 1, with the advantage of including *neutral* primes so that these possibilities can be distinguished. Furthermore, we used a staircase procedure to establish optimal priming parameters individually for each participant.

Method

Participants. Participants were members of the University of Oregon community who were paid for their participation in the study. All 21 participants were naïve to the hypothesis, and reported normal or corrected-to-normal vision, and fluency in English. Experiment duration was approximately 1.5 hours.

Apparatus. The equipment used was identical to that in Experiment 1.

Picture stimuli. The picture pairs for the change blindness portion of the task were identical to those used in Experiment 1.

Staircase procedure. In order to elicit approximately equal numbers of explicit and implicit responses from each participant, a thresholding task was performed in which the prime presentation time was held constant while the grey-value luminance was varied (background grey-value was 45%, where 0% = black and 100% = white). During the staircase, participants were presented with a mask of 10 random, black uppercase letters for 100 ms, then a prime word for 100 ms, and finally the same 10-letter mask for 200 ms. The initial colour of the prime words was black. In subsequent trials, if the participant could read the prime aloud, the prime was made lighter (i.e., closer to the grey level of the background), and if not, it was made darker. Using this procedure, the grey-value was varied with each reversal until it reached the threshold at which the participant could read the prime word aloud in approximately half of the trials, and responded "X" in the other half. After doing the staircase task twice, the grey-value threshold closest to the background colour was used as the prime colour for the change detection task. The entire staircasing procedure took approximately 10–15 min.

Procedure. Each pair of pictures was associated with the same helpful and misdirecting primes as in Experiment 1. In addition, each pair could also be associated with a neutral prime (a string of semirandom numerals). If participants detected a word as a prime, they were to read the word aloud. If a string of numerals was detected they were to respond "numbers"; otherwise, they were to respond "X" to an unreportable prime.

The picture groups from Experiment 1 were rearranged to create three groups of pictures. Each participant was presented with 40 picture pairs preceded by helpful primes, 40 preceded by misdirecting primes, and 20 preceded by neutral primes. As with Experiment 1, the presentation order was randomized for each participant. Because of the individual staircasing of prime luminance, it was expected that for each participant approximately half of the primes would be read (explicit) and half would not be read aloud (implicit). Between subjects, each picture pair was randomized to occur equally often after helpful or misdirecting primes, and half as often after neutral primes.

The overall change blindness trial layout for Experiment 2 was nearly identical to Experiment 1, with a slight modification to the prime parameters. Participants were presented with a black fixation cross for 500 ms which was followed by a 100 ms black premask of random letters, then the prime word for 100 ms (with a luminance determined by the earlier staircase procedure) followed by a 200 ms black postmask of the same random letters. A 500 ms blank delay period followed as participants attempted to read the prime word aloud (or say "numbers" to a neutral prime or "X" if the prime could not be read). After this delay, pictures from a particular picture pair began alternating in an ABAB fashion as in Experiment 1. Participants were requested to press a button as soon as they noticed a change. Prime report and change detection responses were recorded manually by the experimenter.

Results

Again, the four trials that contained no change did not elicit any false alarms, and were eliminated from subsequent data analyses. The remainder of the trials were coded according to whether the participant could read the prime aloud (explicit) or not (implicit). If a helpful prime was misread, it was recoded as misdirecting, as were neutral primes that were reported as words. Some participants consistently read aloud words that were not at all related to the actual prime presented and as a result had very few trials coded as implicit, and few explicit trials coded as neutral or helpful. Because these participants (n = 6) seemed to be using a different response criterion in the change blindness task than in the thresholding task, their data was discarded.

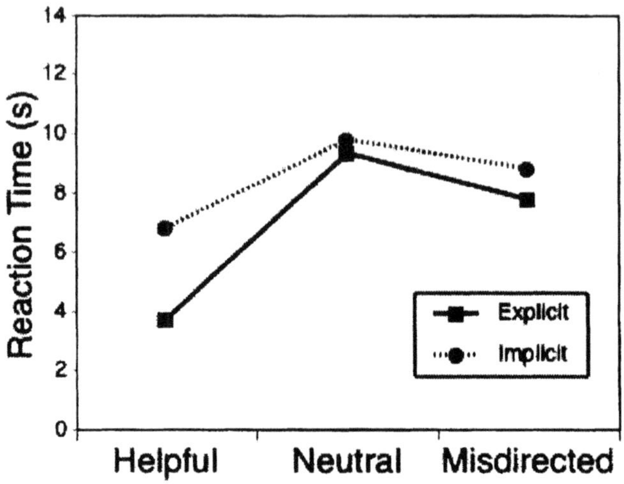

Figure 5. Experiment 2 results. Across both explicit and implicit trials, change detection times were significantly shorter with helpful primes compared to neutral, while there were no differences with misdirecting and neutral primes.

A repeated measures ANOVA of detectability (explicit/implicit) by prime type (helpful/misdirecting/neutral) was performed on the median reaction time data for each condition. Overall there was a significant main effect of prime type, $F(2, 28) = 8.569$, $p < .01$. Planned contrasts indicated that helpful primes significantly differed from the neutral primes ($p < .005$), but that misdirecting primes did not differ significantly from neutral primes (Figure 5). Neither the main effect of detectability, nor the interaction of detectability and prime type reached significance, indicating that the effect of the primes was similar across both the explicit and the implicit trials.

Discussion

Experiment 2 replicated the helpful/misdirecting prime difference found in Experiment 1 in both the explicit and implicit trials. In addition, because of the addition of neutral primes, the results of Experiment 2 indicate that helpful primes do indeed facilitate reaction times relative to neutral primes, and show that misdirecting primes do not significantly hinder reaction times. It seems, then, that the perseveration hypothesis put forth in the Introduction of Experiment 2 does not explain the difference in reaction times between helpful and misdirecting primes. Instead, it appears that the helpful primes allowed participants to quickly orient to the changing object, which decreased reaction times. Misdirecting primes undoubtedly guided attention, but participants did not seem to spend any more time with attention on this inaccurate location than they would have during a search with a neutral prime (in which they would have looked at *some* location first).

EXPERIMENT 3

While the results of Experiments 1 and 2 demonstrate that Helpful primes facilitate change detection times relative to misdirecting and neutral primes even when subjects were unable to read the prime aloud, it is possible that these primes were not truly implicit. Although some studies have found evidence for implicit priming at the word level, other studies suggest that implicit priming may even occur at the level of word fragments. Abrams and Greenwald (2000) repeatedly presented participants with explicit targets (such as ''smut'' and ''bile''), and then found that subjects were slower at categorizing a target word as happy when preceded by an implicit prime (such as ''smile'') which shared some parts with the negatively valenced words already experienced. However, this effect was only seen when participants viewed targets many times. Our prime stimuli were not repeated in this manner, yet it does seem plausible that participants might have been capable of explicitly detecting fragments of the prime word, for example the ''ch'' out of ''chair''. While this information would not have allowed them to read the entire word aloud (and thus they might say ''X'' and the trial would be coded as Implicit), the explicit fragment ''ch'' might still have allowed enough information to later intuit that the prime must have been ''chair'' when later presented with a scene containing a chair, but no chimneys, chains, or chisels. In order to rule out the possibility of change detection mediated by explicit fragment recognition in this way, we replicated the previous experiments, with the addition of a forced-choice task at the end of each trial to verify that the primes were indeed implicitly processed, never reaching explicit awareness.

Method

Participants. Participants were University of Oregon students who took part in exchange for credit in an undergraduate psychology course. All 16 participants were naïve to the hypothesis, and reported normal or corrected-to-normal vision, and fluency in English. Experiment duration was approximately 1.5 hours.

Apparatus. The equipment used was identical to that in Experiment 1.

Picture stimuli. The picture pairs were identical to those used in the previous experiments.

Staircase procedure. The staircase procedure was identical to that in Experiment 2.

Procedure. The four groups of picture pairs were identical to those used in the previous experiments. As in Experiment 1, we used helpful and misdirecting, but not neutral, primes. Each picture pair thus had both a helpful and a misdirecting prime associated with it, though any individual subject was presented with only one of these at the onset of the trial. During each trial, the prime presentation (either helpful or misdirecting), picture presentation, and report of change object/location were identical to the procedure in Experiment 2. However, after the participant provided a verbal description of the observed change, a computer-generated voice read aloud the two prime words associated with that picture pair (one helpful and one misdirecting, with presentation order randomized). Participants then pressed the left of two buttons on the button-box if the first computer-generated word was the previously seen prime word, or the right button if the second word was the previously seen prime word.

Results

Experiment 3 replicated the priming effect found in the previous experiments. Overall, helpful primes elicited faster change detection times than misdirecting primes. On average, changes preceded by helpful primes were detected 3.6 s faster than those preceded by misdirecting primes. A repeated-measures ANOVA of detectability (explicit/implicit) by prime type (helpful/misdirecting) found a significant main effect of prime type, $F(1, 15) = 8.741, p < .05$. Neither the detectability main effect nor the Detectability × Prime type interaction term reached significance, suggesting that priming effects were similar for both explicit and implicit conditions (Figure 6).

An analysis of the forced-choice responses following implicit primes found that participants' ability to correctly identify primes was not significantly different from chance, 52% correctly detected; average sensitivity $d' = 0.109$, $SE = 0.07$, $t(15) = 1.711$, n.s. In contrast, explicit primes elicited correct responses in the forced-choice paradigm 88% of the time, average $d' = 1.624$, $SE = 0.166$, $t(15) = 10.132$, $p < .001$. The relatively low forced-choice identification rate in the explicit condition was primarily due to those participants who occasionally incorrectly read aloud a word that had no semantic similarity to the presented prime. While these trials were treated as explicit in our analysis, neither option in the forced choice corresponded to the word that was read aloud. If these misread primes were omitted, the average correct forced-choice response for explicit primes was 95% (average $d' = 1.868$, $SE = 0.164$, $t(15) = 11.747$, $p < .001$.

Discussion

It seems reasonable to assume that if participants could extract some partial explicit information during the initial prime presentation, then that information should be available for both the change detection task as well as the subsequent

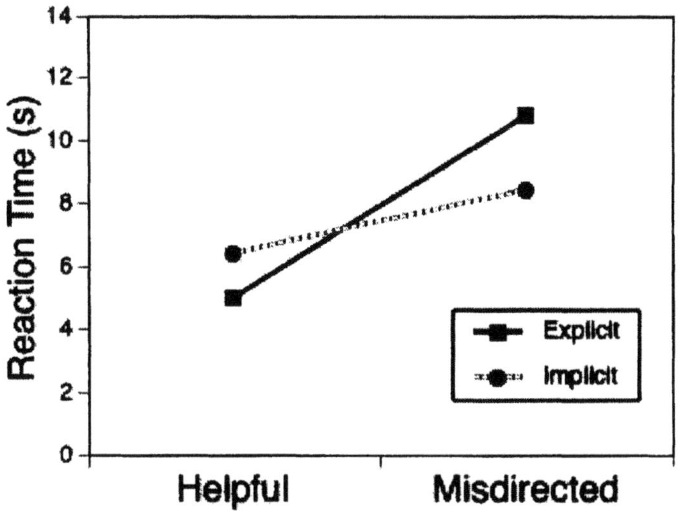

Figure 6. Experiment 3 results. Significant priming effects were found for both explicit and implicit priming conditions, in spite of the fact that participants were no better than chance at reporting the implicit primes in a two-alternative forced-choice task.

forced-choice response. For example, if a participant explicitly detected the "ch" from the prime "chair" then she should be able to use that partial information in the two-alternative forced-choice task to determine that "chair" was the correct choice rather than, for example, "window". In other words, if fragments of primes were being explicitly detected, then the participants should have responded above chance during the forced-choice response phase. The fact that their forced-choice response scores did *not* differ significantly from chance suggests that the primes were indeed processed implicitly.

If one considers visual processing as occurring along a continuum, it appears that primes could fall into three somewhat overlapping categories. At one extreme, stimuli that are effectively masked would yield no measurable priming effects. At the other extreme, an ineffective mask would yield an explicitly recognizable prime. Only between these extremes would one expect stimuli to be processed enough to yield implicit, but not explicit effects. This pattern of results is termed "indirect-without-direct-effects" (Merikle & Daneman, 1998) and has been used as the benchmark for true implicit processing. Previous studies of word–word priming (Draine & Greenwald, 1998) have found evidence for implicit priming in the absence of explicit awareness. In addition, studies investigating the semantic link between words and pictures have also found implicit priming effects. For example, Ferrand, Granger, and Segui (1994) found that implicitly presented primes facilitated picture-naming latencies.

In general, it seems that primes will have the greatest implicit effects when they approach (but do not surpass) the threshold required for explicit processing. For experiments that are designed to test implicit priming effects, this argues against relying on stimulus parameters that keep the explicit detection rate at zero, as parameters too far from threshold would not be expected to produce implicit processing. However, using prime parameters near this explicit detection threshold will necessarily allow a proportion of stimuli to be explicitly detected. This suggests that studies that attempt to demonstrate "indirect-without-direct-effects" should determine on a trial-by-trial basis whether a prime has been explicitly detected.

GENERAL DISCUSSION

Although the experimenter in the present study did not indicate that there was any relation between the prime and the picture pair, most participants reported that they naturally oriented more quickly toward an object related to the word that they had just read aloud. If the prime was helpful, they were able to detect the change more quickly. Most interesting are those trials in which participants were *not* able to read the primes aloud. In spite of the inability to consciously detect the prime, participants were still faster at detecting changes to helpfully primed objects. This finding, in conjunction with the finding that subsequent prime recognition was at chance, seems to be clear evidence that implicit semantic information can be used to increase the efficiency of visual processing when viewing a natural scene.

Natural scenes are interesting experimental stimuli because of their ecological validity, their overall complexity, and the fact that the items within the scene have related and overlapping semantic meanings associated with them. Unlike an array of unrelated objects, real-world scenes offer the chance to identify the semantic mechanisms at work in everyday scene perception. Our visual systems seem to be tuned for processing scenes, rather than unrelated arrays of objects. Just as acoustically presented words are processed more rapidly when presented as part of a structured sentence, our visual world seems to confer a processing advantage on coherently organized scenes (Biederman, 1981).

Some have argued that the visual world is segmented and comprehended though a two-stage process involving first the formation of a preattentive "primal sketch" assessing the overall layout and gist of the scene (Biederman, 1981; Schyns & Oliva, 1994), followed by a continued sweep of attention to detect specific features as needed during subsequent fixations. The primes in the present study theoretically could have affected reaction times by increasing the efficiency of either the primal sketch formation or the subsequent guidance of attention during the search, or both. However, the pattern of results demonstrated here suggests that most, if not all, of the effect occurred during the guidance of attention. First, the magnitude of the reaction time savings (a reduction of 2–4 s when a helpful prime was presented) is too great to have been derived from only a reduction in the primal sketch formation, which itself has been demonstrated to occur in a half-second or less (Potter, 1975, 1976). Second, there is no reason to expect that the helpful primes would lead to a faster determination of a scene's gist than would the misdirected primes, since both primes provided at least some semantic cues to the scene's general content. Thus, these findings indicate that implicit and explicit semantic information can be used by the brain to guide attention within a complex scene.

REFERENCES

Abrams, R., & Greenwald, A. (2000). Parts outweigh the whole (word) in unconscious analysis of meaning. *Psychological Science*, *11*(2), 118–124.

Aginsky, V., & Tarr, M. J. (2000). How are different properties of a scene encoded in visual memory? *Visual Cognition*, *7*(1/2/3), 147–162.

Bates, T. C., & D'Oliveiro, L. (2003). PsyScript: A Macintosh application for scripting experiments. *Behavior Research Methods, Instruments, and Computers*, *35*(4), 565–576. [Computer software] retrieved from http://www.maccs.mq.edu.au/~tim/psyscript/index.html

Biederman, I. (1981). On the semantics of a glance at a scene. In M. Kubovy & J. R. Pomerantz (Eds.), *Perceptual organization* (pp. 213–253). Hillsdale, NJ: Lawrence Erlbaum Associates, Inc.

Bruner, J., & Potter, M. C. (1964). Interference in visual recognition. *Science*, *144*, 424–425.

Carr, T. H., McCauley, C., Sperber, R. D., & Parmelee, C. (1982). Words, pictures, and priming: On semantic activation, conscious identification, and the automaticity of information processing. *Journal of Experimental Psychology: Human Perception and Performance*, *8*(6), 757–777.

Chun, M. M., & Nakayama, K. (2000). On the functional role of implicit visual memory for the adaptive deployment of attention across scenes. *Visual Cognition*, *7*(1/2/3), 65–81.

Draine, S. C., & Greenwald, A. G. (1998). Replicable unconscious semantic priming. *Journal of Experimental Psychology: General*, *127*(3), 286–303.

Ferrand, L., Grainger, J., & Segui, J. (1994). A study of masked form priming in picture and word naming. *Memory and Cognition*, *22*(4), 431–441.

Hollingworth, A., Williams, C. C., & Henderson, J. M. (2001). To see and remember: Visually specific information is retained in memory from previously attended objects in natural scenes. *Psychonomic Bulletin and Review*, *8*(4), 761–768.

Huettel, S., Güzeldere, G., & McCarthy, G. (2001). Dissociating the neural mechanisms of visual attention in change detection using functional MRI. *Journal of Cognitive Neuroscience*, *13*(7), 1006–1018.

Klein, R. M., Kingstone, A., & Pontefract, A. (1992) Orienting of visual attention. In K. Rayner (Ed.), *Eye movements and visual cognition: Scene perception and reading* (pp. 46–65). New York: Springer.

Levin, D. T., Momen, N., Drivdahl, S. B., & Simons. D. J. (2000). Change blindness blindness: The metacognitive error of overestimating change-detection ability. *Visual Cognition*, *7*(1), 397–341.

Merikle, P., & Daneman, M. (1998). Psychological investigations of unconscious perception. *Journal of Consciousness Studies*, *5*, 5–18.

O'Regan, J. K., Deubel, H., Clark, J. J., & Rensink, R. A. (2000). Picture changes during blinks: Looking without seeing and seeing without looking. *Visual Cognition*, *7*(1/2/3), 191–211.

Potter, M. (1975). Meaning in visual search. *Science*, *187*, 965–966.

Potter, M. (1976). Short term conceptual memory for pictures. *Journal of Experimental Psychology: Human Learning and Memory*, *2*, 509–522.

Rensink, R. A. (2000). The dynamic representation of scenes. *Visual Cognition*, *7*(1/2/3), 17–42.

Rensink, R. A. (2002). Change detection. *Annual Reviews of Psychology*, *53*, 245–277.

Rensink, R. A., O'Regan, J. K., & Clark, J. (1997). To see or not to see: The need for attention to perceive changes in scenes. *Psychological Science*, *8*(5), 368–373.

Schyns, P. G., & Oliva, A. (1994). From blobs to boundary edges: Evidence for time- and special scale dependent recognition. *Psychological Science*, *5*, 195–200.

Smith, M. C., Meiran, N., & Besner, D. (2000). On the interaction between linguistic and pictorial systems in the absence of semantic mediation: Evidence from a priming paradigm. *Memory and Cognition*, *28*(2), 204–213.

Werner, S., & Thies, B. (2000). Is ''change blindness'' attenuated by domain-specific expertise? An expert–novices comparison of change detection in football images. *Visual Cognition*, *7*(1/2/3), 163–173.

VISUAL COGNITION, 2005, 12 (6), 1143–1160

Eye movements in search in visual neglect

Sara Forti and Glyn W. Humphreys
Behavioural Brain Sciences, School of Psychology, University of Birmingham, Birmingham, UK

Derrick G. Watson
Department of Psychology, University of Warwick, Coventry, UK

We report data on eye movements in a patient (MP) with unilateral visual neglect, in a task of searching a real-world scene. We demonstrate that MP missed some contralesional targets after fixating on them. On such "miss" trials, MP's eye movements were similar to those on objects in target-absent trials trials, where MP was cued to search for a target that was not actually present. In addition, MP showed little evidence of memory for the locations of objects that remained across a trial, though poor memory fails to explain performance on trials where targets were fixated. We suggest that, in this case, neglect reflected the poor uptake of information from the contralesional side of space, so that even fixated targets sometimes failed to match a "template" held in his memory.

Visual neglect is characterized by an impaired ability to detect stimuli presented on the side of space contralateral to the lesion (Heilman, Watson, & Valenstein, 1993). Recent studies indicate that several different factors may contribute to neglect, including: "Magnetic" orienting of attention to the ipsilesional side (Làdavas, Petronio, & Umiltà, 1990; Shalev & Humphreys, 2000), poor coding of spatial and object-based representations of stimuli (Bisiach & Luzzatti, 1978, Humphreys & Riddoch, 1994), poor ability to sustain attention and/or arousal (Robertson & Manly, 1999), impaired "weighting" of information on the affected side when spatial selection is required (Duncan, Bundesen, Olson, Humphreys, Chavda, & Shibuya, 1999), impaired visuospatial working memory (so that patients may fixate consistently back to the ipsilesional side; see Husain, Manna, Hodgson, Wojciulik, Driver, & Kennard, 2001; Husain, Mannon, Mort, Hodgson, Driver, & Kennard, 2002). In the present study we attempt to

Please address all correspondence to: Sara Forti, Behavioural Brain Sciences, School of Psychology, University of Birmingham, Birmingham B15 2TT, UK. Email: s.forti@bham.ac.uk
This work was supported by grants from the MRC and the Stroke Association to the second author.

DOI:10.1080/13506280444000689

decompose the neglect syndrome by analysing eye movement behaviour in a neglect patient as he performed a real-life search task, looking for objects on a table. Our data suggest that at least two factors contributed to the patient's performance (and particularly to his misses of targets that were present), one of which reflected impaired scanning and fixations of targets, and the other impaired perceptual identification of targets that were fixated. At least in this case, neglect cannot be fully attributed to the processes involved in scanning and the control of overt attention (including impaired visuospatial working memory).

EYE MOVEMENTS IN VISUAL NEGLECT

There have now been several studies where eye movements patterns have been monitored in patients showing visual neglect. These studies demonstrate that neglect patients can make saccades to stimuli on the left side of space (Karnath, Neiemeier, & Dichgans, 1998; Làdavas, Zeloni, Zaccara, & Gangemi, 1997), but these saccades can be relatively infrequent and there can be a biased deviation of the exploratory gaze. Fixations fall more frequently and earlier during search to the right relative to the left (Behrmann, Watt, Black, & Barton, 1997; Karnath, 1988).

Husain and co-workers (2001, 2002; Kennard, Husain, Mannan, Hodgson, Wojciulik, & Driver, 2001) have proposed that poor visuospatial working memory may play a role in this abnormal pattern of exploration. If patients fail to code the locations of objects already inspected in visuospatial working memory, then they may be liable to reexplore the ipsilesional side of space, especially if the problem is coupled with a spatial bias in attention. However, in at least some studies, the neglect symptoms shown by the patient cannot be fully attributed to poor exploratory scanning. For example, Walker and colleagues (Walker, Findlay, Young, & Lincoln, 1996; Walker & Young, 1996) reported data from patient RR, who appeared to present with object-based neglect (e.g., neglecting the left half of an object even when it was placed in his right visual field). Although most of his fixations fell to the right sides of stimuli, some fixations did fall on the left sides, even on trials in which neglect was recorded. Thus neglect could be dissociated from some of the factors leading to impaired exploration of space through eye movements. These aspects of neglect in patient RR, even when he fixated the stimulus, are reminiscent of some errors found with normal observers under condition of "change blindness". For example, O'Regan, Deubel, Clark, and Rensink (2000) measured eye movements when observers explored scenes while looking for a change introduced by the experimenters. They found that subjects failed to detect up to 40% of the changes made to objects being fixated. Similar observations have been made by Ballard, Hayoe, and Peltz (1995) and Hayoe, Besinger, and Ballard (1998). Such results suggest that stimuli can fail to be detected if they are not attended during a fixation, in which visual perception of change in an object occurs only when that object is given focused attention (Rensink, O'Regan, & Clark, 1997; Henderson & Hollingworth, 1999; Hollingworth & Henderson, 2002; see also Mack & Rock, 1998 for evidence from the phenomenon of "inattentional blindness"). Finally Kim, Schwartz, and Heilman (1999) reported a physiological confirmation of this result via the study of pupillary dilatation, which may be used as a behavioural index of stimulus-induced arousal. In their study two patients with neglect were aware of left side stimuli and looked at them, but they did not show any pupillary dilatation; in contrast, pupil dilatation occurred when the patients looked at stimuli on the right side of space. It may be, then, that neglect can reflect inattention to a stimulus even when it is fixated. To assess this last possibility we had patient MP search for a target object in an array of real objects and monitored his eye movements simultaneously. We assessed the characteristics of MP's eye movements as a function of where targets fell in the field, and as a function of whether targets were detected or missed. The data reveal that at least two factors may contribute to impaired search in MP's case: One relating to poor fixation of targets, but the other occurring even on trials where target fixations took place. The results indicate the utility of measuring eye movements in such cases.

CASE HISTORY

MP (born January 19, 1947) is a left-handed former tool worker. He suffered an aneurysm in the right middle cerebral artery in 1992, resulting in the right middle cerebral artery occlusion and infarct. There was a mild left hemiparesis for the upper limb. An MRI scan showed a large lesion including the right frontotemporoparietal junction, the superior and inferior frontal gyri, the superior temporal gyrus, and the postcentral gyrus (see Edwards & Humphreys, 1999; Humphreys & Riddoch, 2001). MP was formerly left handed. Verbal intelligence, assessed by the National Adult Reading Test (NART; Nelson & Willison, 1991), predicted a full-scale IQ of 90. Neuropsychological screening showed reduced memory skills, both for verbal material (prose recall at WMS-R: 24/50 for immediate recalling—53% of age matched controls; 13/50 with 30 minutes delay—23%), and for nonverbal material (visual reproductions at Wechsler Memory Scale Revised; WMS-R; Wechsler, 1987): 12/41 immediate—bottom 2% of age-matched controls; 5/41 delayed—bottom 1% of controls). His digit span was four. There were also poor mental arithmetic abilities and impairments in executive function: Verbal fluency at COWAT = 22; nonverbal fluency at TRAILS (Reitan, 1992) A = 198 s, TRAILS B = 36 (in these tests MP performed more than 2 *SD*s below the mean for controls). Also, some problems in face processing were present (Warrington test of face memory = 25/50; recognition of famous faces = 7/14: Recognition Memory Test; Warrington, 1984). In contrast, reading abilities were spared: Using the Psycholinguistic Assessment of Language Processing in Aphasia (Kay, Lesser, & Coltheart,

1992) he scored 30/30 on regular words and 29/30 on irregular words. Similarly object recognition was good (70/76 on the long naming test of the BORB; Riddoch & Humphreys, 1993). Clinical neglect was apparent in the standardized Behavioural Inattention Test (BIT; Wilson, Cockburn, & Halligan, 1987). On line crossing MP scored 28/36, missing items in the final left column; in the star cancellation task MP omitted all of the target stars on the far left and cancelled 9/19 stars in the next left column; in a line bisection task with lines placed randomly on a page, he omitted all items on the left and showed an average shift of 3% towards the right; he identified the gender of the left side of male–female chimerics on just 5/20 trials responding in all other trials to the gender of the right side face.

METHOD

Eye movements were measured while MP was asked to execute a search task locating a named target. MP was seated at about 15 cm from a table (120 × 120 cm) where a number of everyday objects were randomly placed. On each trial MP was asked to point to the object named by the experimenter. The experiment consisted of five blocks of trials, with an average of 15 objects presented on the table at a time (5 left, 5 central, 5 right), all common household stimuli that varied in volume from 64 cm^2 (address stamps: 7.5 × 8.5 cm) to 5980 cm^3 (kettle; 20 × 23 × 13 cm). Objects were randomly assigned to the positions tested on a trial.

In one block, MP was asked to search for a particular object on the table, with any object being cued as a target no more than once per block. In addition to the verbal cues given to the target objects present on the table, there were also target-absent trials, where he was given a distractor name (corresponding to an object that was not present—on average 5.2 distractor names were given across a trial block). The order of the targets and distractor names was randomly determined, as was the order of the targets cued as a function of their position on the table. MP was told to say "no" when he could not find the object cued. He was told about the presence of distractor names as well as target cues, to encourage him to make "no" responses (misses) on some trials where the cued target was present.

The objects on each trial were drawn from a set of 24 common artefacts. No objects were repeated on a trial and the objects present on a trial were equally distributed across the areas of space (left, centre, and right from MP's midline). The objects chosen for each trial were randomly assigned to the different positions.

Apparatus

Eye movements were measured from the right eye using a SensoMotoric Instruments (GmbH) HED corneal reflection based eyetracker. This system consists of a head-mounted device with a scene camera that captures the participant's field of view and an eye camera that records an image of the eye. After calibration the system produces a video image of the scene with a superimposed cursor that represents the position that the participant is fixating. Because the relationship between the scene camera and eye camera is fixed, the participant is free to move their head. The video output was digitized at 25 frames/s for subsequent offline frame-by-frame analysis using a software based DVD player. The system was calibrated before each block of trials by asking MP to fixate each of five markers placed on the table surface (at the centre and towards the four corners). During analysis a fixation was defined as a sequence of at least two video fields (equivalent to 80 ms) during which the pupil was stationary and within the boundary of the same object. The experiment was conducted in the Department of Psychology, University of Warwick.

RESULTS

The data on eye movements were broken down to illustrate separately three main steps involved in target detection: (1) Scanning of visual field and attentional orienting; (2) uptake of information during fixation; and, finally (3) memory for object locations across the trial blocks.

Scanning of visual field and attention orienting

It has been shown (Behrmann et al., 1997; Chèdru, Leblanc, & Lhermitte, 1973) that subjects with neglect often start searching from a lateral right position, irrespective of the target's location, suggesting a tendency to orient attention automatically to the ipsilesional (right) side (Làdavas et al., 1990; Shalev & Humphreys, 2000). Accordingly, we would expect MP to direct attention at the beginning of the trial to the ipsilesional side, with objects on the right being fixated first. Moreover, if MP showed a defect in disengaging attention from the right side (cf. Posner, Walker, Friedrich, & Rafal, 1984), then only few objects on the left would be fixated, and fixations on right-side objects would be longer than fixations to objects on the left.

Only some of these predictions were confirmed. First of all, MP did show a tendency to orient attention to the right first: Pooling together the data across all trial types (target-detected, target-missed, and target-absent trials), the first fixation fell on a right-side object on 59 occasions, compared to 26 trials for objects on the left and 16 for objects in the centre; the effect of field on the location of the first fixation was reliable, $\chi^2(2) = 13.4, p < .001$; see Figure 1). This also did not alter as a function of the position of the target, $\chi^2(6) = 4.3, p > .05$.

However, the bias in orienting attention towards the right was not exacerbated by a deficit in disengaging attention from that side: MP scanned across objects on the left side of the display on all trials, and left-side fixations were as frequent as those to the right. In order not to have performance confounded by

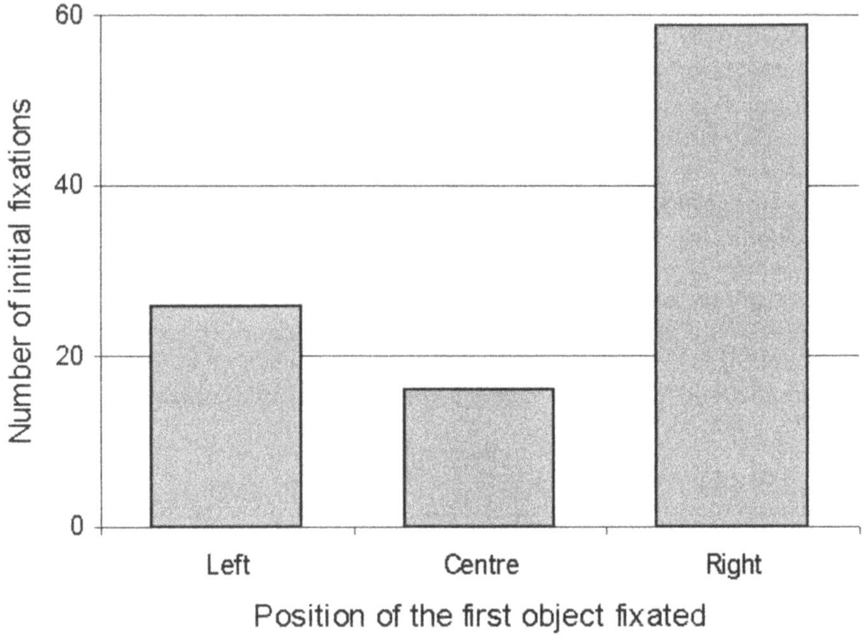

Figure 1. Number of initial fixations made on objects as function of their lateral position.

target detections, we analysed the distribution of fixations during target-absent trials: MP fixated 88% (131/149) of objects located on the right and 83% (129/156) of objects on the left, with no difference between the two, $\chi^2(1) = 1.7$, $p > .05$. Interestingly, MP showed a singular behaviour for objects in the centre, with only 55% (59/107) of the stimuli being directly fixated; this, in turn, led to an overall effect of object position on the proportion of fixated to nonfixated objects, $\chi^2(2) = 42.3$, $p < .001$. This result is consistent with central objects being relatively easy to reject as possible targets, and MP did not rely on direct fixation.

Moreover, the data on the duration of first fixations also confirmed that there was no marked problem in attentional disengagement. We analysed separately data from trials in which targets were detected, and data from target-absent trials (when target recognition processes were not involved[1]). For target-absent trials, we selected the data for objects that matched the location and the serial position of the targets in the search order on a trial. Fixations on target-absent trials

[1] The data from the 12 missed targets were not evenly distributed across space and so were not considered in this analysis.

(made to distractors) were clearly faster overall than those to targets, $F(1,55) = 38.7$, $p < .001$, but there was no effect of stimulus location and no interaction between stimulus location and type of stimulus (objects on target-absent trials vs. detected target) ($F < 1.0$): MP did not make longer fixations to right-side objects, which would be expected if there were impaired disengagement of attention from them (Figure 2).

Additionally, we assessed whether MP showed a persistent scanning pattern across trials, moving from objects on the right to those on the left. If this were the case we would expect more distractors to be fixated when targets fell in the contralesional (left) field. Instead, MP made on average 10 fixations on other objects before detecting a right target, 12 before detecting a left target, and 8 before detecting a central target; the effect of field was not reliable ($F < 1.0$). This indicates that, although he typically started search from the right side, the scanning path varied across trials. Overall there was evidence of a bias in visual orienting, consistent with a distorted gradient of attention across the field, but no evidence for a spatial impairment in attentional disengagement.

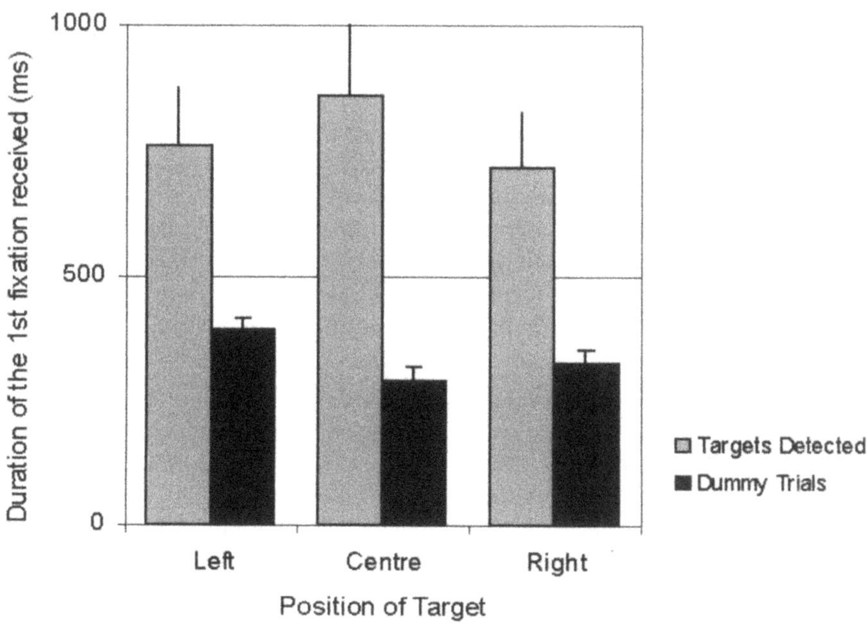

Figure 2. Duration of the first fixation both on targets detected and on objects fixated during target-absent trials, as a function of their position.

Uptake of information at fixations

A second aspect of visual neglect may be poor uptake of visual information on the contralesional side, despite fixating upon objects (e.g., O'Regan et al., 2000; Walker et al., 1996). In this case misses may occur because, on those trials, insufficient target information is acquired to trigger recognition. It would then follow that eye movements on miss trials should be similar to those on target-absent trials—where target information is not available. For example, in both of these cases the first fixations may be shorter than the first fixations on trials where the targets are detected, because recognition processes are not engaged on miss and target-absent trials. However, it is also possible that target detection occurs at a relatively late stage during scanning, in which case effects of target recognition will emerge only on later fixations. Furthermore, if there is poor uptake of visual information on the left side, then left-side targets will accrue more fixations than right-side targets even when they are detected. We describe these points separately.

Failures in detection despite fixating objects. MP sometimes missed targets that were fixated. He failed to detect 16/74 objects. Of these, only 4 (2 in the centre, 2 on the right) were not fixated; the remaining 12 targets (8 on the left, 2 in the centre, 2 on the right) were missed despite being fixated. Interestingly, all 29 targets on the left were fixated, but 8 were still missed. In contrast, MP failed to fixate, and to detect, some objects in the centre and right locations. The difference between the proportions of items missed but fixated on the left, centre and right sides of space was reliable (8/8 left versus 4/8 centre + right; Fisher's exact probability = .04). It is reasonable to assume that, at least in the cases where fixations were made to left-side stimuli, MP may have oriented attention towards the usually neglected side, in an endogenous fashion. This contrasts with a tendency not to fixate central items (see above) and misses due to not fixating targets in the centre and right. This in turn suggests a difference between a failure in scanning items in the centre and on the right, and neglect even when objects are fixated (particularly for objects on the left).

First fixations on miss, dummy, and target detection trials. The first fixations on target-miss trials were, on average, 455 ms for the eight left-missed targets (all fixated), 198 ms for the two fixated but missed central targets, and 453 ms for the two fixated but missed right-side targets. This compares with durations of 760, 857, and 715 ms for first fixations to left, central, and right targets that were detected, and durations of 363, 280, and 450 ms to objects in equivalent locations on target-absent trials. Although there was a trend for first fixations to be shorter on target-miss than on target detection trials, this was not reliable, $F(1, 27) = 0.5$, $p < 1.0$. Also, there was no difference, $F(1, 7) = 4.3$, $p >$ .05, between the first fixation durations to targets missed on the left and those to

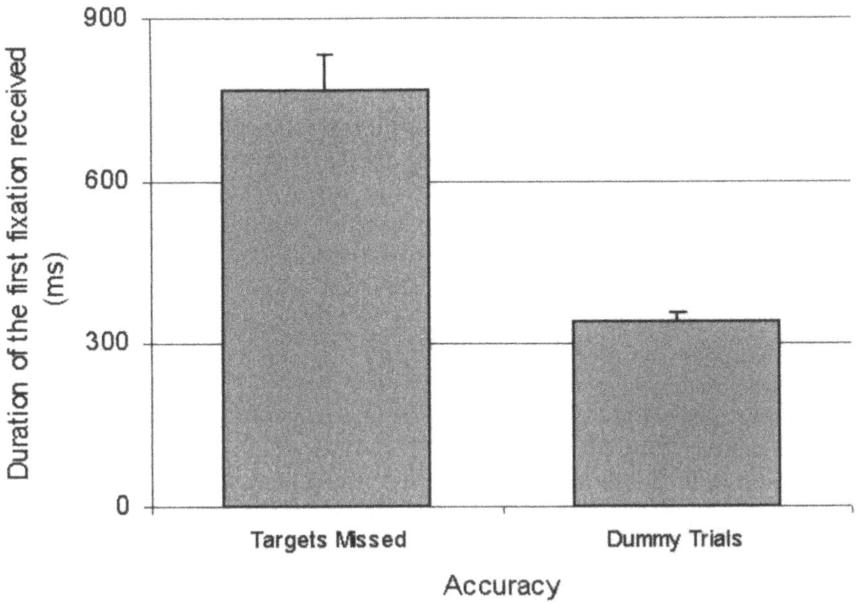

Figure 3. Duration of the first fixation on targets missed on the left and distractors on the left fixated during target-absent trials.

left-side objects on target-absent trials (matching the stimuli for serial position and location within trial) (see Figure 3).

Duration of second fixations. The second fixation may represent a further chance to process stimuli that were not detected on their first fixation. Accordingly, objects receiving a very short first fixation should be more likely to be refixated, and to be refixated longer on the second occasion.

To assess this, the probability of refixation was examined in relation to the duration of the first fixation. Detected targets showed a significant correlation between first fixation duration and the probability of being refixated (Spearman's $\rho = -.29$, $p < .05$): Objects were more likely to be refixated when the first fixation was short (mean first fixation when the object was not refixated = 851 ms; mean first fixation when the object was refixated = 545 ms).

The duration of the second fixation was also compared between detected-targets, missed-targets, and objects fixated during target-absent trials. There was a reliable main effect of the type of item, $F(2, 241) = 19.6$, $p < .001$. Post hoc tests with Bonferroni's α revealed that this effect was due to the difference between target-detected and target-absent trials, but there was no difference between the duration of the second fixations on targets that were detected and targets that were missed (*M* detected = 728 ms; *M* missed = 554 ms; *M* target-absent trials = 328 ms). The contrast with target-absent trials suggests that the second fixation acts as a verification check, on detection trials, or a reaffirmation that the object was not a target, on miss trials (Figure 4).

Finally we compared the duration of the second fixation to the duration of the first fixation, for targets detected after the first fixation (9 on the left, 3 in the centre, 4 on the right), targets missed although being fixated more than once (7 on the left, 2 in the centre, 2 on the right), and for objects on target-absent trials that were fixated more than once. The data were analysed in a mixed-design ANOVA with the factors being: ''Serial order of fixation'' (first fixation, second fixation) and ''trial type'' (targets detected, targets missed, and objects on target-absent trials); the data were not sufficient for spatial position to be considered too. There was no reliable effect of serial order, $F(1, 245) = 0.1, p > .05$; with *M* first fixation = 355 ms, *M* second fixation = 358 ms). However, there was a reliable effect of trial type, $F(2, 245) = 29.2$, $p < .001$. Post hoc comparisons with Bonferroni's α revealed differences in the fixation durations on trials where targets were detected, on target-absent trials, and on trials where targets were missed (*M* detected = 636 ms; *M* missed = 538 ms; *M* objects on target-absent trials = 327 ms). Interestingly, a significant interaction between

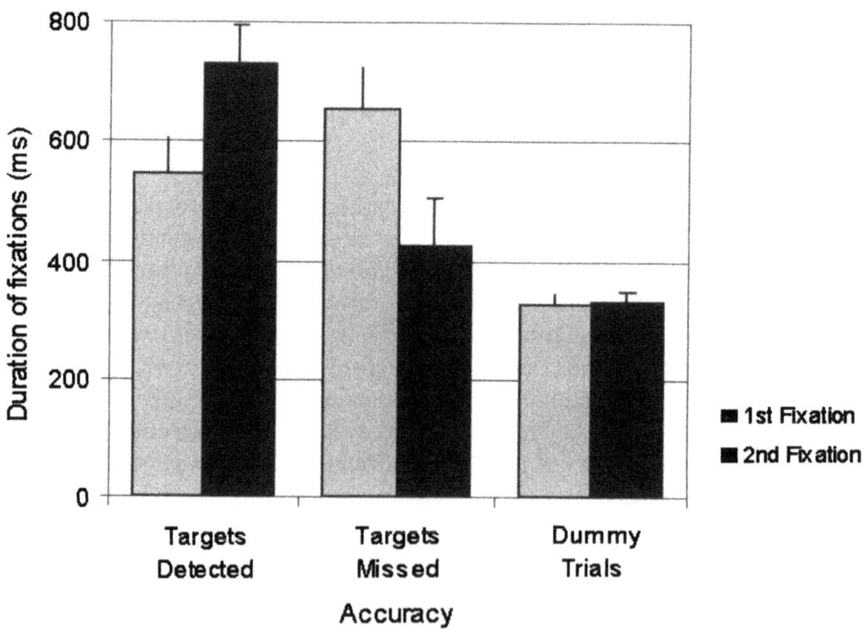

Figure 4. Comparison between the first and the second fixation received by targets detected, targets missed, and objects on target-absent trials.

"serial order" and "trial type" was found, $F(2, 245) = 4.5$, $p < .05$. When targets were detected, the second fixation was longer than the first fixation (*M* first = 545 ms, *M* second = 727 ms); when targets were missed, the second fixation was shorter than the first (*M* first = 651 ms, *M* second = 425 ms). There was no difference between first and second fixation durations on target-absent trials (*M* first = 326 ms, *M* second = 328 ms).

Number of fixations to left- and right-side targets. We have seen that objects were more likely to receive a second fixation when the first fixation was too short to lead either to target-detection or to a target-rejection. However, if a deficit in the uptake of information at a first fixation is linked to visual neglect, we would also expect fixations to be repeated particularly to left-side objects. In contrast, recent research has suggested a link between unilateral neglect and impaired visuospatial memory (Husain et al., 2001, 2002; Kennard et al., 2001), primarily based on patients making repeated fixations on stimuli in their ipsilesional field. In order to evaluate this trend irrespective of target selection processes, we analysed refixations on target-absent trials. On these, each object was fixated 1.8 times on average. In order to assess the likelihood of refixations, we considered, for each position, the proportion of objects fixated only once and the proportion of objects refixated. This association was significant, $\chi^2(2) = 7.7$, $p < .05$: Opposite to an account of impaired visual memory for stimuli on his left field, MP was more likely to refixate objects in left space than objects in the centre and right space (101/129, 78% of objects refixated on the left; 42/59, 71% in the centre; 82/131, 63% on the right). This is consistent with there being poor take-up of visual information for left objects, rather than poor visuospatial memory for items already searched on the ipsilesional side.

We repeated this analysis for fixations on distractors objects for trials where targets were present (either detected or missed). Performance was assessed as a function of both distractor position and target position. A log linear model examined the factors: "Number of fixations and refixations" (objects fixated once, objects refixated), "target position" (left, centre, right), "distractor position" (left, centre, right), and "accuracy" (targets detected, targets missed). The model created, L.R.$\chi^2(24) = 21.5$, $p = .6$, had three bivariate associations: Refixation × Accuracy, Refixation × Target position, and Accuracy × Distractor position. The interaction between the number of refixations and the accuracy of detection arose because, on trials where the target was missed, there were increased numbers of refixations relative to single fixations on distractors; more interestingly, the interaction between the number of refixations and target position was due to MP making more repeated fixations rather than single fixations when searching for left targets, whilst there were more single object fixations when the target was in the centre compared to the right. In addition, the interaction between accuracy and distractor position occurred because, on trials where the targets were missed, there was a relative increase in the number of fixations and refixations to distractors on the left, as if MP was endogenously directing his attention to the affected side (see Figure 5).

Memory for object locations

Memory across trial blocks. The objects presented remained the same, and in the same positions, throughout a trial block (over an average of 15.4 target-present trials and 5.2 target-absent trials). It was possible that MP could memorize the positions of the objects, facilitating search on later trials. This would fit with evidence from normal participants developing detailed visual object representations in long-term memory (Hollingworth & Henderson, 2002). Hence targets prompted later in the block should be more likely to be detected with fewer fixations on distractors, and on their first fixation, compared with targets probed early in a block. In addition, the first fixations on "late-occurring" targets should be shorter than on "early occurring" targets the one on the first objects of the block, since the memory track from previous trials may prime the information uptake.

We conducted split-half analyses on the data from target detection trials, according to whether the target was probed on the first or the second half of the trials within a block. On 28 trials, targets were probed and detected in the first half of the block (8 on the left, 16 in the centre, and 14 on the right) and, on 30 trials, targets were probed and detected in the second half of the block (13 on the left, 8 in the centre, 9 on the right). The number of other objects fixated before detection was compared between targets probed on early trials and targets probed on late trials, as a function of the position of the target. There was no effect of serial position in the block on the number of distractors fixated before detection; also, there was no effect of position of target and no interaction (all *F*s < 1.0). Also, we compared the number of targets detected at the first fixation against the number of targets detected after two or three fixations, for targets probed in the first and the second half of the block. A significant association was found, $\chi^2(1) = 4.8$, $p < .05$. The data are presented in Table 1. Targets probed on "late" trials were less likely to be detected on first fixation than targets probed on "early" trials. This may be due to several factors: It could reflect a fatigue effect toward the end of the trials, or poor sustained attention; it might also be due to previously probed trials attracting attention. Whichever the case, the data suggest that there was little memory of items across probe trials, and certainly MP failed to direct his first fixation immediately to a target on "late" trials. This was also supported by an analysis of the duration of the first fixation on the target, in early compared to late trials (first half vs. second half of the block)—the duration of the first fixation was not affected by the serial position the target was cued in the block, $t(56) = -0.56$, $p > .05$; *M* early trials = 763 ms; *M* late trials = 771 ms).

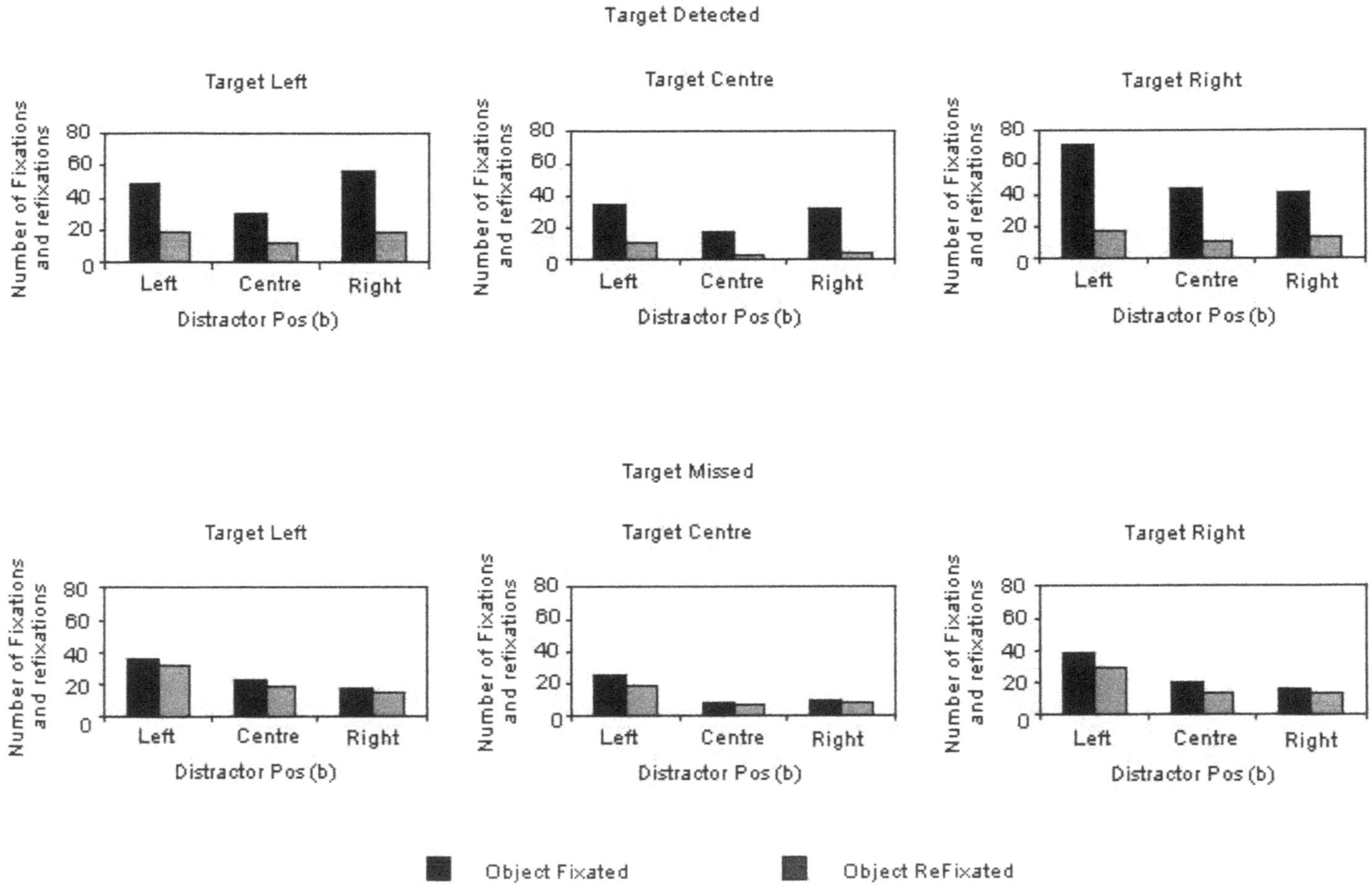

Figure 5. Objects fixated and refixated as a function of target and distractor position, for targets missed and targets detected.

TABLE 1
Fixations for detection in function of the early or late probing of the item within the block

Total number of fixations on target at detection	*First half of the block (early trials)*	*Second half of the block (late trials)*	*Total*
1	24	18	42
2 or 3	4	12	16

Memory within trial blocks. We have described that MP often refixated previously inspected objects even within the same trial block. This result is consistent with the research by Horowitz and Wolfe (1998) in that memory for previously inspected locations is not a determinant variable for visual search efficiency, and observers often return on fixated objects. However, in order to compare MP with normal subjects, we also calculated the ''memory metric'' used by Gilchrist and Harvey (2000). This places the observed refixations in between of two extremes of expected refixations: 0% is given by a perfect-memory model, in which no items would be refixated; 100% is given by a no-memory model, in which items may be refixated irrespectively of previous fixations received in the same trial. Considering target-absent trials, MP scored 60.3%, which is at the extreme of normal observers' performance as reported by Gilchrist and Harvey (ranging from 12.2% to 60.2%).

DISCUSSION

Our analysis of eye movements in patient MP reveals several interesting aspects of his performance. The major findings were as follows.

MP tended to make his first fixation to an object on the right side of space, irrespective of whether the target fell to the right, left, or in the central area around MP's midline. This is consistent with the link between right temporoparietal damage, left neglect, and the tendency to orient attention automatically to the ipsilesional (right) side (Làdavas et al., 1990; Shalev & Humphreys, 2000). However, although his first fixation was to the right, MP was able to scan across to other objects on the left side of the display on all trials. Hence, impaired scanning into the left side does not appear to be the main cause of neglect. Indeed, there was also no effect of field on the duration of the first fixation, which runs against the idea that MP had problems in disengaging attention from an ipsilesional stimulus (cf. Posner et al., 1984).

MP often failed to fixate central objects on target-absent trials, when cued to find a target that was not present. This suggests that fixation on an object was not a prerequisite to the object being processed and rejected as a distractor, at least when it fell in an area of relatively high acuity, at the start of a trial.

The positions of objects in the field also affected the likelihood that a target was detected on its first fixation. Targets were more likely to be detected on their first fixation when they fell to the right of MP's body relative to when they fell to the centre or to the left of his body. This suggests that the position of the object in space tended to affect the uptake of visual information from the object, even when it was fixated. If the uptake was slower/noisier for objects in left space, then a second fixation was more likely to be required before detection took place. This was also confirmed by an analysis of MP's performance on target-absent trials (cued to a target not on the table), and on trials where targets were present but missed. On target-absent trials MP tended to refixate objects on the left, consistent with him needing longer to acquire the information required to evaluate the stimulus (e.g., as a distractor, on target-absent trial).

Misses on fixated objects

Perhaps most interestingly, MP did miss targets even after fixating upon them. This was primarily when targets fell on the left, while misses to central and right side targets were sometimes associated with failures to fixate those stimuli. It may be then, that there are different components to MP's problems: There is some difficulty in controlling scanning, which can affect stimuli even in the central and right sides of space. On top of this, however, is a problem in detecting stimuli on the left even after scanning and fixation. This dissociation between eye movements and left neglect has been noted in other cases of neglect (e.g., Walker et al., 1996). It suggests that fixating a stimulus is not sufficient to ensure full attention on that stimulus (see also O'Regan et al., 2000). Indeed, in general, MP's eye movements to left targets on miss trials resembled those on target-absent trials, when stimuli were rejected as distractors. He tended to make repeated fixations on both missed targets and distractors on target-absent trials, and these stimuli were more likely to be refixated than targets that were detected. Also, on both miss trials and target-absent trials the duration of the first fixation tended to be reduced relative to when the fixation landed on a target that was eventually detected. And, when a second fixation occurred, the duration of this differed according to whether it took place on an object that was detected as a target or, alternatively, on an object that was missed or on a distractor (on target-absent trials). Second fixations tended to be longer than first fixations when they fell on a target that was detected successfully; in contrast, second fixations tended to be shorter than first fixations when they fell on distractors or on targets that were missed. From the short durations of both first fixations, and the even shorter second fixations, we conclude the following: When fixations fell on a distractor or on targets that were missed, MP responded to some mismatch between the perceptual information being acquired and his "template" for the target. These early mismatches led to short fixations and refixations that appeared to be confirmations of the mismatch (and hence were shorter than first fixations). In contrast, after a first fixation fell on a target, a second fixation was sometimes made to confirm the match to template. This second fixation, supporting the match, was then prolonged as the match took place.

Eye movements and memory for objects

There was remarkably little evidence for MP's search being affected by memory for objects present and the locations they occupied across a trial block. For example, there was no effect on the duration of the first fixation according to whether a target was probed early in a trial block relative to when it was probed late. There was also no change in the duration of the first, relative to the second fixation. Given that the stimuli remained in the same location across a block of trials, we might have expected fixation durations to decrease when targets had to be detected late in a trial block. In fact, targets probed late in a trial block were less rather than more likely to be detected on the first fixation. These data are consistent with the argument that patients with right temporoparietal damage and neglect can have poor visuospatial memory (Husain et al., 2001). However, our results also contradict the idea that poor visuospatial memory within a trial is the cause of neglect, since left neglect in MP occurred even when fixations fell on targets. Indeed, MP showed a proportion of refixations within a trial comparable to controls. Rather than poor visuospatial memory being causal here, we suggest that neglect stemmed from poor uptake and matching of visual information from left side targets. The failure to detect targets probed late in a trial block may reflect the influence of further, nonspatial factors on performance, including fatigue, reduced arousal and/or poor sustained attention (cf. Robertson & Manly, 1999). Such nonspatial effects may exacerbate any deficits in perceptual processing in the first place.

If our account of impaired perceptual processing in neglect is correct, then the data raise questions about the nature of the visual codes that are affected. For example, a critical factor here appeared to be the position of objects with respect to MP's body midline, not their position with respect to other body coordinates such as the eye or the head (since objects could be missed even when fixated). It may be that the process of matching a stimulus to any template, for search, is based on coding in terms of body coordinates. The importance of the body coordinate system for understanding MP's performance is also indicated in earlier work that demonstrated effect of "action coding" on search (Humphreys & Riddoch, 2001). In this earlier work, performance was affected by whether objects were oriented towards MP's body, suggesting representations sensitive to the positions of objects with respect to the effectors used in action.

In summary, the data are consistent with the view that MP could miss targets in search for at least two reasons. One relates to his ability to direct fixations to all the stimuli present. This problem was nonspatial and occurred with targets in the centre and on the right side of space. A second problem was more spatial in nature and particularly affected objects on the left side of MP's body. He could miss here targets even after fixating item and, similarly to when MP fixated a distractor, his fixations tended to be brief and to be repeated. We attribute this second problem to impaired uptake of visual information from stimuli on the impaired (contralesional) side of space. The study emphasizes that neglect cannot be attributed simply to poor control of eye movements, and that it is unlikely to reflect a single factor in patients with large lesions.

REFERENCES

Ballard, D. H., Hayoe, M. M., & Peltz, J. B. (1995). Memory representations in natural tasks. *Journal of Cognitive Neuroscience, 7*, 66–80.

Behrmann, M., Watt, S., Black, S. E., & Barton, J. J. S. (1997). Impaired visual search in patients with unilateral neglect: An oculographic analysis. *Neuropsychologia, 35*, 1445–1458.

Bisiach, E., & Luzzatti, C. (1978). Unilateral neglect of representational space. *Cortex, 14*, 129–133.

Chédru, F., Leblanc, M., & Lhermitte, F. (1973). Visual searching in normal and brain-damaged subjects: Contribution to the study of unilateral inattention. *Cortex, 9*, 94–111.

Duncan, J., Bundesen, C., Olson, A., Humphreys, G., Chavda, S., & Shibuya, H. (1999). Systematic analysis of deficits in visual attention. *Journal of Experimental Psychology: General, 128*, 450–478.

Edwards, M. G., & Humphreys, G. W. (1999). Pointing and grasping in unilateral visual neglect: Effect of on-line visual feedback in grasping. *Neuropsychologia, 37*(8), 959–973.

Gilchrist, I. D., & Harvey, M. (2000). Refixation frequency and memory mechanisms in visual search. *Current Biology, 10*, 1209–1212.

Hayoe, M. M., Besinger, D., & Ballard, D. H. (1998). Task constraints in visual working memory. *Vision Research, 38*, 125–137.

Heilman, K. M., Watson, R. T., & Valenstein, E. (1993). Neglect and related disorders. In K. M. Heilman & E. Valenstein (Eds.), *Clinical neuropsychology* (3rd ed., pp. 279–336). New York: Oxford University Press.

Henderson, J. M., & Hollingworth, A. (1999). The role of fixation position in detecting scene changes across saccades. *Psychological Science, 10*(5), 438–443.

Hollingworth, A., & Henderson, J. M. (2002). Accurate visual memory for previously attended objects in natural scenes. *Journal of Experimental Psychology: Human Perception and Performance, 28*(1), 113–136.

Horowitz, T. S., & Wolfe, J. M. (1998). Visual search has no memory. *Nature, 394*, 575–577.

Humphreys, G. W., & Riddoch, M. J. (1994). Attention to within-object and between-object spatial representations: Multiple sites for visual selection. *Cognitive Neuropsychology, 11*(2), 207–241.

Humphreys, G. W., & Riddoch, M. J. (2001). Detection by action: Neuropsychological evidence for action-defined templates in search. *Nature Neuroscience, 4*(1), 84–88.

Husain, M., Manna, S., Hodgson, T., Wojciulik, E., Driver, J., & Kennard, C. (2001). Impaired spatial working memory across saccades contributes to abnormal search in parietal neglect. *Brain, 124*, 941–952.

Husain, M., Mannon, S., Mort, D., Hodgson, T., Driver, J., & Kennard, C. (2002). Impaired spatial memory contributes to unilateral neglect. *Journal of Neurology, Neurosurgery, and Psychiatry, 73*(2), 221.

Karnath, H.-O. (1988). Deficits of attention in acute and recovered visual hemi-neglect. *Neuropsychologia, 26*, 27–43.

Karnath, H.-O., Neiemeier, M., & Dichgans, J. (1998). Space exploration in neglect. *Brain*, *121*, 2357–2357.

Kay, J., Lesser, R., & Coltheart, M. (1992). *Psycholinguistic assessments of language processing in aphasia (PALPA)*. Hove, UK: Lawrence Erlbaum Associates Ltd.

Kennard, C., Husain, M., Mannan, S., Hodgson, T., Wojciulik, E., & Driver, J. (2001). Abnormal visual search in parietal neglect: A defect of spatial working memory. *Annals of Neurology*, *50*, S39.

Kim, M., Schwartz, R. L., & Heilman, K. M. (1999). Pupillographic findings in neglect. *Journal of Neurology, Neurosurgery, and Psychiatry*, *67*, 82–85.

Làdavas, E., Petronio, A., & Umiltà, C. (1990). The deployment of visual attention in the intact field of hemineglect patients. *Cortex*, *26*(3), 307–317.

Làdavas, E., Zeloni, G., Zaccara, G., & Gangemi, P. (1997). Eye movements and orienting of attention in patients with visual neglect. *Journal of Cognitive Neuroscience*, *9*, 67–74.

Mack, A., & Rock, I. (1998). *Inattentional blindness*. Cambridge, MA: MIT Press.

Nelson, H. E., & Willison, J. (1991). *The National Adult Reading Test* (2nd ed.). Windsor, UK: NFER-Nelson.

O'Regan, J. K., Deubel, H., Clark, J. J., & Rensink, R. A. (2000). Picture changes during blinks: Looking without seeing and seeing without looking. *Visual Cognition*, *7*(1/2/3), 191–211.

Posner, M. I., Walker, J. A., Friedrich, F. J., & Rafal, R. D. (1984). Effects of parietal injury on covert orienting of attention. *Journal of Neuroscience*, *4*, 1863–1874.

Reitan, R. M. (1992). *Trail Making Test: Manual for administration and scoring*. Tucson, AZ: Reitan Neuropsychology Laboratory.

Rensink, R. A., O'Regan, K. O., & Clark, J. J. (1997). To see or not to see: The need for attention to perceive changes in scenes. *Psychological Science*, *8*(5), 368–373.

Riddoch, M. J., & Humphreys, G. W. (1993). *BORB: Birmingham Object Recognition Battery*. Hove, UK: Psychology Press.

Robertson, I. H., & Manly, T. (1999). Sustained attention deficits in time and space. In G. W. Humphreys, J. Duncan, & A. M. Treisman (Eds.), *Attention, space, and action: Studies in cognitive neuroscience* (pp. 297–310). Oxford, UK: Oxford University Press.

Shalev, L., & Humphreys, G. W. (2000). Biased attentional shift associated with unilateral left neglect. *Cognitive Neuropsychology*, *17*(4), 339–364.

Walker, R., Findlay, J. M., Young, A. W., & Lincoln, N. B. (1996). Saccadic eye movements in object based neglect. *Cognitive Neuropsychology*, *13*, 569–615.

Walker, R., & Young, A. W. (1996). Object based neglect: An investigation of the contributions of eye movements and perceptual completion. *Cortex*, *32*, 279–295.

Warrington, E. K. (1984). *Recognition Memory Test*. Windsor, UK: NFER-Nelson.

Wechsler, D. (1987). *Wechsler Memory Scale—Revised: Manual*. San Antonio, TX: Psychological Corporation.

Wilson, B., Cockburn, J., & Halligan, P. W. (1987). *Behavioural Inattention Test*. Bury St. Edmunds, UK: Thames Valley Test Company.

VISUAL COGNITION, 2005, 12 (6), 1161–1176

A computational model of visual stability and change detection during eye movements in real-world scenes

Fred H. Hamker
Allgemeine Psychologie, Psychologisches Institut II, Westfälische Wilhelms-Universität, Münster, Germany

Visual cognition and change detection critically depend on attention and gaze. A central question is the exact involvement of attention in change detection. This paper proposes a detailed computational model to illuminate the nature of visual stability and change detection during eye movements. Consistent with experimental findings the model predicts that change detection at the saccade endpoint is relatively easy. According to our model the planning of an eye movement provides an enhanced sensory processing of an object at the saccade target. This facilitates the memorization of that object. A comparison of the memorized target template with the representation after saccade allows for change detection. We hypothesize that the spatial reentry signal that emerges from planning an eye movement could be a crucial neural basis of visual stability.

Change blindness describes the difficulty of humans to notice changes in scenes. Most studies agree that focused attention overcomes this blindness to changes (for an overview see Rensink, 2002). Henderson and Hollingworth investigated change detection to saccade targets and found a high detection rate for saccade target deletions. According to the saccade target theory of visual stability the object at the saccade target is special because its location and features have been selected and memorized prior to the saccade (Currie, McConkie, Carlson-Radvansky, & Irwin, 2000; McConkie & Currie, 1996). The saccade target theory has been contrasted to extraretinal based theories which assume that an extraretinal eye position signal is used to remap all retinal locations. In extension to the saccade target theory, Henderson and Hollingworth suggested a visual memory theory to explain scene memory accross saccades. Regarding change detection they suggested the following processes: Attention is allocated to the saccade target. Therefore, sensory

Please address all correspondence to: F. H. Hamker, Allgemeine Psychologie, Psychologisches Institut II, Westfälische Wilhelms-Universität, Fliednerstrasse 21, 48149 Münster, Germany. Email: fhamker@uni-muenster.de

This research was supported by DFG HA2630/2-1 and in part by the ERC Program of the NSF (EEC-9402726). I thank M. Lappe for helpful comments.

DOI:10.1080/13506280444000698

processing of the saccade target is gated and an internal representation of the stimulus in short-term memory is facilitated. During the saccade the sensory activity declines. Once the eyes reach the target vision starts with the process of comparing the stored representation with the new incoming signals. If no match between both occurs an error signal is generated.

In this paper we investigate visual stability and change detection by means of a computational approach. In the above-mentioned theories the term ''attention'' is not clearly defined and the specific role of attention in change detection has not been described in detail. For example, attention in the saccade target theory is important to identify the saccade target as special. It is well known that the eye movement is associated with attention (Deubel & Schneider, 1996; Kowler, Anderson, Dosher, & Blaser, 1995; Rizzolatti, Riggio, & Sheliga, 1994) but it is not clear if attention is a separate but linked process or if attention emerges due to an extraretinal eye position signal. If the latter is the case the saccade target theory would be less different to the extraretinal based theories.

Several computational models have emphasized different aspects of eye movement control and some of them have been demonstrated in real-world scenes (Backer, Mertsching, & Bollmann, 2001; Findlay & Walker, 1999; Itti & Koch, 2000; Krieger, Rentschler, Hauske, Schill, & Zetzsche, 2000; Rao, Zelinsky, Hayhoe, & Ballard, 2002). In all of these models the process that defines gaze is separated from the process that analyses a scene near gaze point.

Our model (see Appendix) is unique with respect to how perception is influenced by the planning of an eye movement (Hamker, 2003). It has been successfully shown to be consistent with electrophysiological and behavioural findings in attention tasks (Hamker, 2003; Hamker, 2004a) and recently extended to natural scene perception (Hamker, 2005; Hamker & Worcester, 2002). According to our model occulomotor areas responsible for planning an eye movement, such as the frontal eye field, influence perception prior to the eye movement. The activity reflecting the planning of an eye movement reenters the ventral pathway and sensitizes all cells within the movement field. We use this model to simulate a change detection task that is comparable to the one used by Henderson and Hollingworth (2003). Consistent with the saccade target theory the model predicts that a global remapping is not necessary for visual stability. However, since the spatial reentry signal, which is crucial for the emergence of attention in our model, can be defined as an extraretinal eye position signal, our model suggests that the saccade target theory is in part related to the extraretinal based theories.

THE MODEL

The model (Figure 1) consists of visual areas V4, inferotemporal (IT) cortex, prefrontal areas that contain the frontal eye field (FEF) for saccade planning, and more ventrolateral parts for implementing functions of working memory

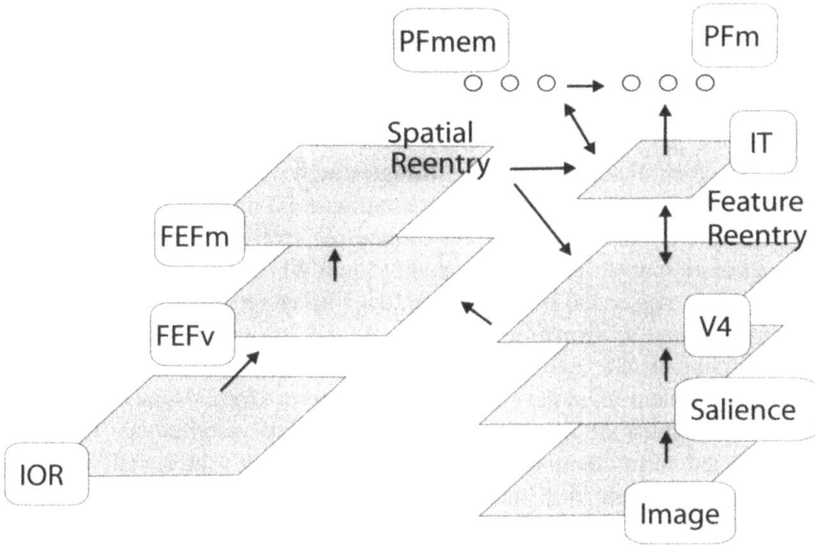

Figure 1. Model for visual perception of natural scenes. First, information about the content and its low level stimulus-driven salience is extracted. This information is sent further upwards to V4 and to IT cells that are broadly tuned to location. The target template is encoded in PF memory (PFmem) cells. PF match (PFm) cells indicate by comparison of PFmem with IT whether the target is actively encoded in IT. Feedback from PFmem to IT increases the strength of all features in IT matching the template. Feedback from IT to V4 sends the information about the target downwards to cells with a higher spatial tuning. FEF visuomovement (FEFv) cells combine the feature information across all dimensions and indicate salient or relevant locations in the scene. The increasing activity of a FEF movement (FEFm) cell is sent into extrastriate visual areas and facilitates the processing within its movement field. The IOR map memorizes recently visited locations and inhibits the FEF visuomovement cells.

(Baddeley, 1986). In addition, we simulate a map for the inhibition of return (IOR) which might be located in posterior parietal areas.

We investigate the putative interplay of the saccade system onto visual perception and memory in a change detection task. Change detection and transsaccadic scene perception requires the storage of visual information in visual short-term memory (VSTM; Irwin, 1992) and the use of the memorized pattern to indicate the change.

If we present a visual scene to the model, features such as colour, intensity, and orientation are computed in different V4 maps. The fact that features that are unique in their environment ''pop-out'' is to a first degree achieved by computing center-surround differences. V4 cells compete for representing their encoded stimuli. This competition is biased by feature-specific reentry from IT cells and spatial reentry from the frontal eye field movement cells. The growing receptive filed size along the processing hierarchy requires that a number of V4 cells encoding identical stimuli at different locations project to a single IT cell. For simplicity the type of features in V4 and IT is identical. IT receives feature-specific reentry from the prefrontal memory cells and location-specific reentry from the frontal eye field movement cells (Hamker, 2003).

Initially, the prefrontal memory is empty. The model is set to memorize a pattern defined by the cells with the strongest activity in IT just prior to each saccade. At this time the dynamic representation in IT is strongly influenced by the occulomotor system, since the movement cell activity provides a delayed reentry signal to extrastriate areas such as V4 and IT. The assumption of such a reentry signal is supported by the finding that FEF microstimulation modulates V4 activity (Moore & Armstrong, 2003).

The planning of the eye movement is implemented as follows. The FEF visuomovement neurons receive convergent afferents from V4 and IT and add up the activity across all dimensions (colour, intensity, orientation). The information from the target template hold in PF memory cells additionally enhances the locations that result in a match between target and encoded feature. This allows the biasing of specific locations by the joint probability that the searched features are encoded at a certain location. The firing rate of FEF visuomovement cells represent the saliency and task relevance of locations. The effect of the FEF visuomovement cells on the FEF movement cells is a feedforward excitation and surround inhibition. Thus, by increasing their activity slowly over time FEF movement cells compete for the selection of the strongest location. If a FEF movement cell exceeds a threshold, an eye movement is indicated towards the location of the centre of gravity of movement cell activity.

There is currently no clear indication where cells that implement a dynamic memory for an inhibition of return are located. We assume that each eye movement triggers a charge of IOR cells at the location of saccade target. This causes a suppression of the recent saccade target in the visuomovement cells. IOR cells slowly decay.

The model "prefrontal cortex" serves for two major functions, memorizing a pattern in PF memory cells and indicating a match of the incoming pattern with the memorized pattern in PF match cells. Cells that memorize patterns are known to exist in prefrontal areas. In studies using a delayed match-to-sample task elevated firing rates within specific populations have been reported during the delay (Miller, Erickson, & Desimone, 1996). The activity of such populations in the prefrontal cortex is robust to interfering stimuli. Other prefrontal cells indicate a match detection in a delayed match-to-category task (Freedman, Riesenhuber, Poggio, & Miller, 2002), which is similar to our PF match cells. The capacity of transsaccadic memory has been shown to comprise about 3–4 items and its content is slow-decaying (Irwin, 1992). An item seems to be an integrated object representation, which comprises identity, colour, and location (Irwin, 1992). The present model capacity is only a single item. This simplification does not allow us to make predictions about the circumstances that may define the size of memory. However, it does allow us to simulate perception of an object at the saccade target.

This model of visual perception has been developed to be consistent with a range of electrophysiological findings. The model "V4 area" (Hamker, 2004b) has been demonstrated to quantitatively account for receptive field competition in V4 (Reynolds, Chelazzi, & Desimone, 1999). A simplified version of the proposed systems model (Hamker, 2003) has been shown to match the time course of IT and V4 activity in visual search (Chelazzi, Duncan, Miller, & Desimone, 1998). The model is also consistent with findings in the FEF (Schall, 2002) and psychophysical data (Hamker & vanRullen, 2002).

We do not explicitly model the properties that emerge from the physical eye movement. During this period vision is impaired. V4 cells would receive input from a smeared retinal image and after the eyes land the new retinal image would mask the most recent representation. In an overt visual search task with a homogeneous background V4 cells have been shown to persist activity after saccade onset but the firing rate declines within 100 ms (Chelazzi, Miller, Duncan, & Desimone, 2001). Due to the large receptive fields and late position in the hierarchy, IT cells maintain activity during the movement and increase their rate after the eyes reach the target (Chelazzi et al., 1998). It seems that perception during a saccade is not necessarily interrupted. Although once the eye started sweeping across the scene the image in early visual areas is smeared and masked with the postsaccadic image, late processing areas could maintain the representation that is present just prior to the saccade in order to support a continuous percept.

The high foveal resolution of the object at the target location after the saccade allows a preferred processing (Rolls & Tovee, 1995). Any form of spatial attention that might be present initially after the eyes land cannot be dissociated from this fovea effect. Since we do not model the retinal resolution and cortical magnification we use the decline phase of the reentry signal to simulate the condition after eye movement. The decreasing reentry signal enhances sensory processing of the object at the target location for a brief period until it vanishes.

EXPERIMENT

Henderson and Hollingworth (2003) instructed participants to observe complex scenes to prepare for a later memory test. Participants had been told that they were shown two scenes: One original and a slightly modified version. They would have to indicate whether the original and the modified scene differ. More important is the additional instruction according to which they have to immediately report object changes while studying the original scene. Here we are only interested in the deletion condition. Objects that were deleted had been placed in an uncluttered region of the scene. If a subject's gaze crossed a boundary around such an object it was removed by displaying a changed scene.

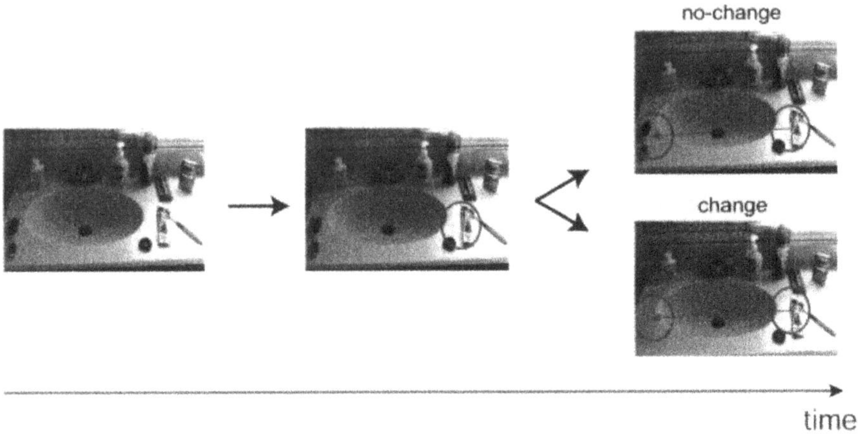

Figure 2. Experimental paradigm to simulate change detection of objects at the saccade target. A natural scene is presented to the model. Initially, the selection of the saccade target location is only guided by stimulus-driven saliency. The most salient location is at the toothpaste. In the present simulations we use the second saccade, which is directed towards the hairbrush to investigate change detection. At the time the FEF movement cells indicate the onset of a saccade the scene either remains identical (no-change condition) or the hairbrush is removed (change condition). The activity within the model is not reset during the change of the scenes.

This experiment is comparable to a free viewing task, in which the model's gaze is only driven by the stimulus saliency. The model scans the scene by visiting one object after another (Hamker & Worcester, 2002). We set up the model to memorize the strongest IT pattern in PF memory cells just prior to each saccade. We use this model's behaviour for change detection (Figure 2). Once a location is selected for an eye movement we either do not change the scene or we remove the object at the saccade target. Thus, we receive simulation data for no-change and change trials. The PF match cells indicate if the model finds the memorized pattern in the scene.

RESULTS

We first demonstrate how the model operates in a free viewing task and then demonstrate change detection. Figure 3 shows the FEF visuomovement and movement cell activity around the time the model selects the hairbrush. FEF visuomovement activity can be interpreted as a measure of the strength of processing at each location since visuomovement cells receive their main input from V4 and IT and average across all dimensions. Prior to each saccade all locations with objects are processed in parallel and represent potential targets. The movement cells read out the information from the visuomovement cells and select the target location for the eye movement. Thirty ms prior to saccade, the

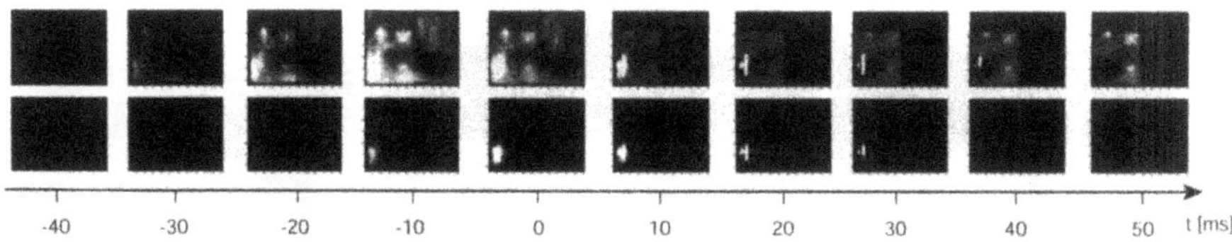

Figure 3. Activity of the FEF visuomovement (top) and movement (bottom) cells around the time of saccade begin (0 ms). Each image is a snapshot of the activity at the time indicated on the axis. The space within each image refers to the scene shown in Figure 2. The planning of the saccade is a dynamic process. Visuomovement cells reflect areas of high processing in V4 and IT and thus potential target locations for saccadic eye movements. The area around the location of the most recent eye movement is inhibited by cells from the IOR map. Movement cells determine the saccade target and provide a reentry signal to enhance sensory processing in V4 and IT. Spatial reentry emerges earlier than 30 ms prior to the saccade and is most prominent at the beginning of the eye movement.

model shows a preferred processing around the planned target location. This effect emerges from the reentry of movement cell activity into V4 and IT. After target selection the IOR cells get charged by the movement cells and inhibit visuomovement cells with receptive fields around the target location. The movement cell activity gets suppressed by FEF fixation cells, that are activated after the eye movement.

We have explained the mechanisms of overt and covert attention in the model and demonstrated the models performance. Regarding change detection the model has to recognize whether the object at the target location is removed or not. As we have explained in the model section the PF memory cells memorize the strongest pattern active in IT just prior to saccade selection. Objects at the target location are most likely to be memorized since the reentry signal enhances their sensory processing (Hamker, 2004a). Once a pattern is in short-term memory it biases competition among IT cells by a feature-specific reentry signal. This form of feature-based attention represents a search process for a specific pattern in the scene. PF match cells indicate by their activity if the search template hold in PF memory cells is currently encoded in the scene (Figure 1). We expect that the removal of an object results in a decline of activity in IT cells which encode properties of this object. Figure 4A shows this for the intensity channel in the model. The dark areas of the hairbrush located in the lower left part of the scene (Figure 2) are represented by cells on the left side of the feature space (cells 2–3 in Figure 4A). Although the receptive fields of IT cells cover the full scene the model dynamics at the time of the eye movement ensures that the removal of the hairbrush results in a decline of representative IT cells, even if the scene contains other dark areas elsewhere. The higher firing of IT cells (Figure 4A) in the no-change condition relative to the change condition is reflected by light areas in the difference signal. Thus, the visual persistence during the eye movement encoded by IT cells reflects the recently attended item.

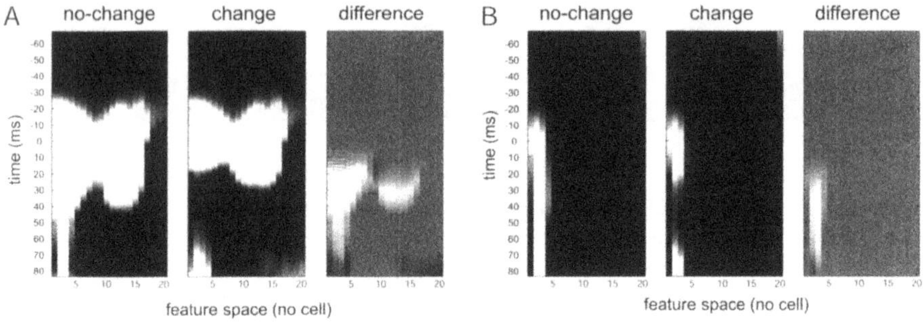

Figure 4. Comparison of IT and PF match cell activity in the no-change and change condition. (A) IT cell activity in the intensity channel. The model encodes intensity information by a population of 20 cells. Cells with lower numbers are selective for dark patches and cells with high numbers for light areas. We simulate nine populations of IT cells with different receptive fields (RF). In order the present IT activity across the full scene we only plot the maximal activity regardless of RF location. 50 ms prior to saccade IT activity is very low since inhibitory cell populations with a larger time constant suppressed IT activity after a strong firing due to the most recent eye movement. IT cells increase again their activity and initially encode the scene in parallel. Due to the spatial reentry signal they focus their processing at the saccade target by suppressing IT cells with RFs that do not contain the saccade target (not shown). The removal of the object results in a decline of the activity of cells encoding the objects properties. By taking a difference between the no-change and change conditions we see that the no-change condition is reflected by a stronger activity of several IT cells (grey indicates no difference). (B) PF match cell activity in the intensity channel. Despite the obvious difference in IT activity, change detection is performed by searching for a target template hold in PF memory cells. In this case the model selects the dark patches of the hairbrush as a search template since the IT cells encoding this feature fired most prominent prior to the eye movement. After the eye movement the model continues to search for this pattern. PF match cells indicate a successful search by their activity since they only fire by the joint activity of a PF memory and respective IT cell.

The detection of a change, however, has to be an active process, which requires the match of IT activity with the memorized target template hold in PF memory cells. In the intensity channel the model has memorized a pattern that represents the dark spots of the hairbrush. Thus, in the no-change condition the PF match cells (Figure 4B) continue to fire around the time of the eye movement but in the change condition their activity declines.

PREDICTIONS

First of all the model predicts that an object change at the saccade target location can be well detected. The reentry signal that necessarily occurs at the saccade target location enhances sensory processing of objects in its vicinity and increases their probability to be memorized. The model also predicts that object properties can be reported when the scene disappeares before the eyes land at the target location. Irwin & Zelinsky (2002) found a high accuracy of report even when the object had never been foveated. Thus, object identification begins before the eyes actually fixate on the object.

Visual stability might be achieved through continuity. Before the eye movement spatial reentry generates a representation that prepares for the one that can be expected after the eye movement. Processing capacity is restricted to the same issues just prior, during and just after the eye movement. Thus, parallel changes elsewhere in the scene are likely to go unnoticed.

The model also predicts that change detection is an active process that requires to actively maintain a search template while the eyes sweep across the scene. Thus, eye movements recruit resources that could interfere with other tasks to be done in parallel. For example, it has been shown that mental rotation (Irwin & Brockmole, 2000) and shifts between global and local judgements (Brockmole, Carlson, & Irwin, 2002) are impaired during saccadic eye movements.

For simplicity the model searches for the same pattern it has memorized prior to the saccade. One can only speculate about the exact templates used for identification. It is likely that the brain generates a prediction instead of a copy. This makes it difficult to predict the search template used in experiments. Henderson and Hollingworth (2003) investigated the effect of deletion as well as type and token changes. The detection rates for type and token changes decrease with the length of the saccade. Since the spatial resolution is very poor in the periphery, and predictions are difficult to make based on very little information, we would expect a slower decline if changes do not affect low frequency components and the sign of contrast.

DISCUSSION

The perception of a visually stable environment during eye movements is an ongoing debate. Although perception during eye movements is strongly impaired we do not perceive the world in snapshots. Von Helmholtz (1896) argued that image motion during the eye movement could be sensed to compensate for the shift in image position. Extraretinal theories postulate an extraretinal signal that remaps the full scene during each saccade in order to achieve a space-constant representation. It was suggested that the motor-related signal of saccades can be sent back as a "corollary discharge" (Sperry, 1950) or "efference copy" (von Holst & Mittelstaedt, 1950). However, the origin and type of such an extraretinal signal has remained unclear. Perisaccadic mislocalization experiments in total darkness lead to a hypothetical eye-position signal that starts to rise 200 ms before the saccade begins and ends after the eyes have landed (Schlag & Schlag-Rey, 2002). Remapping of receptive fields has been found in LIP (Kusunoki & Goldberg, 2003), the superior colliculus (Walker, Fitzgibbon, & Goldberg, 1995), and the frontal eye field (Umeno & Goldberg, 1997). However, it is not clear if this predictive shift of receptive fields contributes to visual stability or primarliy updates the occulomotor system.

Alternatively to the visual remapping hypothesis a link between the representation before and after the saccade could be established by the saccadic target object or nearby landmarks (Currie et al., 2000; Deubel, Schneider, & Bridgeman, 2002; McConkie & Currie, 1996). Locating this object at the onset of the new fixation would be sufficient for relating the contents of the two representations with each other (Bridgeman, van der Heijden, & Velichkovsky, 1994).

Our model is consistent with the saccade target theory of visual stability. However, since the spatial reentry signal from the movement cells can be defined as an extraretinal eye position signal our model indicates a possible relation of the saccade target theory to extraretinal theories. We predict a modulatory signal that is directed to the saccade target. This signal could be responsible for the reported mislocalization of briefly flashed stimuli towards the endpoint of the saccade in illuminated rooms (Lappe, Awater, & Krekelberg, 2000; Ross, Morrone, & Burr, 1997).

In extension to a mere theory the presented model helps to reveal the underlying processes and illuminates those effects often referred to as attention. Early perception is primarily feedforward driven and meets high-level expectations while late perception emphasizes the features of interest and implements focal processing (Hamker, 2004a). We suggest that competition and cooperation implements a dynamic filter that allows the brain to connect planning processes with the physical world. The neural correlate of visual stability across eye movements could have its origin in spatial reentry. Perception just prior to a saccade is focused onto the saccade target and its vicinity and allows to make a hypothesis about the situation after saccade. The testing of this hypothesis links cognition before and after the saccade with each other which might be crucial for perceiving stability. This definition of vision as an active process does not require a stable space-based representation around the eye movement.

Regarding change detection the saccade target theory refers to attention for mediating the processes of object selection and memorization. Our model provides a testable explanation about the underlying neural processes. Spatial reentry (e.g., from the FEF movement cells) provides an enhanced sensory processing of the item at the saccade target. It becomes the object of interest, it is more likely to be memorized and its representation can be used to make an estimate of the situation after the eye movement. After the saccade the generated search template allows a comparison with a high-level representation such as in area IT. Similar as suggested by the coherence theory of Rensink (2000) our model demonstrates that an all-purpose representation is not necessary. The representation needed for the task at hand is dynamically created.

This idea receives further support from IT cell recordings during search for familiar objects in natural scenes. Sheinberg and Logothetis (2001) compared trials in which a familiar object was present but the eye movement was directed elsewhere with trials in which the monkey directed his eye towards the familiar object. They found significantly different target-selective responses 95 ms before the eyes move. This provides strong evidence for our assumption that object analysis occurs even before the object is fixated. Present familiar objects that are not the goal of the next fixation show no perisaccadic activity. Our model explains this effect primarily based on the spatial reentry signal, which is under normal conditions directed to the saccade target location.

Limitations of the model

We do not model the properties that emerge through the physical movement of the eyes and the decrease of spatial resolution from the fovea to the periphery. Thus, we are unable to make predictions about the period of time from the beginning of the eye movement to the new fixation and ignore this period in the simulation. We think that our assumptions about the conditions after saccade are plausible. The objects which appear in the fovea will receive an enhanced sensory processing such that an hypothesis testing is possible without engaging in a detailed search to locate the object of interest.

Our search template contains simple object features. Thus, change detection is only possible if these features are unique in the near environment of the saccade. The grouping of complex parts into some object description is not implemented in the present version. However, we have demonstrated that the dynamics of the model suppress the same features elsewhere in the scene. The limitation due to the simple feature spaces used does not seem to affect the line of arguments.

Studies suggest that approximately four visual objects or six spatial locations can be stored in VSTM (Luck & Vogel, 1997; Pashler, 1988). The present model makes very little assumptions about memory. We have implemented a memory with a single item capacity. However, even sophisticated models of spatial working memory are limited to a single entry (Compte, Brunel, Goldman-Rakic, & Wang, 2000). This limitation is inherent to simple continuous attractor neural networks that implement a population-based winner-take-all process (Amari & Arbib, 1977; Deneve, Latham, & Pouget, 1999). The assumption of a fixed number of storage places with a single item capacity is probably too simple, since the capacity varies due to experimental conditions. Alternatively short-term memory could store patterns with nested oscillations in which memory patterns repeat on each low-frequency (5–12 Hz) oscillation (Lisman & Idiart, 1995). If we would extend the model to have a capacity of four items it would still predict that the saccade target is special. However, it would allow us to investigate more subtle aspects of visual salience and task relevance of objects at other locations. Another factor that limits present neural models of working memory is that we did not understand sufficiently well how the brain defines an object as compared to a collection of features. Working memory seems to encode integrated objects (Luck & Vogel, 1997). Although our model of working memory is relatively simple it is sufficient to study the interaction of perception and memory. In future versions it would be interesting to extend the model in order to provide predictions for the controversy about the amount of information that can be maintained during consecutive eye movements (Henderson & Hollingworth, 2003; Irwin & Zelinsky, 2002). Despite these limitations the presented model is a novel and very comprehensive computational account for explaining the role of covert and overt attention in change detection. Earlier computational models have typically focused on either recognition and attention or memory alone. Our model combines these three aspects of perception. It helps to explore the underlying neural mechanisms on the systems level that cannot be addressed at the level of more abstract theories.

REFERENCES

Amari, S., & Arbib, M. A. (1977). Competition and cooperation in neural nets. In J. Metzler (Ed.), *Systems neuroscience* (pp. 119–165). San Diego, CA: Academic Press.

Backer, G., Mertsching, B., & Bollmann, M. (2001). Data- and model-driven gaze control for an active-vision system. *IEEE Transactions on Pattern Analysis and Machine Intelligence, 23*, 1415–1429.

Baddeley, A. D. (1986). *Working memory.* Oxford, UK: Oxford University Press/Clarendon Press.

Bridgeman, B., van der Heijden, A., & Velichkovsky, B. (1994). A theory of visual stability across saccadic eye movements. *Behavioral and Brain Sciences, 17*, 247–292.

Brockmole, J. R., Carlson, L. A., & Irwin, D. E. (2002). Inhibition of attended processing during saccadic eye movements. *Perception and Psychophysics, 64*, 867–881.

Chelazzi, L., Duncan, J., Miller, E. K., & Desimone, R. (1998). Responses of neurons in inferior temporal cortex during memory-guided visual search. *Journal of Neurophysiology, 80*, 2918–2940.

Chelazzi, L., Miller, E. K., Duncan, J., & Desimone, R. (2001). Responses of neurons in macaque area V4 during memory guided visual search. *Cerebral Cortex, 11*, 761–772.

Compte, A., Brunel, N., Goldman-Rakic, P. S., & Wang, X. J. (2000). Synaptic mechanisms and network dynamics underlying spatial working memory in a cortical network model. *Cerebral Cortex, 10*, 910–923.

Currie, C. B., McConkie, G. W., Carlson-Radvansky, L. A., & Irwin, D. E. (2000). The role of the saccade target object in the perception of a visually stable world. *Perception and Psychophysics, 62*, 673–683.

Deneve, S., Latham, P. E., & Pouget, A. (1999). Reading population codes: A neural implementation of ideal observers. *Nature Neuroscience, 2*, 740–745.

Desimone, R., & Duncan, J. (1995). Neural mechanisms of selective attention. *Annual Review of Neuroscience, 18*, 193–222.

Deubel, H., & Schneider, W. X. (1996). Saccade target selection and object recognition: Evidence for a common attentional mechanism. *Vision Research, 36*, 1827–1837.

Deubel, H., Schneider, W. X., & Bridgeman, B. (2002). Transsaccadic memory of position and form. *Progress in Brain Research, 140*, 165–180.

Findlay, J. M., & Walker, R. (1999). A model of saccade generation based on parallel processing and competitive inhibition. *Behavioral and Brain Sciences, 22*, 661–674.

Freedman, D. J., Riesenhuber, M., Poggio, T., & Miller, E. K. (2002). Visual catagorization and the primate prefrontal cortex: Neurophysiology and behavior. *Journal of Neurophysiology, 88*, 929–941.

Hamker, F. H. (2003). The reentry hypothesis: Linking eye movements to visual perception. *Journal of Vision, 11*, 808–816.

Hamker, F. H. (2004a). A dynamic model of how feature cues guide spatial attention. *Vision Research, 44*, 501–521.

Hamker, F. H. (2004b). Predictions of a model of spatial attention using sum- and max-pooling functions. *Neurocomputing, 56C*, 329–343.

Hamker, F. H. (2005). How the detection of objects in natural scenes constrains attention in time. In L. Itti, G. Rees, & J. Tsotsos (Eds.), *Neurobiology of attention*. Amsterdam: Elsevier.

Hamker, F. H., & VanRullen, R. (2002). The time course of attentional selection among competing locations [Abstract]. *Journal of Vision, 2*(7), 7a.

Hamker, F. H., & Worcester, J. (2002). Object detection in natural scenes by feedback. In H. H. Bülthoff, S.-W. Lee, T. A. Poggio, & C. Wallraven (Eds.), *Biologically motivated computer vision* (pp. 398–407). New York: Springer.

Henderson, J. M., & Hollingworth, A. (2003). Eye movements and visual memory: Detecting changes to saccade targets in scenes. *Perception and Psychophysics, 65*, 58–71.

Irwin, D. E. (1992). Memory for position and identity across eye movements. *Journal of Experimental Psychology: Learning, Memory, and Cognition, 18*, 307–317.

Irwin, D. E., & Brockmole, J. R. (2000). Mental rotation is suppressed during saccadic eye movements. *Psychonomic Bulletin and Review, 7*, 654–661.

Irwin, D. E., & Zelinsky, G. J. (2002). Eye movements and scene perception: Memory for things observed. *Perception and Psychophysics, 64*, 882–895.

Itti, L., & Koch, C. (2000). A saliency-based search mechanism for overt and covert shifts of visual attention. *Vision Research, 40*, 1489–1506.

Itti, L., Koch, C., & Niebur, E. (1998). A model of saliency-based visual attention for rapid scene analysis. *IEEE Transactions on Pattern Analysis and Machine Intelligence, 20*, 1254–1259.

Kowler, E., Anderson, E., Dosher, B., & Blaser, E. (1995). The role of attention in the programming of saccades. *Vision Research, 35*, 1897–1916.

Krieger, G., Rentschler, I., Hauske, G., Schill, K., & Zetzsche, C. (2000). Object and scene analysis by saccadic eye-movements: An investigation with higher-order statistics. *Spatial Vision, 13*, 201–214.

Kusunoki, M., & Goldberg, M. E. (2003). The time course of perisaccadic receptive field shifts in the lateral intraparietal area of the monkey. *Journal of Neurophysiology, 89*, 1519–1527.

Lappe, M., Awater, H., & Krekelberger, B. (2000). Postsaccadic visual references generate presaccadic compression of space. *Nature, 403*, 892–895.

Lisman, J. E., & Idiart, M. A. (1995). Storage of 7 +/– 2 short-term memories in oscillatory subcycles. *Science, 267*, 1512–1515.

Luck, S. J., & Vogel, E. K. (1997). The capacity of visual working memory for features and conjunctions. *Nature, 390*, 279–281.

McConkie, G. W., & Currie, C. B. (1996). Visual stability across saccades while viewing complex pictures. *Journal of Experimental Psychology: Human Perception and Performance, 22*, 563–581.

Miller, E. K., Erickson, C. A., & Desimone, R. (1996). Neural and mechanisms of visual working memory in prefrontal cortex of the macaque. *Journal of Neuroscience, 16*, 5154–5167.

Moore, T., & Armstrong, K. M. (2003). Selective gating of visual signals by microstimulation of frontal cortex. *Nature, 421*, 370–373.

Pashler, H. (1988). Familiarity and the detection of change in visual displays. *Perception and Psychophysics, 44*, 369–378.

Rensink, R. A. (2000). The dynamic representation of scenes. *Visual Cognition, 7*(1/2/3), 17–42.

Rensink, R. A. (2002). Change detection. *Annual Review of Psychology, 53*, 245–277.

Rao, R. P., Zelinsky, G. J., Hayhoe, M. M., & Ballard, D. H. (2002). Eye movements in iconic visual search. *Vision Research, 42*, 1447–1463.

Reynolds, J. H., Chelazzi, L., & Desimone, R. (1999). Competitive mechanism subserve attention in macaque areas V2 and V4. *Journal of Neuroscience, 19*, 1736–1753.

Rizzolatti, G., Riggio, L., & Sheliga, B. M. (1994). Space and selective attention. In C. Umiltà & M. Moscovitch (Eds.), *Attention and performance XV: Conscious and nonconscious information processing* (pp. 231–265). Cambridges MA: MIT Press.

Rolls, E. T., & Tovee, M. J. (1995). The responses of single neurons in the temporal visual cortical areas of the macaque when more than one stimulus is present in the receptive field. *Experimental Brain Research, 103*, 409–420.

Ross, J., Morrone, M. C., & Burr, D. C. (1997). Compression of visual space before saccades. *Nature, 386*, 598–601.

Schall, J. D. (2002). The neural selection and control of saccades by the frontal eye field. *Philosophical Transactions of the Royal Society of London, Series B, 357*, 1073–1082.

Schlag, J., & Schlag-Rey, M. (2002). Through the eye, slowly: Delays and localization errors in the visual system. *Nature Reviews: Neuroscience, 3*, 191–215.

Sheinberg, D. L., & Logothetis, N. K. (2001). Noticing familiar objects in real world scenes: The role of temporal cortical neurons in natural vision. *Journal of Neuroscience, 21*, 1340–1350.

Sperry, R. W. (1950). Neural basis of the spontaneous optokinetic response produced by visual inversion. *Journal Comparative Physiology and Psychology, 43*, 482–489.

Unemo, M. M., Goldberg, M. E. (1997). Spatial processing in the monkey frontal eye field: I. Predictive visual responses. *Journal of Neurophysiology, 78*, 1373–1383.

Von Helmholtz, Hermann (1896). *Handbuch der Physiologischen Optik.* Hamburg-Leipzig, Germany: Leopold Voss.

Von Holst, E., & Mittelstaedt, H. (1950). Das Reafferenz-Prinzip. *Die Naturwissenschaft, 37*, 464–476.

Walker, M. F., Fitzgibbon, E. J., & Goldberg, M. E. (1995). Neurons in the monkey superior colliculus predict the visual result of impending saccadic eye movements. *Journal of Neurophysiology, 73*, 1988–2003.

APPENDIX

We now give a formal description of the model. The model can be divided into a static preprocessing part that computes stimulus-driven saliency and a dynamic part that simulates the temporal dynamics of perception.

Preprocessing and saliency. Starting from the colour image, we extract orientation maps $O(\sigma, \theta)$ with varying resolution σ and orientation θ, intensity I, red–green $RG = R - G$, and blue–yellow $BY = B - Y$ maps (Itti, Koch, & Niebur, 1998).

The conspicuity of each feature is implemented by centre-surround operations "$\ominus$". We construct orientation contrast $\mathcal{O}(c, s, \theta)$, intensity contrast $\mathcal{I}(c)$, as well as red–green $\mathcal{RG}(c)$ and blue–yellow $\mathcal{BY}(c)$ double opponency (Itti et al., 1998). For each feature, we combine the feature information into an attribute $\mathbf{V}$ and its corresponding contrast value into a gain factor P of a population code. This dual coding principle is a very important characteristic. A feature is presented by the location of cell activity, and the conspicuity of this feature is represented by the strength of activity. At each location x_1, x_2 we construct a space, whose axes are defined by the represented features and by one additional conspicuity axis. The population is then defined by a set of neurons $i \in N$ sampling the feature space, with each neuron tuned around its preferred value $\mathbf{u}_i$. For each neuron y_i we obtain an activity value:

$$y_i = P \cdot g(\mathrm{u}_i - \mathrm{V}) \tag{1}$$

Specifically we use a Gaussian tuning curve with the selectivity parameter σ_g:

$$g(\mathbf{u}_i - \mathbf{V}) = \exp\left(-\frac{\|\mathbf{u}_i - \mathbf{V}\|^2}{\sigma_g^2}\right) \tag{2}$$

To apply the same range of selectivity parameters $\sigma_g^2 \in \{0.05 \ldots 0.2\}$ for all channels we normalize the feature values $\mathrm{V} \in \{I, RG, BY, \theta, \sigma\}$ of each channel between zero and one. The cell activity within the population should typically lie within the range of zero and one. Thus, we also normalize the contrast values to $\widetilde{\mathcal{I}}, \widetilde{\mathcal{RG}}, \widetilde{\mathcal{BY}}, \widetilde{\mathcal{O}}$. We finally receive the populations for each channel with scale c at each location x:

$$\begin{aligned}
y_i^I(c,\mathrm{x}) &= \widetilde{\mathcal{I}}(c,\mathrm{x}) \cdot g(u_i - I(c,\mathrm{x})) \\
y_i^{RG}(c,\mathrm{x}) &= \widetilde{\mathcal{RG}}(c,\mathrm{x}) \cdot g(u_i - RG(c,\mathrm{x})) \\
y_i^{BY}(c,\mathrm{x}) &= \widetilde{\mathcal{BY}}(c,\mathrm{x}) \cdot g(u_i - BY(c,\mathrm{x})) \\
y_i^{\theta}(c,\mathrm{x}) &= \max_{\theta}\left(\widetilde{\mathcal{O}}(c,\theta,\mathrm{x}) \cdot g(u_i - \theta)\right) \\
y_i^{\sigma}(c,\mathrm{x}) &= \max_{\theta}\left(\widetilde{\mathcal{O}}(c,\theta,\mathrm{x}) \cdot g(u_i - \sigma)\right)
\end{aligned} \tag{3}$$

We now have $\#c$ maps, where $\#c$ is the number of center scales, with a population at each point x for a different centre scale c. To combine these maps across space into one map with the lowest resolution (highest c) we use a maximum operation ($\max_{c,\mathrm{x}' \in RF(\mathrm{x})}$).

Temporal dynamics. Each map in the model represents a functional area of the brain. It contains at each location x a population of i cells encoding feature values (Equation 4), with the exception of the maps in the frontal eye field and IOR that only encode space ($i = 1$). In addition V4 and IT have separate maps for different dimensions d (RG, BY, etc.). The population of cells is driven by its input $y^{\uparrow}_{d,i,\mathrm{x}}$. Feedback implements an input gain control to enhance the representation of certain features and biases the competition (Desimone & Duncan, 1995) among active populations. Feature-specific reentry operates within the models "ventral pathway" and enhances cell populations whose input matches the feedback signal. Spatial reentry arrives from the frontal eye field and boosts features at a certain location, generally the target of the next soccade. $I^{f\,inh}$ *induces competition among cells and* I^{inh} causes a normalization and saturation. Both terms have a strong short range and weak long range inhibitory effect.

$$\tau \frac{d}{dt} y_{d,i,\mathrm{x}} = y^{\uparrow}_{d,i,\mathrm{x}} + I^L + I^G - y_{d,i,\mathrm{x}} \cdot I^{inh}_{d,\mathrm{x}} - I^{f\,inh}_{d,\mathrm{x}} \tag{4}$$

The following maps use implementations of the general equation quoted above (Equation 4).

V4. Each V4 layer receives input from a different dimension (d) in the feature conspicuity maps: $y^{\theta}_{i,\mathrm{x}}$ for orientation, $y^{I}_{i,\mathrm{x}}$ for intensity, $y^{RG}_{i,\mathrm{x}}$ for red–green opponency, $y^{BY}_{i,\mathrm{x}}$ for blue–yellow opponency and $y^{\sigma}_{i,\mathrm{x}}$ for spatial frequency. V4 cells receive feature specific feedback from IT cells ($I^L = I^L(y^{IT})$) and spatial reentry from the frontal eye field ($I^G = I^G(y^{FEFm})$).

IT. The populations from different locations in V4 project to IT, but only within the same dimension. We simulate a map containing nine populations with overlapping receptive fields. We do not increase the complexity of features from V4 to IT. Thus, our model IT populations represent the same feature space as our model V4 populations. The receptive field size, however, increases in our model, so that several populations in V4 converge onto one population in IT: $y^{\uparrow}_{i,d,\mathrm{x}} = w^{\uparrow} \max_{\mathrm{x}' \in RF(\mathrm{x})} y^{\mathrm{V4}}_{i,d,\mathrm{x}'}$. IT receives feature-specific feedback from the prefrontal memory ($I^L = I^L(y^{PFmem})$) and location-specific feedback from the frontal eye field ($I^G = I^G(y^{FFEm})$).

PFmem. The memorization of pattern is achieved through recurrent excitation. Whether a pattern should be memorized depends on the task. Thus, we define a variable $I^{\text{store}}(t) \in \{0, 1\}$ that controls the inhibition and determines that a pattern is memorized prior to each saccade. The first and strongest pattern closes the gate for other patterns.

$$
\begin{aligned}
I_d^{inh}(t) &= w_{inh} \sum_j y_{d,j}(t) \\
I_d^{f\,inh}(t) &= (0.7 - 0.6 \cdot I^{\text{store}}) \sum_j y_{d,j}(t) \\
y_{d,i}^{\uparrow}(t) &= \sigma(\Gamma_{mem} - w^{\text{cue}} \max_j y_{d,j}(t)) \max_{\mathrm{x}} \sigma(y_{d,i,\mathrm{x}}^{\text{IT}}(t) - C); \\
\sigma(a) &= max(a, 0)
\end{aligned}
\tag{5}
$$

PFm. To determine whether a pattern in the visual scene is similar to the pattern in memory we define PF match cells which compare the current pattern in IT cells with those in PF memory. Activity rises in PF match cells only if both populations match.

$$
\begin{aligned}
I_d^{inh}(t) &= w_{inh} \sum_j y_{d,j}(t) \\
I_d^{f\,inh}(t) &= w_{f\,inh} \sum_j y_{d,j}(t) \\
y_{d,i}^{\uparrow}(t) &:= w^{\uparrow} y_{d,i}^{\text{PFmem}}(t) \max_{\mathrm{x}} y_{d,i,\mathrm{x}}^{\text{IT}}(t)
\end{aligned}
\tag{6}
$$

FEFv. The visuomovement neurons receive convergent afferents from V4 and IT $y_{\mathrm{x}}^{\uparrow a} = w^{\text{V4}} \sum_d \max_i y_{d,i,\mathrm{x}}^{\text{V4}} + w^{\text{IT}} \sum_d \max_{i,\mathrm{x}' \in RF(\mathrm{x})} y_{d,i,\mathrm{x}'}^{\text{IT}} + w^{\text{FEFm}} \; y^{\text{FEFm}}$. The information from the target template additionally enhances the locations that result in a match between target and encoded feature $y_{\mathrm{x}}^{\uparrow b} = w^{\text{PFmem}} \prod_d \max_i y_{d,i}^{\text{PFmem}} \; y_{d,i,\mathrm{x}}^{\text{V4}}$ at all locations simultaneously. This allows the biasing of specific locations by the joint probability that the searched features are encoded at a certain location.

FEFm. The effect of the visuomovement cells on the movement cells is a feedforward excitation and surround inhibition: $y_{\mathrm{x}}^{\uparrow} = w^{\text{FEFv}} \; y_{\mathrm{x}}^{\text{FEFv}} - w^{\text{FEFf}} y^{\text{FEFf}} - w_{inh}^{\text{FEFv}} \sum y_{\mathrm{x}}^{\text{FEFv}}$. Thus, by increasing their activity slowly over time premotor cells compete for the selection of the strongest location.

FEFf. After the FEFm cells exceed a threshold Γ_o^{FEF} and initiate an eye movement at the time t_0, we assume that the FEFf cells receive input for a brief period T^{SAC} and then decay.

$$
\begin{aligned}
\tau \frac{d}{dt} y(t) &= w^{\text{FEF}} \; I^{\text{FEF}}(t) - y(t) \\
I^{\text{FEF}}(t) &= \begin{cases} 1 \; if & y^{\text{FEFm}}(t_0) > \Gamma_o^{\text{FEF}} \quad t < t_0 + T^{\text{SAC}} \\ 0 \; else & \end{cases}
\end{aligned}
\tag{7}
$$

IOR. There is currently no clear indication where cells that ensure an inhibition of return are located. We regard each location x as inspected, dependent on the selection of an eye movement at $y_{\mathrm{x}}^{\text{FEFm}}(t_e) > \Gamma_o^{\text{FEF}}$ or when a match in the PFm cells is lost. In this case the IOR cells are charged at the location of the strongest FEFm cell for a period of time T^{IOR}. This causes a suppression of the recently attended location in the FEFv map. IOR cells get slowly discharged by decay with a low weight w_{inh}.

$$
\begin{aligned}
\tau \frac{d}{dt} y_{\mathrm{x}}^{\text{IOR}} &= (1 - y_{\mathrm{x}}^{\text{IOR}})(w^{\text{FEFm}} \; I_{\mathrm{x}}^{\text{FEFm}} - w_{inh} y_{\mathrm{x}}^{\text{IOR}}) \\
I_{\mathrm{x}}^{\text{FEFm}} &= \begin{cases} \exp - \dfrac{(\mathrm{x} - \mathrm{x_m})^2}{0.01} & if \quad t < t_e + T^{\text{IOR}} \\ 0 & else \end{cases} ; \; y_{\mathrm{x_m}}^{\text{FEFm}} = \max_{\mathrm{x}} (y_{\mathrm{x}}^{\text{FEFm}})
\end{aligned}
\tag{8}
$$

VISUAL COGNITION, 2005, 12 (6), 1177–1198

Infants' object category formation and use: Real-world context effects on category use in object processing

Gwenden L. Dueker
Grand Valley State University, Allendale, MI, USA
Amy Needham
Duke University, Durham, NC, USA

Three experiments investigated category formation based on real-world encounters with objects and how that category knowledge functioned as a top-down influence on infants' scene processing. 5-month-old infants received various experiences with exemplars of an object category and then were shown a display containing 2 adjacent novel objects, 1 of which was a novel exemplar of the category. If infants recognized this novel exemplar, they could use category information to determine the boundary between the 2 objects. Only infants who experienced object exemplars in situations that closely mimicked everyday infant experiences with objects successfully parsed the display. Results suggest that regularities in the contexts where infants experience similar objects help infants link their experiences and form object categories that can be used to segregate novel scenes. These results shed light on the real-world process of infant knowledge base formation and on how infants use their knowledge base to segment real-world scenes.

Recognition of familiar objects or types of objects in a novel scene helps infants and adults to organize and segregate complex visual information. Evidence for this top-down effect of prior knowledge on scene perception exists in both the adult (e.g., Bravo & Farid, 2003; Goldstone, Lippa, & Shiffrin, 2001) and infant (e.g., Dueker, Modi, & Needham, 2003; Needham & Baillargeon, 1998)

Please address all correspondence to: Gwenden Dueker, Assistant Professor of Psychology, Grand Valley State University, 1 Campus Drive, 2118 AuSable Hall, Allendale, MI 49401, USA. Email: duekerg@gvsu.edu

This research was supported by FIRST grant HD32129 and grant HD37049 from the NICHD to Amy Needham. We would like to thank Tracy Barrett, Eric Bracey, Stephanie DiGuiseppi, Brian Maxey, Ruth Ormsbee, Susan Ormsbee, Lubna Zaeem, and the undergraduate students (especially all the willing Boxer Ones) working in the Infant Perception Lab at Duke University for their help with the data collection. We would also like to thank the many parents and infants who generously spent their time participating in the studies reported here.

DOI:10.1080/13506280444000706

literatures. Recognizing objects facilitates segmentation of both known and unknown objects, because (1) subtracting out the known objects can simplify a scene and (2) where known and unknown objects intersect, recognizing where the familiar object ends can indicate where an unfamiliar object begins. This top-down effect on visual processing depends upon development of a knowledge base that includes, among other things, knowledge about object categories. Knowledge about the ''kinds'' or categories of objects that exist in the world is vital in assisting infants' parsing of complex visual scenes, but how do infants acquire the necessary knowledge base?

One way would be to retain an exhaustive catalogue of all the objects an infant has seen before, something that would tax even the memory system of an adult. Nonetheless, it appears that infants can remember having seen an object 24 hours before and can use that information to help them segment a novel scene containing that object (Dueker et al., 2003; Needham & Baillargeon, 1998). A knowledge base that would be both more useful and more viable than one that requires the representation of each object the infant has seen is one in which object *categories* are represented. Such a knowledge base would allow infants to generalize object boundary information not just to novel scenes involving a familiar object, but to scenes containing novel exemplars of a familiar category as well. There is also evidence that category-based knowledge about object boundaries remains available to support infant interpretations of novel scenes longer than experience with one of the actual objects present in the scene (Dueker et al., 2003). So, although we know that infant scene processing can benefit from the top-down effects of object category knowledge, what we do not yet know is how infants form object categories from their real-world experience of objects to use for this purpose.

INFANT CATEGORY FORMATION

Despite recent calls to shift the focus of infant categorization research from the presence or absence of categories at different ages to the process by which categories are formed (e.g., Cohen, 2003; Oakes & Madole, 2000, 2003), scientists still know relatively little about this process. Many studies have demonstrated that after seeing multiple examples of a similar kind of thing, even very young infants begin to treat those kinds of things equivalently (the definition of categorization; Mervis & Rosch, 1981) under highly controlled and simplified circumstances (e.g., Behl-Chadha, 1996; Bomba & Siqueland, 1983; Hayne, Rovee-Collier, & Perris, 1987; McGurk, 1972; Oakes & Spalding, 1997; Quinn & Johnson, 2000; Younger & Cohen, 1986) and that infants can use their category knowledge to guide the types of inferences they make in novel situations (e.g., Mandler & McDonough, 1996, 1998; Needham, Dueker, & Lockhead, 2005a). Typically the stimuli used are not real objects, but are pictures, models, or line drawings of objects, and the stimuli are usually presented in isolation. In essence, the researchers have done much of the work for the infants by choosing similar objects and grouping them together. It is doubtful that such experiences are common in the real-world experiences of infants. It is quite possible that these tightly controlled laboratory procedures might be under- or overestimating infants' performance in a real-world situation.

In the research presented here we study infants' visual scene perception using an object segregation task. We chose this particular task for two reasons. First, this task uses three-dimensional objects, making it similar to infants' everyday experiences of objects. Second, prior studies have established that infant perception of a novel test display can be affected by the infant's prior experiences with objects (Needham, 2001; Needham & Baillargeon, 1998; Needham et al., 2005a). However, as in other category research, all of those past studies presented these prior experiences with objects in highly artificial laboratory situations. The point of this set of studies is to learn whether infants could benefit from prior experiences with objects in more realistic situations/settings. First we will briefly summarize the prior research and then we will describe the ways we adjusted the procedures to better mimic infant real-world experiences.

Object segregation and prior experience

Object segregation is the process of determining where one object begins and another object ends. The object segregation task used in the present research presents infants with a display containing two objects, a tall thin blue box and a curved yellow cylinder, that are adjacent and touching one another (see Figure 1). The question of interest is whether infants perceive this display as one or two objects. Infants' segregation of the novel display is assessed by giving infants physical evidence about how many objects are part of the display and observing infant looking responses. The logic is that if the physical evidence of the number of objects in the display (one or two) contradicts infants' interpretation of that display, they will look longer at the unexpected event than at an analogous event depicting the expected composition.

Prior research with this particular display established that for 4- to 6-month-old infants, the composition of the display is ambiguous. That is, after seeing the stationary display, infants showed no clear expectation about whether there were one or two objects present (Needham, 1998; Needham & Baillargeon, 1998). However, a short prior exposure to either individual piece of the display facilitated infant segregation of the test display. Further, while prior experience with *one* object that was similar to an object in the display did not facilitate infant segregation of the test display (Needham, 2001), prior experience with a set of *three* objects similar to one of the objects in the test display did facilitate infant parsing of the test display (Needham et al., 2005a), even when that prior experience happened 72 hours before test (Dueker et al., 2003). Thus, prior experience allowed infants to add information to their knowledge base, in this

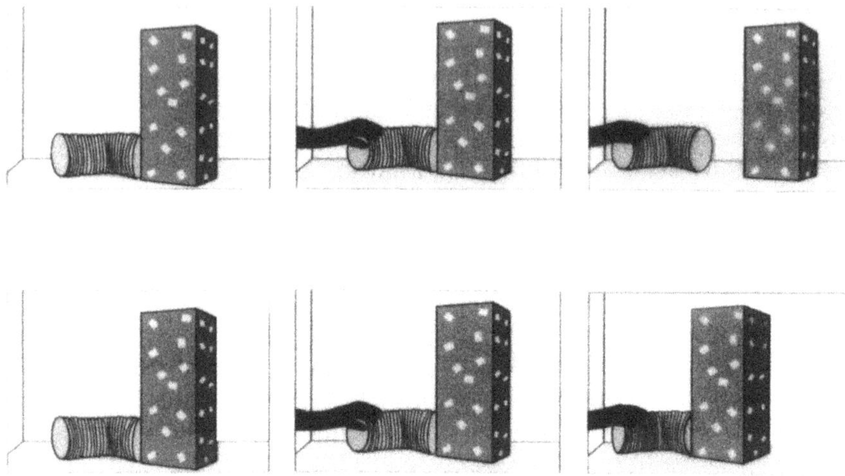

Figure 1. Test display and test events. Top: Move-apart event. Bottom: Move-together event.

case what the boundaries of a particular object or kind of object were, and to use that information later to segment the test display (Needham & Baillargeon, 1998; Needham & Modi, 1999; Needham et al., 2005a). These last sets of results indicate that infants can form object categories and generalize what they know to novel objects in a complex visual scene.

The object familiarization experience that infants were given in the previous experiments was quite artificial—all three similar objects were present in the same place at the same time in an experimental context where all other objects and distractions were excluded. However, it seems unlikely that infants often encounter such situations in their every day lives. Rather, it seems more likely that infants often encounter similar objects singly in different physical contexts. For example, an infant might see a dog in the front yard in the morning, visit her grandmother and see another dog, and still later encounter a dog while on a walk with her father. Prior results suggest that after seeing a few real dogs in a group (Needham et al., 2005a) or many pictures of dogs (Quinn & Eimas, 1996; Quinn, Eimas, & Rosenkrantz, 1993), infants would be able to form a category in the lab, but is this the same process that operates in infants' daily experience? The three experiments below represent our successive approximations of infants' real-world experiences of objects. By manipulating the timing and the presentation of similar objects we learned about the real-world formation of an object category that supported infants' top-down segmentation of a complex visual scene.

Experiment 1 examines the impact of non-simultaneous experiences—can infants form a category when presented with three similar objects individually rather than simultaneously? Experiment 2 examines the impact of background contextual variability, and Experiment 3 examines the impact of contextual regularities embedded in a changing background context.

EXPERIMENT 1

In Experiment 1, infants were presented with the three similar objects (three boxes similar to the blue test box) sequentially rather than simultaneously as had been done in prior research. Recall that after simultaneous exposure to this set of three boxes, infants recognized the test box as a member of a category that they knew something about (namely, what the boundaries of this kind of box were), and they used this knowledge to parse the novel test display into two objects. The simultaneous presentation allowed for easy comparison between the boxes, and comparison is one of the processes known to facilitate infants' and children's category formation (Gentner & Namy, 2000; Namy & Gentner, 2002; Oakes, 2001). There is also evidence that events in which everything is visually available are easier for infants to reason about than are events that involve the infant remembering information and comparing later events to what is stored in memory (Baillargeon, 1992). This suggests that category formation might be more difficult for infants if they were only shown one exemplar at a time. With this experience, which seems more like the kind of experience that infants would have outside the lab, infants would have to hold each box in memory and perform comparisons across remembered objects.

The question is whether simultaneous exposure to all three exemplars is necessary in order for infants' object segregation to be facilitated. If this physical context of simultaneous presentation of the ''set'' is necessary (perhaps because it makes comparison easier) then we would expect that after sequential experience with the boxes, infants' object segregation would not be facilitated. On the other hand, most infants' experiences with similar objects are probably sequential, and there is evidence with older infants that they have conceptual categories based on every day experience with objects like cars and keys (e.g., Mandler & McDonough, 1996) and that these categories can affect their object segregation expectations (Needham, Ormsbee, & Cantlon, 2005b). From this we might expect that sequential presentation would be useful for infants and that their object segregation would in fact be facilitated. Therefore, infants were shown the same three boxes as used in prior studies, but the boxes were presented individually, and infant expectations about the composition of the test display were assessed. Infants saw each box on its own for 10 s rather than all three together for 5 s as had been the case in previous studies (e.g., Needham et al., 2005a). We expected that this change would probably make it more likely that infants could benefit from prior experience with the boxes and it also produced a procedure that was more amenable to later real-world variations.

Method

Participants

The participants were 16 healthy, full-term infants (8 females, 8 males) ranging in age from 4 months, 8 days to 5 months, 25 days (M = 4 months, 28 days; SD =13.8). Half of the infants saw the move-apart test event (M = 5 months, 0 days; SD =14.2) and half saw the move-together test event (M = 4 months, 26 days; SD =14.1). Data from four additional infants was collected and excluded, one due to computer related malfunctions, two due to infant fussiness, and one due to experimenter error.

Infants' names in this experiment and the following experiments were obtained from the Durham County (North Carolina) vital records office and families were contacted via letter and follow-up phone call. Parents were reimbursed for their travel but not compensated for their participation.

Apparatus

The puppet-stage like apparatus consisted of a wooden cubicle 200 cm high, 106 cm wide, and 49.5 cm deep. The infant faced an opening 56 cm high and 95 cm wide in the front wall of the apparatus. The floor of the apparatus was covered with pale blue (No. 2.5 PB 8/2)[1] cardboard with a clear Plexiglas cover that allowed the felt-bottomed objects to move smoothly and quietly across the apparatus floor. The side walls were painted white and the back wall was covered with a brightly patterned contact paper. The infants were tested in a brightly lit room, with four clip-on lights attached to the back and side walls of the apparatus to provide additional light. Two wooden frames, 200 cm high and 69 cm wide and covered with blue cloth, stood on either side of the apparatus to isolate the infants from the experimental room. At the end of each trial, a white curtain was lowered to cover the opening to the apparatus.

There were two objects used in the test trials, a yellow cylinder and a blue box. The cylinder was 22 cm long and 10 cm in diameter. It consisted of a section of clothes dryer vent hose that was stuffed with Styrofoam so that it was rigid and formed a modified ''C'' shape with its ends curved slightly forward. The left end of the cylinder was covered with cardboard; the right end was covered with a thin metal disc. The entire cylinder and the ends were painted bright yellow (No. 2.5Y 8/10). The box was 35 cm high, 13 cm wide, and 13 cm deep. It was covered with bright blue (No. 2.5PB 7/6) contact paper decorated with small (approximately 2 cm on a side) white squares. One corner of the box faced the infants. The left rear wall of the box (not visible to the infants) held a

[1] Numbers in parentheses after colour descriptors refer to Munsell colour chip matches. These matches were made with the objects in their positions on the apparatus, so that lighting conditions would be the same as during the experiment.

magnet. The cylinder lay on the floor of the apparatus with its metallic end set against the box's magnet. This made it possible for the box to move along with the cylinder when the cylinder was pulled by the experimenter's hand. The bottom surfaces of the cylinder and the box were covered with felt so they both slid smoothly and quietly across the floor. The front 2.5 cm of the cylinder's right end protruded from the box's left corner; this protrusion was designed to make clear to the infants that the cylinder and box were adjacent. In its starting position, the box was 17.5 cm from the front edge of the apparatus and 31.5 cm from the right wall; the cylinder was 28 cm from the front edge of the apparatus and 33.5 cm from the left wall. Together, the cylinder and box subtended about 30 degrees (horizontal) and 27 degrees (vertical) of visual angle from the infants' viewpoint.

In each test event, the cylinder was pulled to the side by an experimenter's hand wearing a lavender spandex glove that was 59 cm in length. The hand entered the apparatus through a wall opening that was partially hidden by a white curtain; the infant could not see the experimenter's face through this opening.

Infants were familiarized (see procedure) with three boxes that were each similar to the test box. Each was identical to the test box (blue with white squares) in its dimensions and had squares or circles of approximately the same size and in the same positions as the squares on the test box. The colours and texture elements for each of the boxes were as follows: green (No. 2.5G 5/12) with white triangles, blue (No. 2.5PB 7/6) with red (No. 5R 4/12) squares, and purple (No. 7.5P 4/10) with white squares.

Procedure

Each infant sat on his or her parent's lap in front of the apparatus. The infant's head was approximately 63.5 cm from the box. The infant's looking behaviour was monitored by two observers who viewed the infant through peepholes on either side of the apparatus. The observers could not see and did not know which test event the infant was observing. Each observer held a joystick connected to a computer and depressed the trigger whenever the infant attended to the events, keeping the trigger depressed for the duration of looking. Each trial was divided into intervals of 100 ms duration. For each interval, the computer determined whether the two observers agreed on the direction of the infant's gaze. Interobserver agreement was calculated for each trial as the number of intervals in which the computer registered agreement, out of the total number of intervals in the trial. Agreement in the experiments reported here averaged 88% per trial per infant. The input from the primary (more experienced) observer was used to determine the end of the trials.

Each infant first received three brief familiarization trials that featured one of the three different boxes. The boxes were presented in a random order. The caretaker and infant were seated in front of the previously described testing apparatus. The screen was raised and one of the boxes was visible, held by a purple-gloved hand. Each box was held approximately 10 cm above the apparatus floor and tilted alternately to the right and left in a smooth continuous motion to maintain infant interest in the object. When the infant had accumulated 10 s of looking at the box, the trial ended with the screen closing. It then reopened showing the next box. After all three boxes had been presented in this manner, the infants were given a brief familiarization trial with the stationary test display. This trial ended when the infant (1) looked away from the display for 2 consecutive seconds after having looked at it for at least 10 cumulative seconds or (2) looked at the display for 30 cumulative seconds without looking away for 2 consecutive seconds.

Following the familiarization trials, each infant saw either the move-apart *or* the move-together test event[2] (see below for description of test events) on each of six successive trials. Each test trial ended when the infant (1) looked away from the event for 2 consecutive seconds after having looked at it for at least 8 cumulative seconds (the length of one event cycle) or (2) looked at the event for 60 cumulative seconds without looking away for 2 consecutive seconds. Each infant completed the entire set of six test trials. There were no reliable differences in looking times during the stationary familiarization trial for the infants who would see the move-apart (M = 16.04; SD = 3.3) versus the move-together test event (M = 16.22; SD =6.5), $F(1, 14) = 0.005$.

Events

Move-together event. At the start of each test trial, the curtain was raised and the infant could see the experimenter's hand on the floor of the apparatus about half-way between the cylinder and the opening in the left wall. After 1 s, the hand grasped the cylinder (1 s) and pulled it 14 cm to the left at the approximate rate of 7 cm/s (2 s). The cylinder and box moved as a single, rigid unit. The hand paused for 1 s and then pushed the cylinder and the box back to their starting positions (2 s). The hand then resumed its initial position on the apparatus floor (1 s). Each event cycle thus lasted about 8 s. Cycles were repeated without stop until the computer signalled that the trial had ended. When this occurred, a second experimenter lowered the curtain in front of the apparatus.

[2] As in Needham and Baillargeon (1998), and all of the other research conducted on object segregation in our lab, a between-subjects design was used: Each infant saw either the move-apart or the move-together test event. Because we assume that the infants are responding to the test event relative to their initial interpretation of the display during the familiarization trial (and we do not want to open up the possibility of elevated levels of looking being a result of surprise that the nature of the display as one or two objects changed between trials), we believe it is important to show each infant only one of the possible compositions of the display.

Move-apart event. The move-apart event was identical to that just described except that only the cylinder moved and the box remained stationary throughout the trial. To accomplish this, the magnet was removed and a weight was placed inside the box to prevent its movement (see Figure 1).

Results

Preliminary analyses. Preliminary analyses were first conducted to determine whether there was a reliable effect of trial or sex on the infants' looking times at the two test events. A two-factor analysis of variance (ANOVA) was conducted with event as the between-groups factor and trial as the repeated measure, which indicated that there was a reliable effect of trial on the infants' looking times. Infants looked reliably less as the experiment progressed, $F(1,5) = 14.01$, $p < .0001$. There was no differential effect of trial on the infants' looking times at the two test events, $F(1,5) = 0.45$. A second two-factor ANOVA with event and sex as between-groups factors showed no significant effect of the infants' sex on looking times at the two test events, $F(1,12) = 0$. Therefore, data were collapsed across trial and sex for further analyses.

Main analysis. The infants' looking times (see Figure 2) were analysed by means of a one-factor ANOVA with event (move-together or move-apart) as the between-subjects factor. This analysis yielded no significant effect of test event, $F(1,14) = 0.349$, indicating that infants looked about equally at the move-together ($M = 26.7$; $SD = 13.1$) and the move-apart ($M = 30.4$, $SD = 11.9$) test events.

Discussion

It appears that serial presentation of the three box exemplars in this laboratory context does not promote the category formation that would support infants' segregation of the test display. In this context, three exemplars that are similar to the test box and are presented simultaneously appear necessary to facilitate subsequent segregation. This could suggest a number of interesting possibilities. It could be that direct comparison between the three exemplars facilitates category formation, and that infants at this age are not able to hold the non-present exemplars in memory long or well enough for comparison or category formation to occur. However studies of infant memory at this age suggest that this is probably not the case (see Rovee-Collier & Hayne, 2000).

An alternate explanation is that our procedure itself impeded infant category formation. Perhaps comparison between the objects themselves isn't sufficient for category formation. In order to group the objects together, it is necessary for the infants to notice the *similarities* between the objects. Our sequential presentation of the boxes in an unchanging context might have highlighted the differences between each box. In any case, this finding implies that experience

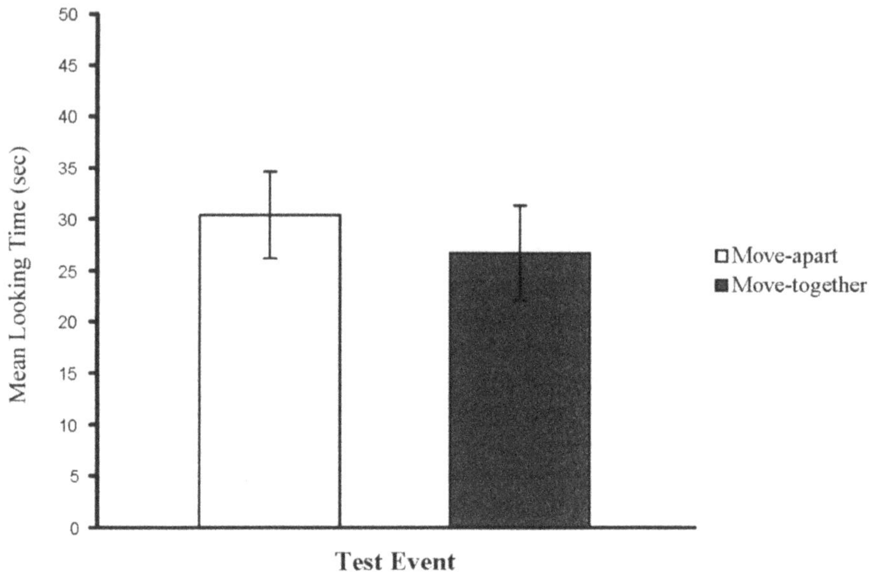

Figure 2. Mean looking time (+*SE*) to test event after immediate prior experience with three similar boxes presented sequentially.

with a small number of similar objects is useful to young infants only if all the exemplars are present at the same time.

However, while our Experiment 1 procedure was more similar to real-world experiences in that it addressed the question of simultaneous versus individual experiences of objects, it was still quite artificial, and it seemed likely that infants could learn from experiences with single exemplars presented sequentially in everyday life. One difference between infants' experience outside the lab and the method of Experiment 1 was the physical context of the experience. In the lab, the physical context was purposefully unchanging. In fact, the only things that changed were the objects themselves. In contrast, outside the lab, sequential experiences with similar types of objects would probably be in varied physical contexts, and there is some infant memory data that suggest that training in varied contexts promotes infants' memory retrieval in a novel context (Amabile & Rovee-Collier, 1991). Despite the fact that this reasoning leads to the (possibly) counterintuitive prediction that changing the context in which infants saw each object would actually facilitate their performance, we sought to create a lab setting that more closely resembled infants' typical experiences outside the lab. Recall the previous example of an infant's exposure to dogs. The infant's three experiences with dogs happened serially, but also occurred in different physical places and separated by longer time intervals than those used in our Experiment 1 procedures.

EXPERIMENT 2

Experiment 2 was an attempt to present the objects of Experiment 1 serially, but to do so in a way that better mimicked the parameters of infants' experiences outside the lab. We designed a version of serial presentation that incorporated longer time intervals between exemplar presentation as well as changes in background physical context in an attempt to learn if this would make the category information more available to infants. If it is the case that varying the background physical context helps infants notice the similarities between the boxes, then we would expect that after this experience, infants would demonstrate facilitated object segregation performance.

Method

Participants

The participants were 16 healthy, full-term infants (8 females, 8 males) ranging in age from 4 months, 9 days to 5 months, 27 days (M = 5 months, 8 days; SD = 15.3). Half of the infants saw the move-apart test event (M = 5 months, 2 days; SD = 10.9) and half saw the move-together test event (M = 5 months, 14 days; SD = 17.2). Data from 11 additional infants was collected and excluded, 2 due to computer related malfunctions, 2 due to infant fussiness, 6 due to intervals between familiarizations and or test being too long (see explanation below), and 1 due to infant failure to habituate.[3]

Apparatus and procedure

The set of stimuli and test events used for these studies was identical to those used in Experiment 1. However, the familiarization procedure with the three similar boxes was changed. Each box—blue, purple, and green—was presented individually to the infant for 10 s as in Experiment 1, but there was a delay of 3.5 min between each presentation. The presenter and location of presentation were changed for each box. In other words, the infant saw three different boxes presented by three different people in three different locations (see Figure 3).

Interval timing was accomplished by the passing of a stopwatch from experimenter to experimenter out of sight of the infant, and data from infants for whom the intervals were not maintained (e.g., it took longer than 3.5 min to reach the lab from the parking lot) were excluded from the analyses (n = 2). The

[3] We do not use specific criteria for habituation in our studies. Instead we show the events to all infants for six trials. We reliably find that infants habituate as the trials progress, regardless of what event they see, indicating that they have processed the event and are beginning to lose interest in it. If an infant looks for the maximum possible time (60 s) for each test trial, there is no indication that the infant is processing the task, and, as such, his or her data are excluded.

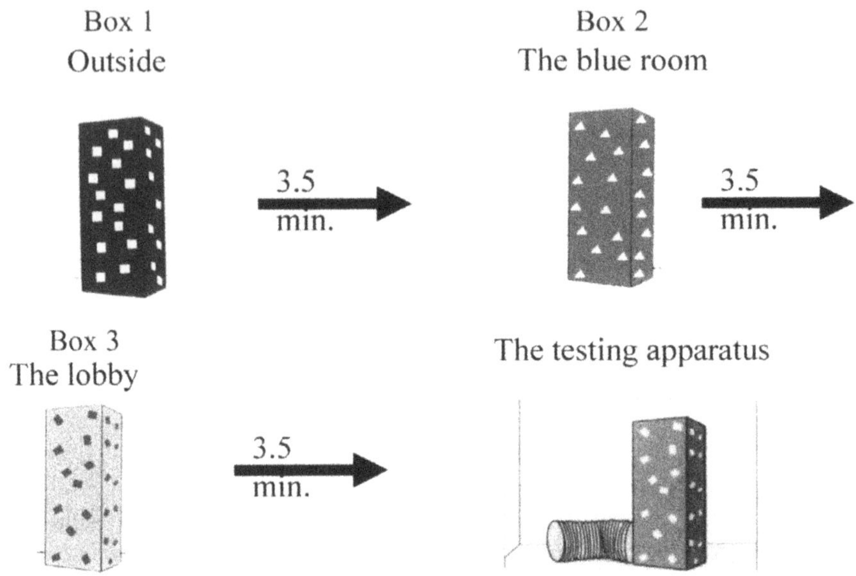

Figure 3. Schematic of procedure for Experiment 2. Box 1: Purple with white squares. Box 2: Green with white triangles. Box 3: Blue with red squares.

order of box presentation was random, and occurred in the following three locations.

Location 1. The first box was presented to the infant outdoors when the caretaker was met at the lab parking space. Experimenter 1 explained what the study was about, received verbal consent from the caregiver to proceed with the study, and showed the infant a box for 10 s, moving the box slowly (in the manner described in Experiment 1) to maintain the infant's attention. The experimenter was standing approximately 60 cm in front of the infant. The box was then hidden, the timer reset, and the caretaker and infant were escorted to the lab.

Location 2. Upon reaching the lab, the caretaker and infant were ushered into a room with walls covered by long blue curtains where Experimenter 2 was sitting. Experimenter 2 chatted with the caretaker and infant until exactly 3.5 min had elapsed from the presentation of the first box, and then presented the second box. After 10 s, the box was hidden and Experimenter 2 left the room.

Location 3. The caretaker and infant were escorted to the lobby of the lab, a room decorated with colourful posters and many toys, where the caretaker finished paperwork and chatted with the lab manager. Three and a half minutes after the presentation of the second box, Experimenter 3 entered the lobby, presented the third box for 10 s, and then left the room.

Location 4. The caretaker and child were then escorted to the previously described testing apparatus in yet another room, and the test procedure was begun approximately 3.5 min after the presentation of the last box. Because the amount of time to get the caretaker and infant settled and ready to begin the test trials was variable, the time between the presentation of the final box and the beginning of the familiarization trial with the stationary display was recorded for each infant. Data from infants whose final interval was longer than 4 min ($n = 4$) were not included in these analyses.

Infants first saw a brief familiarization trial with the stationary test display (described in Experiment 1) and then saw either the move-apart or move-together test event. All infants contributed data for the full set of six test trials. There were no reliable differences in the looking times during the stationary familiarization trial for the infants who would see the move-apart ($M = 20.7$; $SD = 9.7$) versus the move-together test event ($M = 16.6$; $SD = 6.9$), $F(1, 14) = 0.93$. There was also no significant difference in the delay between seeing the last box and the beginning of the stationary familiarization trial for infants who would see the move-apart ($M = 3.40$; $SD = 10.5$) versus the move-together test event ($M = 3.39$; $SD = 23.0$), $F(1, 14) = 0.01$.

Results

Preliminary analyses. The data were analysed as in Experiment 1. Infants looked reliably less as the experiment progressed $F(1, 5) = 4.49$, $p < .01$. There was no differential effect of trial on the infants' looking times at the two test events, $F(1, 5) = 0.28$. There was no significant effect of the infants' sex on looking times at the two test events, $F(1, 12) = 0.14$. Therefore, data were collapsed across trial and sex for further analyses.

Main analysis. The infants' looking times (see Figure 4) were analysed by means of a one-factor ANOVA with event (move-together or move-apart) as the between-subjects factor. This analysis yielded no significant effect of test event, $F(1, 14) = 0.638$, indicating that infants looked about equally at the move-apart ($M = 34.64$, $SD = 10.2$) and move-together ($M = 38.56$; $SD = 9.4$) test events.

Discussion

Infant segregation of the test display was not facilitated by sequential presentation of the three similar boxes prior to seeing the test display, even when those experiences happened in distinct physical locations. This finding was a bit surprising given our intuitions about the types of experience infants are likely to have in daily life. There could be several explanations for this finding (as is the

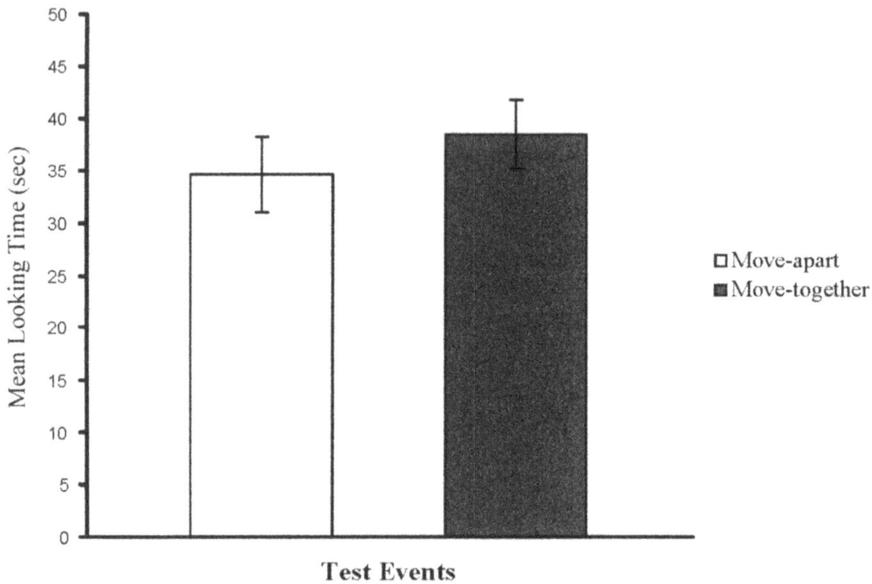

Figure 4. Mean looking time (+*SE*) to test event after prior experience with three similar boxes presented sequentially and in three distinct physical locations.

case with all null findings). First, the task might have just been too difficult for the infants. Previous research has shown that at 6 months old, infants' memory is quite sensitive to changes in the physical context (Rovee-Collier & Gulya, 2000), and for the infants to benefit from their experiences with the objects they certainly needed to remember the various box exemplars. The memory demand could have been too much for infants to overcome.

Reflection on the experiment suggested another alternative. Perhaps the procedure did not accurately capture the essence of infants' everyday experiences. We tried to vary everything except for the objects themselves, with the idea that the background differences would make the similarities among the boxes salient to the infant. We were focusing on how the changing of everything outside of the object could affect infants' understanding of the test display. Although changes in physical context are probably a common experience for infants, it is also the case that similarities often exist among the types of contexts in which specific objects are encountered. Recall the example of an infant's experiences with various dogs. Most dogs wear collars, have leads, and behave in similar ways. Further, the people around the infants may interact with dogs similarly—leaning down to pet them, speaking in ''puppy-voices'' and taking them for walks.

Along with physical contextual variability, there is also some contextual regularity to infants' real-world experiences with similar objects. It could be that infants use these regularities to recognize the similarities between various experiences with objects. In effect, these similarities could help infants link together their experiences in a way that allows them to from a category they can access at a later time. Perhaps it is the interaction of physical contextual variability and regularity that combine to help infants gather category information from their sequential experiences with category exemplars.

EXPERIMENT 3

This experiment is an attempt to include contextual regularity in the infants' physically varied experiences. We wanted to duplicate the procedure of Experiment 2, but include a contextual regularity in all of the infants' experiences with the boxes. This meant we needed a link that could be present during the test events as well. So, the purple glove that was normally present during the test events was worn by each experimenter for all of the trials—familiarization and test, thus providing a mix of variability and regularity in the physical contexts of infants' experiences. If infants need both variation in the physical context (different locations) and contextual regularity (purple glove), we would expect that this prior experience would lead to facilitation in infants' parsing of the test display.

Method

Participants

The participants were 16 healthy, full-term infants (8 females, 8 males) ranging in age from 4 months, 6 days to 5 months, 28 days (M = 4 months, 28 days; SD = 19.0). Half of the infants saw the move-apart test event (M = 4 months, 22 days; SD = 15.3) and half saw the move-together test event (M = 5 months, 5 days; SD = 20.6). Data from 17 additional infants was collected and excluded, 3 due to infant fussiness, 6 due to intervals between familiarizations being too long, 5 due to longer than 4 min between familiarization and test, 2 due to experimenter error, and 1 as a statistical outlier (Z = 3.1).

Procedure and apparatus

The familiarization procedure for Experiment 3 was identical to that of Experiment 2 with the following exception: Each experimenter wore a purple glove (the same one the infants would see during the test trials) on the hand that they were using to present the box during the familiarization trials. The testing procedure was identical to the previous experiments. Each infant first saw a brief familiarization trial with the stationary display, and then saw either the move-apart event or the move-together event. Fifteen infants contributed data for all six test trials and one infant contributed only five trials due to poor visibility for the observers during the last test trial. There were no reliable differences in the looking times during the stationary familiarization trial for the infants who would see the move-apart ($M = 18.64$; $SD = 6.8$) versus the move-together test event ($M = 17.93$; $SD = 5.7$), $F(1, 14) = 0.05$. There was also no significant difference in the delay between seeing the last box and the beginning of the stationary familiarization trial for infants who would see the move-apart ($M = 3.34$; $SD = 14.2$) versus the move-together test event ($M = 3.36$; $SD = 15.9$), $F(1, 14) = 0.09$.

Results

Preliminary analyses. The data were analysed as in Experiment 1. Infants looked reliably less as the experiment progressed, $F(1, 5) = 5.97, p < .001$. There was no differential effect of trial on the infants' looking times at the two test events, $F(1, 5) = 0.62$. There was no significant effect of the infants' sex on looking times at the two test events, $F(1, 12) = 1.86$. Therefore, data were collapsed across trial and sex for further analyses.

Main analysis. The infants' looking times (see Figure 5) were analysed by means of a one-factor ANOVA with event (move-together or move-apart) as the between-subjects factor. This analysis yielded a significant effect of test event, $F(1, 14) = 6.28$, $p < .03$, indicating that infants looked reliably longer at the move-together ($M = 44.09$, $SD = 8.2$) test event than at the move-apart ($M = 30.22$; $SD = 13.3$) test event.

Discussion

After sequential experience with three similar objects in three distinct locations that each included a single contextual regularity, infants showed a clear expectation that the test display was in fact composed of two pieces. This suggests that, given this set of experiences, infants were able to form a category that they could use to disambiguate the test display by generalizing knowledge about the probable boundaries of the test box.

It is important to note that infants could not simply have been attending only to the glove and still have shown facilitation on the object segregation task for two reasons. First, it is unclear how simply recognizing the glove in the test situation would lead infants to attend differently to the two test events, and second, the glove did not act in the same way each time it was seen. For the familiarizations, the glove was present on the hand presenting each box, and at test, the gloved hand reached out and grasped the yellow cylinder—it never touched the test box. In order to have shown facilitated object segregation, infants must have been attending to the boxes as well as the glove. It could be that by including the glove in all of the experiences with the boxes, we merely

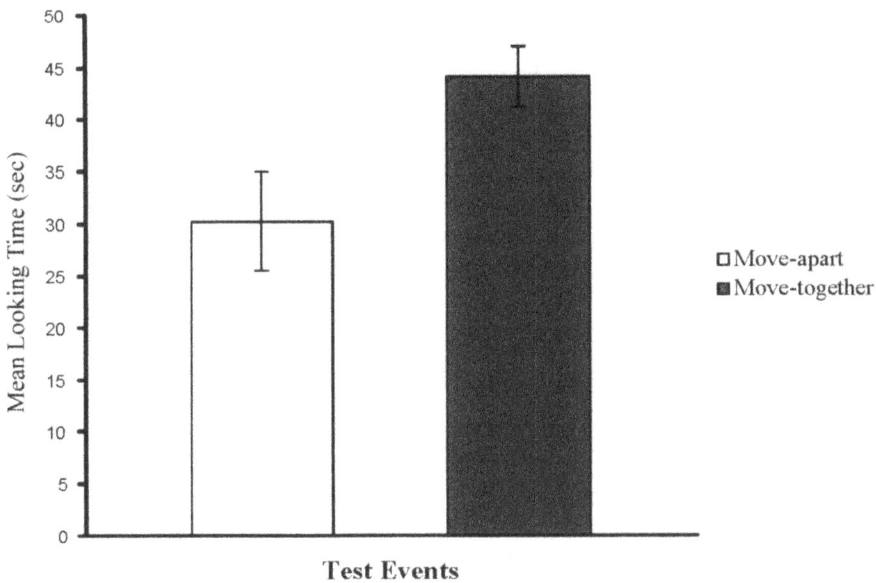

Figure 5. Mean looking time (+*SE*) to test event after prior experience with three similar boxes presented sequentially, in three distinct physical locations with the addition of physical (purple glove) contextual regularity.

increased the overall similarity of the experiences. Alternately, the glove could have served as a salient memory cue that helped infants to access their memories of their previous experiences when the glove was present. In either case, the presence of a physical regularity in the changing physical contexts of infants' experiences with the three boxes seemed to help infants to link those experiences into an object category that facilitated their segregation of the test display.

Cross–experiment analyses. In order to directly compare the results of the three experiments we estimated an overall (Experiment × Condition) ANOVA model. The resulting model was significant, $F(5, 42) = 2.6$, $p < .04$, but there were no significant main effects or interactions in the model. A series of planned comparisons were estimated to clarify this result. The results of Experiments 1 and 3 were significantly different, $F(1, 31) = 4.73$, $p < .04$, and this effect was driven by a significant interaction between experiment and condition, $F(1, 31) = 4.94$, $p < .03$, indicating that while infants looked similarly at the move-apart and move-together events in Experiment 1, in Experiment 3 they looked significantly longer at the move-together event. This set of results supports the interpretation that changes in physical context are beneficial for category formation since contextual regularity (the purple glove) was present in both

Experiments 1 and 3. The results of Experiments 2 and 3 were not significantly different, $F(1,42) = 0.2$, nor was there a significant interaction between experiment and condition, suggesting that infant behaviour during these two experiments was similar. In fact, scrutiny of the means shows that the results were qualitatively similar—in both Experiments 2 and 3 the infants looked longer at the move-together event. This pattern suggests that varying the background physical context might have been helpful to some of the infants, but it was only the combination of the varying physical context with an element of contextual regularity that reliably facilitated infant category formation and parsing of the novel scene.

GENERAL DISCUSSION

Taken together, the pattern of results obtained in these studies suggests that infants' scene perception can benefit from sequential experiences with similar objects when those experiences happen in different physical contexts and contain at least one element of contextual regularity. It appears that it is the combination of these two elements that is important because in the presence of only one, facilitation did not occur. Recall that only contextual regularity (in the form of the purple glove) was present in Experiment 1 when each of the boxes was encountered sequentially in the testing apparatus. In Experiment 2, only physical variability was present, with each experience with a box happening in a distinct location. Neither of these types of experiences was useful for infants. It was the combination of the two in Experiment 3 that enabled infants to notice category information that they could use to segregate the objects in the test display.

These results have many intriguing implications for current theories and future research. First, we know something more about how infants develop a knowledge base that can support the top-down segmenting of visual scenes. By studying category formation in more realistic situations, we have learned about the importance of context to the development of object recognition. This beneficial role for changes in the physical context of experience seems counter to some of the information reported from studies of infant memory performance, but we do not actually believe it is an opposing claim. Similar disparate roles for physical context have been found in adult populations as well (Smith, Glenberg, & Bjork, 1978) and have been discussed as representing effects at different stages of memory. Encoding the same or (we would argue) similar information in different physical contexts provides more cues that might overlap with the current physical context making recovery of that information more likely, while information learned in one context and retrieved elsewhere has fewer cues available.

Second, these findings expand what we know about infants' processing of contextual information. The idea that infants notice and can benefit from regularities present in the environment is not a new one. Mervis and Rosch (1981) argued that the remarkable congruence/similarity in categorization between individuals resulted from the similarities people experienced in their natural environments. This similarity in experience was said to result from exposure (and attention) to the correlatedness of sets of attributes in the physical world and the realization that not all possible combinations of features/attributes are present in the natural world. According to Mervis and Rosch, it is the discontinuity between various sets of attributes that form natural category boundaries. Thus, the correlatedness of attributes within objects supports category formation.

We suggest that it is not just the correlatedness of attributes within similar objects in the real world that support category formation, but that it is also the correlation between specific contextual regularities and specific types of objects that help infants to link together their experiences to form categories. In fact, similar objects typically appear in similar types of contexts or scenes, and this natural correlatedness of object kind and overall context supports category formation. For example, spoons are typically encountered in the kitchen and almost always in the presence of food (Mandler, Fivush, & Reznick, 1987). We argue that these contextual regularities help infants to link together their experiences with various spoons to create a category of spoon. That category knowledge could then help infants to recognize spoons in later encounters and this could be a kind of scene-schema knowledge (e.g., Henderson, 2003) in infants

This argument may imply that infant object knowledge becomes less context-dependent over time because once a category is present it would facilitate recognition of a spoon even in an unusual context. Once infants link together experiences with similar objects to form a category they might be able to access that category information later when they encounter a similar object even if the object is in an atypical setting (e.g., a spoon in the infant's crib). New evidence from our lab supports this reasoning. In keeping with the theme of these studies, which was to try to understand infant cognition in context, we were eager to identify a contextual regularity that might be present in infants' real-world experiences (because purple velvet gloves are, at least for most infants, a rarity). To return to the dog example one last time, beyond the physical regularities like leads and dog kennels, infants could also notice similarities in the way other people interact with the dog as well as the word(s) ''Dog'' or ''Doggie'' commonly used to refer to the dog. We thought that language, specifically the labelling of objects, could be one of the contextual regularities that even young infants might benefit from in daily experiences. In fact, recent data suggest that this is the case (Dueker & Needham, 2003, 2005). After parents label the three boxes during familiarization, infants recognize the test box (which is not named by the parents) in the novel test display. It appears that once the experiences with the original three boxes are linked together to form a category, use of that category is no longer dependent upon the specific contextual link. This suggests that while contextual regularities appear necessary for real-world category formation, once a category is established, infants can use it in contexts that do not contain such contextual regularities.

FINAL COMMENTS

The findings from this set of studies have broad implications for the study of infants' understanding of objects in the world around them. Generally, in studies of infant categorization or object understanding, the physical context is kept constant, and contextual regularities that might be available in everyday infant experience are excluded. For example, many studies have addressed whether and at what age infants can discriminate between the category of dog and of cat (e.g., Quinn et al., 1993), but in these studies infants are presented only with a picture (or two pictures) of a dog on a blank background. Even in studies that use realistic models of dogs (e.g., Mandler & McDonough, 1996), other objects that might help infants link this experience to another experience with a dog (e.g., a lead or dog kennel) were not often available. Our results suggest that by stripping away the trappings of an infant's everyday experiences we might be underestimating her competence because the infant no longer has access to the contextual regularities she could use in a real-world situation to link together her various experiences with similar objects. These studies are just a beginning, but the techniques developed here can be used to learn much more about the origins of real-world scene processing and the processes by which scenes become meaningful to us.

REFERENCES

Amabile, T. A., & Rovee-Collier, C. (1991). Contextual variation and memory retrieval at six months. *Child Development*, *62*, 1155–1166.

Baillargeon, R. (1992). The object concept revisited: New directions in the investigations of infants' physical knowledge. In C. Granrund (Ed.), *Visual perception and cognition in infancy: Carnegie Mellon symposia on cognition* (pp. 265–315). Hillsdale, NJ: Lawrence Erlbaum Associates, Inc.

Behl-Chadha, G. (1996). Basic-level and superordinate-like categorical representations in early infancy. *Cognition*, *60*, 105–141.

Bomba, P. C., & Siqueland, E. R. (1983). The nature and structure of infant form categories. *Journal of Experimental Child Psychology*, *35*, 294–328.

Bravo, M. J., & Farid, H. (2003). Object segmentation by top-down process. *Visual Cognition*, *10*(4), 471–491.

Cohen, L. B. (2003). Unresolved issues in infant categorization. In D. H. Rakison & L. M. Oakes (Eds.), *Early category and concept development* (pp. 193–209). New York: Oxford University Press.

Dueker, G. L., Modi, A., & Needham, A. (2003). 4.5-month-old infants' learning, retention and use of object boundary information. *Infant Behavior and Development*, *26*(4), 588–605.

Dueker, G. L., & Needham, A. (2003). *Labeling facilitates 5-month-old infant's category formation.* Poster presented at the Society for Research on Child Development, Tampa, FL.

Dueker, G. L., & Needham, A. (2005). The impact of object naming on 5-month-old infants' object category formation. *Manuscript submitted for publication.*

Gentner, D., & Namy, L. (2000). Comparison in the development of categories. *Cognitive Development, 14*, 487–513.

Goldstone, R. L., Lippa, Y., & Shiffrin, R. M. (2001). Altering object representations through category learning. *Cognition, 78*, 27–43.

Hayne, H., Rovee-Collier, C., & Perris, E. (1987). Categorization and memory retrieval in 3-month olds. *Child Development, 58*, 750–767.

Henderson, J. M. (2003). Human gaze control during real-world scene perception. *Trends in Cognitive Development, 7*(11), 498–504.

Mandler, J. M., Fivush, R., & Reznick, J. S. (1987). The development of contextual categories. *Cognitive Development, 2*, 339–354.

Mandler, J. M., & McDonough, L. (1996). Drinking and driving don't mix: Inductive generalization in infancy. *Infancy, 37*, 60–96.

Mandler, J. M., & McDonough, L. (1998). Studies in inductive inference in infancy. *Infancy, 37*, 60–96.

McGurk, H. (1972). Infant discrimination of orientation. *Journal of Experimental Child Psychology, 14*, 151–164.

Mervis, C. B., & Rosch, E. (1981). Categorization of natural objects. *Annual Review of Psychology, 32*, 89–115.

Namy, L. L., & Gentner, D. (2002). Making a silk purse out of two sows ears: Young children's use of comparison in category learning. *Journal of Experimental Psychology: General, 131*(1), 5–15.

Needham, A. (1998). Infants' use of featural information in the segregation of stationary objects. *Infant Behavior and Development, 21*, 47–76.

Needham, A. (2001). Object recognition and object segregation in 4.5-month-old infants. *Journal of Experimental Child Psychology, 78*, 3–24.

Needham, A., & Baillargeon, R. (1998). Effects of prior experience in 4.5-month-old infants' object segregation. *Infant Behavior and Development, 21*, 1–24.

Needham, A., Dueker, G., & Lockhead, G. (2005a). Infants' formation and use of categories to segregate objects. *Cognition, 94*, 215–240.

Needham, A., & Modi, A. (1999). Infants' use of prior experiences in object segregation. In H. W. Reese (Ed.), *Advances in child development and behavior* (Vol. 27, pp. 99–133). San Diego, CA: Academic Press.

Needham, A., Ormsbee, S. M., & Cantlon, J. F. (2005b). Seeing a key ring as one whole object: Infants use of category knowledge in object perception at 8.5 months of age. *Manuscript submitted for publication.*

Oakes, L. M. (2001, April). The role of comparison in category formation in infancy. In L. M. Oakes & D. Rakison (Chairs), *Mechanisms of development in category formation in infancy.* Symposium conducted at the meeting of the Society for Research in Child Development, Minneapolis, MN.

Oakes, L. M., & Madole, K. L. (2000). The future of infant categorization research: A process-oriented approach. *Child Development, 71*(1), 119–126.

Oakes, L. M., & Madole, K. L. (2003). Principles of developmental change in infants' category formation. In D. H. Rakison & L. M. Oakes (Eds.), *Early category and concept development* (pp. 132–158). New York: Oxford University Press.

Oakes, L. M., & Spalding, T. L. (1997). The role of exemplar distribution in infants' differentiation of categories. *Infant Behavior and Development, 20*(4), 457–475.

Quinn, P. C., & Eimas, P. D. (1996). Perceptual cues that permit categorical differentiation of animal species by infants. *Journal of Experimental Child Psychology, 63*, 189–211.

Quinn, P. C., Eimas, P. D., & Rosenkrantz, S. L. (1993). Evidence for representations of perceptually similar natural categories by 3-month-old and 4-month-old infants. *Perception, 22*, 463–475.

Quinn, P., & Johnson, M. H. (2000). Global before basic categorization in connectionist networks and two-month old infants. *Infancy*, *1*(1), 31–46.

Rovee-Collier, C., & Gulya, M. (2000). Infant memory: Cues, context, categories and lists. In D. L. Medin (Ed.), *The psychology of learning and motivation* (pp. 1–46). San Diego, CA: Academic Press.

Rovee-Collier, C., & Hayne, H. (2000). Memory in infancy and early childhood. In E. Tulving & F. Craik (Eds.), *The Oxford handbook of memory* (pp. 267–282). New York: Oxford University Press.

Smith, S. M., Glenberg, A., & Bjork, R. A. (1978). Environmental context and human memory. *Memory and Cognition*, *6*(4), 342–353.

Younger, B. A., & Cohen, L. (1986). Developmental change in the infants' perception of correlations among attributes. *Child Development*, *57*, 803–815.

VISUAL COGNITION, 2005, 12 (6), 1199–1234

Influence of scale and orientation on the visual perception of natural scenes

Bruce C. Hansen

Department of Psychological and Brain Sciences, University of Louisville, KY, USA

Edward A. Essock

Department of Psychological and Brain Sciences, and Department of Ophthalmology and Visual Science, University of Louisville, KY, USA

It is well known that the distribution of spatial content with respect to spatial scale in real-world scenes falls in accordance with a $1/f^{\alpha}$ relation. Equally well known is the tendency for an orientation bias in scene content with the predominant bias in content at the horizontal and vertical orientations. This has led to the suggestion of a relationship in which the mechanisms of the human visual system are optimized for processing such regularities. Here we review current literature concerning the measurement of these regularities (via Fourier analysis) of natural scenes in the context of other work that has psychophysically assessed the extent to which visual perception exploits such regularities of spatial content. In addition, 2 psychophysical experiments are presented that extend this literature and argue for the importance of these regularities in perceiving orientation in real-world visual stimuli.

Within the past decade there has been a concerted effort to relate the spatial characteristics of the content contained in real-world scenes to the processing strategies of the visual system. Specifically, it has been posited that the biological processes that underlie visual encoding are *tuned* to optimize the encoding of the spatial relationships present in the content of the visual world (e.g., the spatial scale, orientation, luminance and colour contrast, and texture characteristics that form the structural content of the visual image). Accordingly,

Please address all correspondence to: Edward A. Essock, Department of Psychological and Brain Sciences, 317 Life Sciences Building, University of Louisville, Louisville, KY 40292 USA. Email: essock@louisville.edu

This work was supported by grants nos. N00014-99-1-0516 and N00014-03-1-0224 from the Office of Naval Research and the Kentucky Space Grant Consortium (KSGC-NASA).

DOI:10.1080/13506280444000715

these spatial relationships of the content contained in real-world imagery[1] have been extensively examined and quantified by numerous techniques such as optical and digital Fourier analysis, higher order autocorrelation analyses, principal component analysis, and convolution with various types of gradient kernels (Baddeley & Hancock, 1991; Burton & Moorhead, 1987; Coppola, Purves, McCoy, & Purves, 1998a; Craven, 1993; Howe & Purves, 2002; Keil & Cristóbal, 2000; Oliva & Torralba, 2001; Switkes, Mayer, & Sloan, 1978; Thomson, 2001b; Tolhurst, Tadmor, & Chao, 1992; van der Schaaf & van Hateren, 1996). Regardless of the analysis technique utilized, or the scene specific content in the types natural scene imagery investigated, two general regularities of spatial content are found: (1) The amount of physical energy (or intensity) of any real-world image peaks at the larger scales (i.e., lower spatial frequencies) and decreases with increasing scale (i.e., higher spatial frequencies) of the image content; and (2) for typical scenes, an anisotropy with a bias in image content favouring horizontal and vertical axes (cardinal axes) relative to the oblique axes (45° and 135° oblique). This review will examine these regularities of scale and orientation while emphasizing results obtained with the most frequently used method of measurement, that being Fourier analysis.

Given that such regularities are generally encountered, regardless of the type of environment, several researchers have proposed that it would be to the benefit of the visual system to take advantage of such regularities in encoding image information. However, such statements have left vision researchers with the question of identifying how, if at all, the visual system could make use of such regularities. One effective means for investigating the extent to which visual processing depends on such regularities is through visual performance-based behavioural experiments. Specifically, one could vary the amount of content present across spatial scales, or at specific orientations, or both, in order to determine which conditions are ideally suited for the effective processing of visual information. That is, if visual perceptual-performance depends on the presence of these regularities, any deviation from this will result in less than ideal performance. However, such an expectation is not as clear cut as it seems, and is often dependent on the experimental paradigm utilized.

The vast majority of studies that have concentrated on examining whether or not the visual system is optimally suited to process visual scenes that possess the previously mentioned regularities of image content are still in the early phases of investigation. To date, much of the work that has been done includes studies that measure the regularities of image content and then emphasize parallels between those regularities and certain visual performance phenomena. Others have taken

[1] Specifically, any stimulus possessing semantically meaningful spatial content (whether or not it contains carpentered/manufactured content) at a broad range of scales will be referred to here as a ''real-world'' image/''natural scene''.

this a step further by empirically measuring the extent to which experimental manipulation of the regularities of natural stimuli can impact visually driven performance. It is the latter of the two that will be the focus of this review. Most empirical investigations to date have focused on whether or not the typical distribution of image content across scales found to occur in natural scenes yields maximal performance relative to distributions of image content that deviate from the typical distribution. Only recently has there been a drive to investigate visual processing of natural scene stimuli as the amount of content at different orientations is varied. In this review we will present the current state of research with respect to visually driven performance-based experiments as they relate to the structural regularities most often reported in the literature. In addition, data obtained from recent experiments carried out in our laboratory will be reported as an extension of the research that will be reviewed. It will be shown that some parallels that have been drawn between the observed content regularities of natural scenes and the occurrence of certain visual phenomena are not supported when assessed directly. Before moving into a discussion of the current status of the area of research devoted to measuring visual performance with respect to natural scene content regularities; the following section will describe how these regularities have typically been measured and subsequently reported.

THE CHARACTERISTIC REGULARITIES OF NATURAL SCENES

Content regularities with respect to different spatial scales

It may seem unlikely that, for example, an image of the Grand Canyon would have anything structurally in common with an image of one's own backyard, but there is a great deal of evidence that argues they are quite similar. One particular measurement showing such similarity is the global 2-D Discrete Fourier Transform (DFT). Any complex waveform (in this case, an image is treated as a complex 2-D luminance waveform) can be represented as the sum of sinusoidal waveforms of different amplitudes, frequencies, orientations, and phases. The amplitude plotted as a function of spatial frequency and orientation is often referred to as the amplitude spectrum and the phase of the waveforms as a function of orientation and spatial frequency is called the phase spectrum (Bracewell, 2000; Shapley & Lennie, 1985). The amplitude (or power/energy) spectra of different natural scene images have been measured in a number of studies (e.g., Oliva & Torralba, 2001; Simoncelli & Olshausen, 2001; Tolhurst et al., 1992; van der Schaaf & van Hateren, 1996). A typical method of analysis involves examining global spectral properties as a function of spatial frequency. This involves averaging the amplitudes across orientation at each spatial frequency and plotting the results on logarithmic axes. The typical plot peaks at the lowest spatial frequencies and falls with increasing frequency, f, in accor-

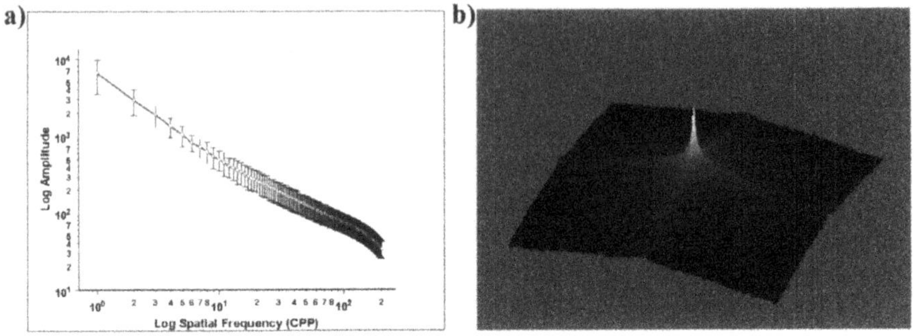

Figure 1. (a) Plot of the typical $1/f^{\alpha}$ fall-off of amplitude (averaged across orientation) as a function of spatial scale (cycles per picture). This particular example is the average of 70 natural scene images; error bars are ±1 *SD*. (b) Two-dimensional representation of an amplitude spectrum (spatial frequency increases with radius, orientation changes with theta) obtained by averaging the same 70 spectra used in (a); note the concentration of amplitude is along the horizontal and vertical axes.

dance with the $1/f^{\alpha}$ relationship in logarithmic coordinates (see Figure 1a). The exact exponent, α, that characterizes natural scenes has been the subject of much debate, but the general consensus has been that of an observed range from $1/f^{0.6}$ to $1/f^{1.6}$ (Field & Brady, 1997).

Characteristic slopes for different scene content

The issue of determining the exact exponent, α, that describes the fall-off of amplitude with increasing image scale is not a trivial one. As will be discussed in later sections of the current review, the exponent of content-amplitude fall-off can determine the scale at which the visual system is most sensitive, as well as determine the ability of humans to discriminate between different content in different natural scenes (Knill, Field, & Kersten, 1990; Párraga & Tolhurst, 2000; Párraga, Troscianko, & Tolhurst, 2000; Tadmor & Tolhurst, 1994; Tolhurst & Tadmor, 1997, 2000; Webster & Miyahara, 1997). In the literature devoted to measuring the distribution of content in large samples of real-world imagery, the image sets investigated generally consisted of imagery ranging from purely *naturalistic* content (i.e., woodlands, meadows, general shrubbery, etc.) to images of purely *carpentered* structures (both indoor as well as outdoor structures) (Burton & Moorhead, 1987; Keil & Cristóbal, 2000; Switkes et al., 1978; Tolhurst et al., 1992; van der Schaaf & van Hateren, 1996). While the α distributions from those studies did not reveal any obvious multimodal tendencies, it could very well be the case that the primary reason behind the large range of α values mentioned earlier was due to the broad range of image content present in the imagery. Our contention is that, in a given type of environment, the range of α values encountered may be very small. This is not a novel idea however, Oliva and Torralba (2001) and Torralba and Oliva (2003) have argued that the ''spectral signature'' (i.e., the nature in which power (squared amplitude) is distributed in a given spectrum across spatial frequencies and orientations) of real-world imagery depends very much on type of environment depicted in the imagery. While it wasn't the focus of their analyses, the reported α values obtained from within different orientation bandwidths were found to depend on scene category (i.e., ''forest'', ''field'', ''coast'', ''street'', ''portrait'' ''high building'', etc.). Specifically, α values were reported (Torralba & Oliva, 2003) for cardinal and oblique orientations for imagery of purely naturalistic and purely carpentered environments. When these values are averaged across orientation within each category type, the difference is negligible: For naturalistic scenes, $\alpha = 1.43$, and for carpentered environments, $\alpha = 1.44$ (α values have been corrected to reflect amplitude spectra fall-offs). However, while the image sets were large, the size of the imagery was limited to 256 $\times$ 256 pixels and were gathered from variable sources, which could introduce numerous unspecified and potentially confounding influences on their analyses (i.e., images in the Corel stock photo library have likely been subjected to numerous image processing algorithms in order to enhance the aesthetic appeal of the imagery). An additional confound could have also been present in that the authors do not indicate whether or not the imagery had been fully linearized (i.e., appropriately gamma corrected) before subjecting the imagery to content analysis.

Given the limitations of the imagery discussed above, we were interested in determining whether α values ''describing'' the orientation-averaged amplitude spectra of real-world imagery do in fact depend on the type of content present in the imagery. We have amassed an image library of over 1800 natural scene images ranging in content from purely naturalistic content to images containing only indoor or outdoor carpentered materials (see Hansen, Essock, Zheng, & DeFord, 2003 and Hansen & Essock, 2004 for image acquisition and content analysis specifications and procedures). Since we wanted to conduct an analysis of imagery most likely to be encountered, we grouped imagery into three very general categories, specifically: (1) Purely naturalistic, (2) mixed naturalistic and carpentered content, and (3) purely carpentered content (see Figure 2a for examples). Our image database contained 131 images of carpentered environments, and thus served as the limiting factor in terms of the number of scenes selected for equal n comparisons. One-hundred and thirty-one images were randomly selected to comprise the other two image sets (all images were cropped to 1024 $\times$ 1024 pixels for analysis). As shown in Figure 3a, there is a clear multimodal distribution of α values, with each peak corresponding to a different scene category and considerable overlap between all categories. Torralba and Oliva (2003) reported that the ''scene-scale'' (i.e., perceived distance of the primary structures that make up a given scene) of different imagery within their categories had an impact on the corresponding spectral signatures.

Figure 2. (a) Far left: Typical image in the naturalistic-carpentered content image set; middle: Example of a typical outdoor image from the carpentered image content set; right: Typical indoor image from the carpentered image content set. (b) From left to right, typical close-range, mid-range, and far-range naturalistic-content imagery.

Since we did not control for scene-scale with our image database, such an influence could be a potential confound in the distributions plotted in Figure 3a. Unfortunately, the number of carpentered and mixed naturalistic/carpentered images in our database is limited, thereby discouraging any meaningful analyses with respect to different scene-scales. However, our naturalistic content image set exceeded 900 images, thus we were able to split this set into three sets (i.e., scene scale sets): Close-up images, mid-range images, and far-range images (refer to Figure 2b for examples). This process involved placing obviously close-range imagery and far-range imagery into their respective categories; if there was any doubt with respect to how close or far the range of content in an image was, it was automatically placed into the mid-range category. Admittedly, while we were as objective as possible, all possibility of experimenter bias in this selection process cannot be ruled out. This categorization process resulted in 201 far-range naturalistic images, which served as the limiting sample size as there were 305 images classified as close-range, and 475 mid-range images. We therefore randomly sampled 201 images from those sets in order to make equal n comparisons between the different naturalistic scene-scale image sets. As shown in Figure 3b, this categorization procedure yielded very similar α values; also the scene-scale categorizations differ very little from the sample of naturalistic images depicted in Figure 3a.

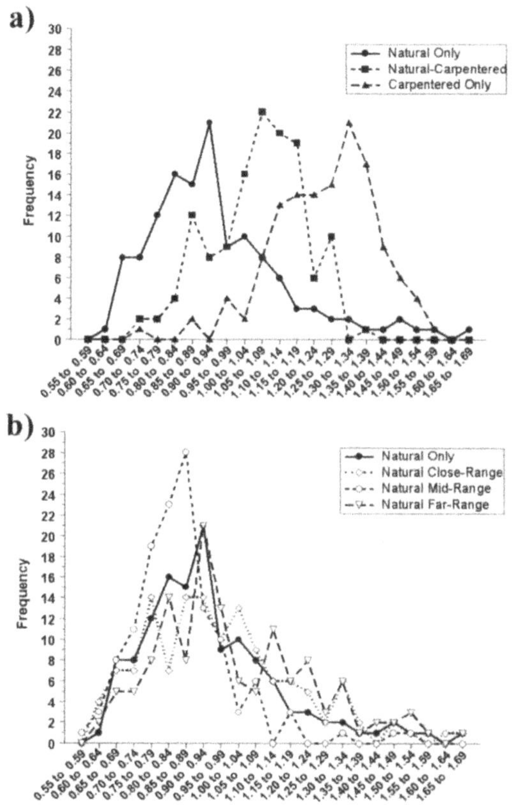

Figure 3. (a) Frequency curves showing the distribution for each of the three different image sets' α values (naturalistic, naturalistic-carpentered mix, and carpentered image sets). On the abscissa is plotted the full range of α values observed in the sample, where each tick mark represents a small range (~.05) in which the frequency of occurrence was measured. (b) Frequency curves showing the distribution for each of the three different image range sets' α values (close-range, mid-range, and far-range sets). The solid curve is replotted from (a) to serve as a reference to the original naturalistic-content sample.

The overall range of α values reported in Figure 3 (~0.59 to ~1.66) is very much in agreement with the typical range reported in the respective literature (i.e., ~0.60 to ~1.6). However, this range is greatly reduced when considering general environments one may encounter (e.g., Figure 2a). As mentioned earlier, the exponent of content-amplitude fall-off can determine the scale at which the visual system is most sensitive, as well as determining the ability in which humans can discriminate between different objects in different natural scenes. This issue will be addressed further in the sections devoted to reviewing empirical studies where visually driven performance was measured under conditions employing stimuli with variable α values.

Content regularities with respect to orientation

The first type of regularity discussed in the preceding section involved analyses of the orientation-averaged amplitude spectra of real-world imagery. However, such an account ignores the variation of amplitude as a function of orientation. As mentioned in the introduction, there have been numerous studies that set out to measure the distribution of content in real-world imagery, and many of these investigations focus on the distribution of content with respect to orientation, and utilize a number of methods to measure this, including: Fourier analysis (Hansen & Essock, 2004; Keil & Cristóbal, 2000; Oliva & Torralba, 2001; Switkes et al., 1978; Torralba & Oliva, 2003; van der Schaaf & van Hateren, 1996), second and higher order autocorrelation analyses (Baddeley, 1997; Thomson, 1999, 2001a, 2001b), principle components analysis (Baddeley & Hancock, 1991; Hancock, Baddeley, & Smith, 1992), and a variety of convolution kernels (Coppola et al., 1998a; Craven, 1993). As mentioned earlier, typical natural scenes are found to be anisotropic with relatively less content at oblique orientations and the most content at orientations near horizontal and vertical (see Figure 1b). The first study to address this, Switkes et al. (1978), used optical Fourier analysis and found that a prominent bias of power at the cardinal axes exists, and also observed the anisotropy regardless of whether the scene content was purely natural outdoor scenery or scenes of carpentered content. Studies of magnitude and direction of gradients in imagery (Coppola et al., 1998a) and of Fourier energy/amplitude examined as a function of orientation within weighted or sectored regions of images (Hansen & Essock, 2004; Keil & Cristóbal, 2000; Oliva & Torralba, 2001; Torralba & Oliva, 2003; van der Schaaf & van Hateren, 1996) support the anisotropic distribution of image content.

A question that will arise out of reviewed research in later sections of this paper is whether or not there is, on average, more horizontal content relative to vertical content. Unfortunately the answer has not been clear. The primary reason for this lies in the fact that of the studies that do report magnitudes for the different orientations, the image sample size was very limited, or the method in which measurements were made likely suffer from sampling errors induced by the discrete structure of the amplitude spectrum obtained via the digital DFT process (refer to Hansen & Essock, 2004 for a discussion on digital sampling errors of amplitude spectra). We will return to the issue of orientation biases in the sections that review human perception of orientation in natural images.

It is well established that visual sensitivity varies as a function of stimulus orientation. There is a long history of research devoted to examining how human (e.g., Jastrow, 1893) and animal (e.g., Lashley, 1938) performance on a variety of tasks varies with stimulus orientation. The typical finding is that performance is superior for horizontal and vertical orientations (cardinal axes), relative to 45° and 135° oblique orientations (oblique axes). This general perceptual bias in visual processing has been labelled the oblique effect (Appelle, 1972). However, it is important to realize that there are two different sources of anisotropic "oblique effect" behavioural performance: An orientation bias in the basic functioning of the visual system (e.g., contrast sensitivity and acuity), and a bias of later visual processing, presumably beyond V1 obtained on more cognitive tasks such as the naming, encoding, and memory of orientation, in which oblique orientations are more confused with each other than are stimuli of cardinal orientations. To help avoid confusing these two oblique effects in the literature, they have been termed *Class 1* and *Class 2* oblique effects, respectively, distinguishing them on the basis of the type of task on which they occur (Essock, 1980). That the two types of oblique effect are fundamentally different is readily apparent by considering that the bias of contrast sensitivity (i.e., Class 1) is fixed to retinal coordinates if the viewer's head or body is rotated, but that the oblique effect of orientation memory (i.e., Class 2) is often labile, dependent upon the observer's sense of "up". Furthermore, the memory anisotropy is readily obtained with nonvisual stimuli, whereas the Class 1 oblique effect seems closely tied to a physiological bias in the early visual cortical area.

Several authors (Annis & Frost, 1973; Coppola et al., 1998a; Keil & Cristóbal, 2000; Ross & Woodhouse, 1979; Switkes et al., 1978; Timney & Muir, 1976) have pointed to a possible phylogenic or ontogenetic relationship between the physical anisotropy in natural scene content and the oblique effect where visual performance with oblique stimuli is poorest relative to stimuli at the cardinal axes. In essence, the conjecture is that since, in the natural environment, there is less information at the oblique axes, then the relative number of cortical units devoted to processing such information should be less when compared to the other, more prominent orientations. In fact there are numerous reports from studies utilizing single-unit recording, visual-evoked potentials, optical imaging of intrinsic signals, and functional magnetic resonance imaging that support a bias in cortical processing of orientation at the cardinal axes (Chapman, Stryker, & Bonhoeffer, 1996; Coppola et al., 1998a; De Valois, Yund, & Hepler, 1982; Furmanski & Engel, 2000; Li, Peterson, & Freeman, 2003; Maffei & Campbell, 1970; Mansfield & Ronner, 1978; Orban & Kennedy, 1980; Sokol, Moskowitz, & Hansen, 1989). While there is indeed a high degree of overlap between the observed orientation biases in natural scenes and the visual phenomenon known as the oblique effect, when tested empirically, this parallel takes on a different interpretation.

HUMAN PERCEPTUAL PERFORMANCE AND THE CONTENT REGULARITIES OF NATURAL SCENES

The following sections review the psychophysical experiments that have examined the extent to which human visual performance is guided by the scale and orientation content regularities described in the earlier sections. In the following sections, the relationship between the value of α of natural scenes and three aspects of human visual processing will be examined. The first aspect of visual processing will deal with adaptation to natural scene and natural scene-like imagery and its subsequent effects on visual sensation. The second will involve the ability of human observers to detect, or discriminate, a change in the spectral α of different natural scenes. Finally, the ability of human observers to discriminate between two different objects or natural scene images that have been progressively blended (e.g., an object in image "A" is gradually "morphed" or "spectrally blended" into an object in image "B"), with the ability to discriminate such changes under conditions where α was varied will be considered. In subsequent sections, we will review recent work from our laboratory where human visual perception of orientation was measured in the context of broad-scale visual noise in which α was varied, as well as in the context of natural scenes containing content biases at different orientations. Lastly, we will present two experiments as logical extensions of the reviewed research in order to provide a more complete description of human visual performance as a function of the content regularities observed in real-world imagery.

Broad-scale visual adaptation of natural scene imagery

Before discussing visual adaptation, it is necessary to first discuss some aspects of human visual sensitivity as defined by the contrast sensitivity function (CSF). Simply put, the CSF is a plot of the minimum contrast required for a pattern to be detected. Contrast sensitivity varies as a function of spatial frequency of the test grating, resulting in a CSF that is band-pass in nature, with a peak at about 3–6 cpd and a cut-off around 60 cpd at 100% contrast. Adaptation experiments first demonstrated that the CSF is not due to a single sensitivity mechanism, but rather is the "envelope" of multiple, relatively independent, narrowly tuned mechanisms (Campbell, Kulikowski, & Levinson, 1966; Pantle & Sekuler, 1968). That is, by adapting an observer to a pattern of a single spatial frequency, sensitivity will be reduced at a relatively narrow band of spatial frequencies rather than at all spatial frequencies. Similarly, the demonstration of these spatial frequency "channels" was soon followed by the demonstration of narrowly tuned channels for different orientations (Blakemore & Campbell, 1969; Blakemore & Nachmias, 1971). When considering natural scenes as stimuli, the single-frequency/single-orientation studies raise the important question regarding the effects of prolonged viewing of natural scenes on the visibility of the various components of natural scenes. Specifically, what affect does the distribution of content (scale and orientation) in typical natural scenes have on the sensitivity to single image components?

Webster and Miyahara (1997) investigated whether or not prolonged viewing of the $1/f^{\alpha}$ distribution of spatial content of natural scenes (Figure 1a) alters sensitivity to individual sinusoidal gratings. The authors set out to investigate this question with a set of natural image stimuli ranging from open-range scenes to close-up scenes containing no artificial structure, as well as a set of visual-noise images possessing amplitude spectra similar to natural images (i.e., characterized by the $1/f^{\alpha}$ fall-off). The visual-noise patterns that were used consisted of six different spectral slopes ranging from $\alpha = 0.0$ to $\alpha = -1.5$, in steps of -0.5. The general procedure involved an initial 5 min adaptation period where the subjects viewed a rapid and random succession of either natural scenes or noise patterns (for the noise pattern conditions, α was constant, with each α tested separately). Following adaptation to a particular image type, observers engaged in a suprathreshold matching task that utilized sinusoidal gratings across a broad range of spatial frequencies. The results were quite striking in that they revealed CSFs with their peaks shifted to higher spatial frequencies for the natural scene stimuli. Given that the αs of the natural scenes that were used were quite similar, there were no apparent differences in the shifting of the adapted CSF peaks for the different scene categories. Considering the large amount of amplitude present at low spatial frequencies, it makes sense that adaptation to such scenes would shift sensitivity away from the lower to middle frequencies. Even more interesting were the CSFs obtained after adaptation to the noise patterns with different αs, in that peak sensitivity for the adapted CSFs seemed to shift to higher and higher spatial frequencies as α increased, with virtually no change in the position of the peak for the patterns with $\alpha = 0.0$.

Thus, the traditional, unadapted, CSF does not characterize an observer's sensitivity to spatial content contained in natural scenes in typical viewing conditions. Just as the bias in the distribution of natural scene content across spatial scale biases sensitivity away from the most prevalent content, our work (e.g., Essock, DeFord, Hansen, & Sinai, 2003; Hansen & Essock, 2004; Hansen et al., 2003) has shown that the orientation bias contained in natural scenes biases the visibility of oriented spatial content at the orientation of the most prevalent content in a given natural scene. When considered concurrently, human sensitivity to the specific content of natural scenes is adjusted by both the scale and orientation of content regularities typically found in natural scenes.

Detection and discrimination of changes in the spectral slope of natural scenes

As mentioned earlier in this review, the typical α for any typical natural scene is approximately 1.0. This was shown in the sampling distribution (i.e., Figure 3b) obtained from randomly sampling natural images from the larger sample of naturalistic images. Each point on the curve represents the number of orientation-averaged amplitude spectra slopes divided into smaller ranges. As can be clearly seen, the majority of images have an α value within the 0.80–1.0 range. Such a plot could be thought of as indicating the number of times one might encounter natural scenes with any particular α for any given day. From that sampling distribution, the likelihood of encountering scenes that deviate far from the 0.80–1.0 range is fairly small. This point has been made by a number of studies that have examined human visual performance in the context of natural scenes, in which it was argued that, since we are more likely to encounter environments exhibiting these αs, our visual systems should function optimally in such environments. The idea here is that during early development (and also long-term development on an evolutionary time scale), such image statistics were likely present and thus, over time, our visual systems were "wired" in a manner best suited to process visual information characterized by the $1/f^{\alpha}$ spatial relationship.

Knill et al. (1990) first set out to examine whether or not our visual systems might be optimally suited to process complex imagery at any given α. In their study, Knill et al. used greyscale visual noise stimuli with varying αs where observers engaged in a 2AFC task in which they were to detect a change in spectral slope. The results revealed that observers were best at detecting a change in α when the standard noise pattern was in the range of 1.4–1.8. It was concluded that the human visual system is "tuned" to better process images with these α values. However, Tadmor and Tolhurst (1994) questioned these results by arguing that the range that was proposed to be best in terms of visual processing was well outside the typical range one might encounter. They also pointed out that while the noise patterns had spectral properties similar to natural scenes, they were still not representative in their content (i.e., the patterns had random phase spectra). Particularly, Tadmor and Tolhurst were interested in determining whether or not the human visual system was best suited to process information in the form of the typical $1/f^{\alpha}$ relationship. They gathered two sets of imagery consisting of either natural scenes (with αs ranging from 0.78 to 1.38) or visual noise patterns with a comparable range of α values. The task they utilized was a spatial 3AFC, where two of the images (either natural scenes or visual noise) were set to one of seven reference αs and the other had a small change in α. The observers were to identify which of the three images was the "odd one out". The results for both of the visual noise and natural scene stimuli consisted of lower thresholds for steeper αs, with the highest thresholds obtained for reference αs near 1.0. The authors interpreted this finding as supporting the fact that the human visual system is optimally suited to process images exhibiting αs in the range of 0.8–1.0. This interpretation may seem odd since thresholds were highest in this range, but consider what a high threshold for discriminating changes in α implies. Tadmor and Tolhurst argue that such a high discrimination threshold for reference images in the typical α range suggests a high degree of "tolerance" for changes in α that might occur within this range.

They reasoned that since we typically encounter scenes with αs near 1.0, slight deviations from this value that might occur, as a result of, for example, shifting our gaze from one area to another within a given environment, or slight accommodative errors due to fixating objects at different distances, would have little impact on our ability to effectively process visual information. Such a tolerance allows us to view the visual system as being set up to "expect" small deviations from the typical α and therefore not waste valuable processing time and resources by signalling to the observer that such changes have occurred. It is in this sense that higher thresholds for images with typical α values can be considered optimal.

Another aspect to consider is that the reference αs that were used as the comparison from which the changes in α were to be discriminated were not the true αs of the original images (also the mean luminance and rms contrast were constrained to very low values). This begs the question of whether or not these results would be obtained if observers were to discriminate changes in α relative to the true αs of the respective natural scene. Thomson and Foster (1997) addressed this question indirectly while examining the role of the phase spectra of natural scenes in detecting changes in α. The task was a temporal 2AFC that utilized natural scenes ranging from texture patterns to actual scenes. For each α category tested (0.7, 1.0, 1.3), thresholds were highest when the natural image's α was closest to its original value. This seems to argue against the idea of an ideal α about which our visual systems are most tolerant. It should be noted that the images with α values deviating from 1.0 (i.e., 0.7 and 1.3) that were used were rather atypical, consisting primarily of texture patterns (utilized to address a different issue), and are still within the typical range of α values mentioned earlier. Their findings are still compelling, and call for a more extensive, highly controlled investigation of this issue.

The role of spectral slope in discriminating between different scene content

In the previous section, the question of interest was whether or not the human visual system is tuned to optimally process natural scenes that have the characteristic $\alpha = 1.0$; and, with the caveat stated, it appeared as though it was. One interesting aspect from the Tadmor and Tolhurst (1994) paper was the rating that participants gave to the highly constrained imagery when presented at different reference α values. Participants were asked to subjectively rate the images in terms of "best perceptive quality", that is, at which reference α did the images appear as having the best quality (these images were considerably "degraded" in terms of having low rms contrast and mean luminance). Regardless of the original images' true αs, all images were rated to have the best perceptual clarity when the reference α was set to roughly $\alpha = 0.8$. The implication here is that when imagery is of substandard quality, the α that yields highest perceptual quality was close to the typical α of natural scenes. It is at this value (when scenes are highly constrained) that people might expect to best perform some visual task. One of the many assumptions here is that when adjusting degraded imagery by some factor (α in this case), observers might be using an internal criterion that could be associated with performing various tasks that require images to be of the best quality. This raises the issue of whether or not there is an ideal α under which people can best perform some visual task. Párraga et al. (2000), and Tadmor and Tolhurst (2000) addressed this issue with two unique visual discrimination tasks that provided some potential answers to this issue.

The task that was devised by Párraga et al. (2000) involved progressively ''morphing'' two natural images that were very different in perceptual meaning (e.g., car and bull or the face of a man and the face of a woman), but had similar salient features. In a modified 2AFC procedure where three images were presented sequentially (with the second image always being the unchanged reference image), observers were instructed to indicate which interval (first or third) contained the stimulus image that was different from the reference (second interval) image. All three images were set to a single α value. Threshold was considered the amount of morphing (applied in progressive steps) that occurred in the test image that allowed observers to discriminate a change. The results indicated that observers were best able to discriminate when an image was being morphed into another image when α was close to 1.0. Tadmor and Tolhurst (2000) had conducted a similar study where the Fourier spectra (amplitude and phase) of two different images were spectrally blended in progressive steps. Again it was found that observers had the lowest discrimination thresholds when the reference α was near 1.0. However, lower thresholds around $\alpha = 1.0$ were not clearly observed when the imagery was highly constrained in the manner discussed earlier; instead discrimination thresholds were lowest across a much broader range. Taken together, the studies mentioned above provide some evidence that the human visual system is best suited to discriminate spatial form when in the context of natural scene imagery possessing the typical α that characterizes natural scenes.

Human perception of orientation in broad-scale stimuli

As mentioned earlier, numerous researchers have drawn parallels between the structural anisotropy observed in real-world imagery and the behavioural oblique effect measured psychophysically, with the implication being that the structural bias my serve as a potential origin of the oblique effect performance bias. However, there exists a limitation in most of the discussion of the development and basis of the oblique effect that has previously been ignored. Specifically, the Class 1 oblique effect is invariably tested with simple line, edge, or grating stimuli. Such simple visual stimuli presented in isolation are quite unlike the visual information humans are actually presented with in typical visual environments. In other words, the Class 1 oblique effect is demonstrated with very simple, rather nonnaturalistic stimuli such as high spatial frequency gratings, whereas discussed earlier, our natural environment is highly broadband (or broad-scale), containing energy at all spatial frequencies and orientations.

In a series of psychophysical studies conducted in our laboratory, we sought to determine whether or not the oblique effect is present when the stimuli consist of broadband structure (Essock et al., 2003). Several sets of stimuli were constructed in the frequency domain and made to have amplitude spectra with α values of 0.0, 0.5, 1.0, and 1.5. The phase spectra of the stimuli were randomly assigned values ranging from $-\pi$ to π, thus the corresponding spatial "patterns" did not possess any meaningful spatial content. The reasoning behind utilizing visual noise patterns was two-fold: First we wanted to examine visual perception of orientation purely as a function of the slope of the amplitude spectra, and, second, we wanted to avoid any possible interaction with the information "carried" by the phase spectra (e.g., the semantic meaning). In order to "embed" broadband oriented structure in these stimuli, a triangle weighting function (with an orientation bandwidth of 45°) was multiplied with the amplitude spectra (centred at each of four different orientations: Vertical, 45° oblique, horizontal, and 135° oblique). Thus the oriented content present in the stimuli was solely defined in terms of oriented amplitude increments. The experimental paradigm involved a standard suprathreshold psychophysical matching procedure (i.e., method of adjustment). Participants were seated in front of a monitor covered by a large circular mask with two circular, laterally displaced windows in which two noise patterns were presented. On the left was a comparison pattern that contained a fixed suprathreshold oriented increment of amplitude (at either 22.5° or 112.5°), and on the right was a test pattern (with an oriented increment of amplitude at one of the four nominal orientations) in which observers were instructed to adjust (via keypress) the magnitude of the content until it appeared to match that contained in the comparison pattern. Surprisingly, the data clearly demonstrated that matches were most accurate for oblique stimuli, least accurate for horizontal, and intermediate for vertical; an effect we termed the "horizontal effect" (see Figure 4a). This effect was found to occur in all α conditions; however, the magnitude of the effect was maximal in α conditions closest to 1.0 (refer to Figure 4b). Worst performance for horizontal was also obtained at threshold in a 2AFC paradigm (see Essock et al., 2003 for further details). Thus, the perception of oriented content in broadband noise patterns with amplitude spectra mimicking that of natural scenes produces performance very different from that found with simple stimuli or that might have been predicted if performance is presumed to match the prevalence of oriented content in typical natural scenes.

An argument could be made that the results were specific to stimuli consisting of broadband visual noise, and that such an effect may not have been

a)
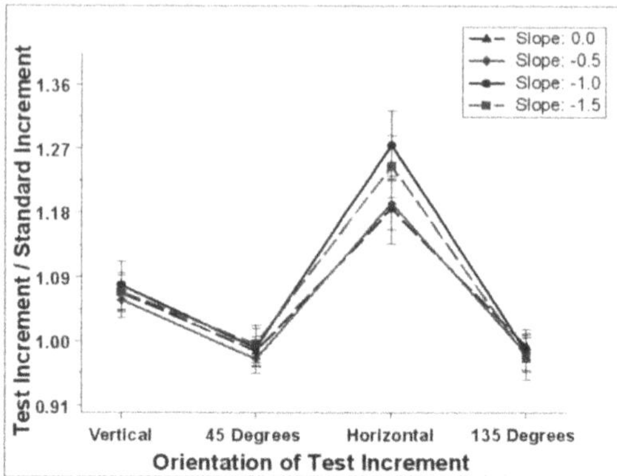

b)
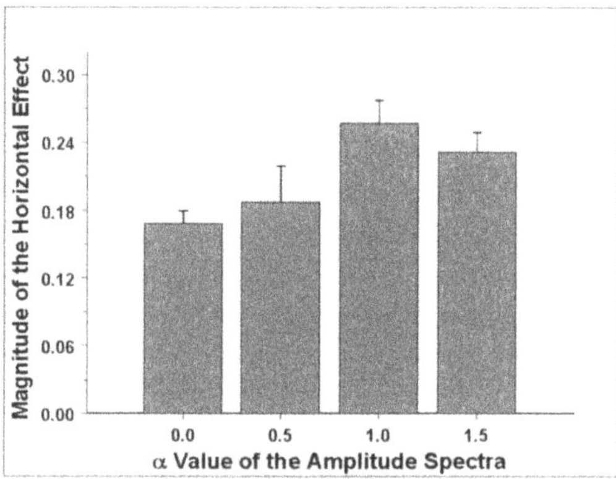

Figure 4. (a) Data replotted from Essock et al. (2003). The paradigm was suprathreshold matching where participants adjusted the magnitude of the broadband increment of amplitude contained in a test pattern in order to match the perceived magnitude of the oriented structure in a standard pattern. On the ordinate is the ratio of the test pattern increment to that of the standard; a ratio of one indicates a perfect match. Note that a ratio above one indicates amplitude at a given orientation had to be *increased* in order to make a perceptual match to the standard. (b) Histogram showing the magnitude of the horizontal effect (i.e., the difference between the average ratio for vertical, 45° and 135° oblique matches and horizontal matches) for each of the α conditions; error bars are +1 *SEM*.

present if the stimuli had consisted of actual real-world scenes. Such an argument reduces to whether or not the phase spectra would interact with the detection of oriented structure defined as broadband increments of amplitude. That is, since it is the phase spectra that specifies how the different sinusoidal components of an image's ''signal'' sum to form a given scene, the amount and localization of the lines and edges that form the meaningful content are determined by each scene's phase spectrum. Whether or not the presence of the phase defined content of an image would produce the same effect observed with visual noise patterns was investigated by Hansen and Essock (2004) and Hansen et al.

(2003). In those studies sets of imagery were gathered that contained naturalistic content biases at one of each of the four nominal orientations as well as imagery containing equal amounts of naturalistic content at all orientations (refer to Hansen et al., 2003 for a detailed description of the image sampling, correction, and selection procedures). The oriented increments of amplitude were applied in the same manner as with the noise patterns, only before weighting the spectra with the triangle filter, the amplitude spectra of the imagery were averaged across orientation in order to eliminate the content biases as defined by the original spectra. This results in maintaining the form (i.e., slope or α) of the spectra while preserving the content biases as defined by the phase spectra of the imagery (refer to Figure 5 for further details). The experimental paradigm was a single interval Yes/No design where on any given trial, observers were required to indicate (via keypress) whether or not the stimulus image contained an oriented increment of amplitude—trials were blocked by orientation of the amplitude increment. The data are replotted in Figure 6 and show the presence of two performance biases. First, regardless of the content bias present in the natural stimulus images, performance for detecting horizontal increments of amplitude was always poor. Second, performance for detecting an oriented

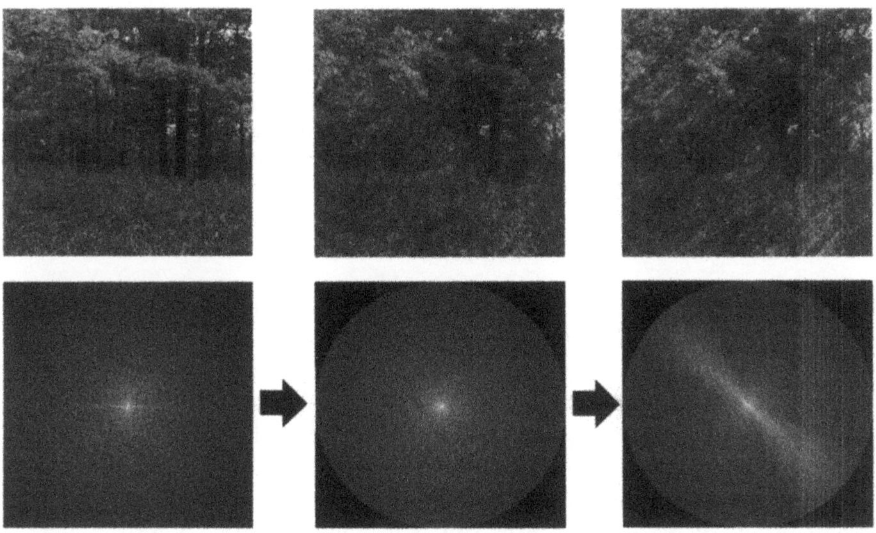

Figure 5. An example of the method by which an oriented increment in amplitude was applied to each image's spectra. The example runs from left to right, with the top row showing the resulting spatial image containing the manipulation in the frequency domain. Starting leftmost with an unaltered image (top left) and the corresponding amplitude spectrum (bottom left); bottom middle: After the spectrum was made isotropic; bottom right: After the amplitude spectrum has been made to contain an oriented increment in amplitude at 135° (note that the coordinates of the Fourier amplitude spectrum are rotated 90°) with the triangle filter referred to in the text.

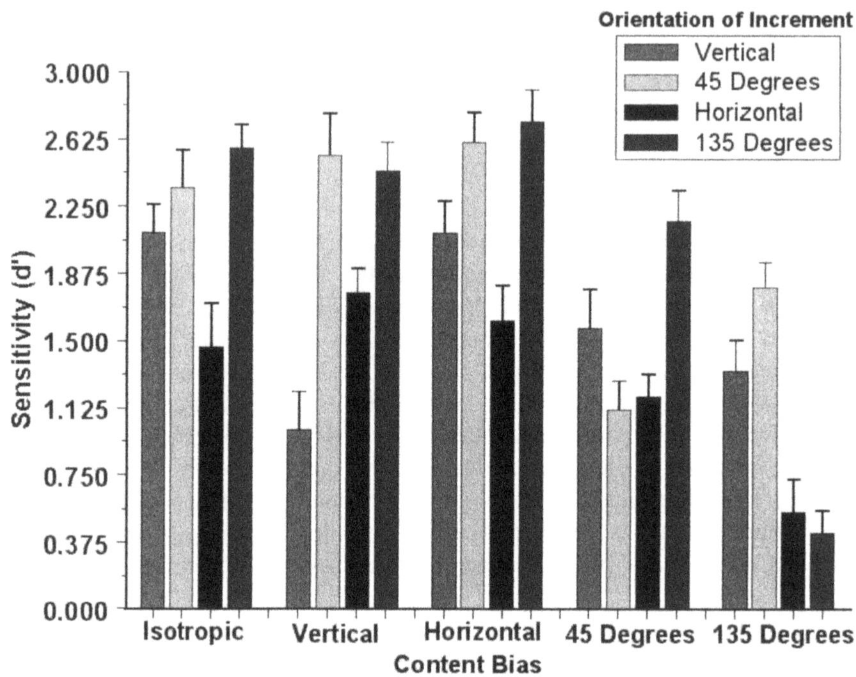

Figure 6. Data replotted from Hansen et al. (2003). On the abscissa is the *averaged* observer data (and the averaged within-subject *SEM*), showing performance for detecting an oriented increment in amplitude at each of the four test orientations for each image content type subset. For example, the grouping labelled *45 degrees* on the abscissa plots averaged observer sensitivity (d') for detecting an oriented increment of amplitude at each of the test orientations with scenes that were dominated by image content at 45°.

increment of amplitude was reduced when the oriented increment was at the same orientation as the phase-defined content bias (i.e., a ''content-dependent effect'').

The association between the behavioural performance horizontal effect observed with broadband natural stimuli and the prevalence of content at the nominal orientations in natural scenes was investigated by Hansen and Essock (2004). It has been demonstrated that the most prevalent content in real-world scenes occurs at the cardinal axes (for both purely natural as well as purely carpentered environments as described in the earlier sections). Determining whether or not there is more content at the horizontal as opposed to vertical orientation is not as easily answered. Specifically, this issue arises out of the fact that whether or not horizontal or vertical content biases are present, and their magnitude, will most likely be determined by the environment from which imagery is sampled. Recently, several reports argue that there is a tendency for a greater bias in favour of horizontal content relative to vertical in typical natural scenes (Baddeley, 1997; Baddeley & Hancock, 1991; Hancock et al., 1992; Hansen & Essock, 2004). Specifically, Hansen and Essock (2004) have shown that the bias of horizontal over vertical content is present in imagery that contains a ground/tree or tree/sky horizon line, contains a predominant receding ground plane, or contains general foliage/shrubbery. This bias was also present in a random sample of imagery obtained from an independent database (e.g., van Hateren & van der Schaaf, 1998). That the content of typical scenes has a horizontal bias is an important finding as we have proposed that the behavioural horizontal effect would have evolutionary utility in such environments (Essock et al., 2003; Hansen et al., 2003). We have suggested that a mechanism that reduces the perceptual salience of the most prevalent content in a scene would have the effect of enhancing the "unique" content in such an image. That is, a mechanism that turns down sensitivity for the "expected" content in a typical scene would serve to relatively enhance the salience of "unexpected", or novel, content at off-horizontal orientations. We have modelled such a mechanism in terms of an anisotropic feedback gain control that adjusts broadband sensitivity for horizontal content, and suggest that the feedback gain control arises out of a larger number of cortical neurons in striate cortex tuned to horizontal orientations (i.e., individual horizontal output signals are normalized by a larger feedback signal, and hence turned down more than output units tuned to other orientations). Such a theory receives much support from single unit recordings and optical imaging of intrinsic signals (Chapman et al., 1996; Chapman & Bonhoeffer, 1998; Coppola, White, Fitzpatrick, & Purves, 1998b; Li et al., 2003).

EXPERIMENT 1: HUMAN PERCEPTION OF ORIENTATION AND THE SLOPE OF THE AMPLITUDE SPECTRA OF NATURAL STIMULI[2]

In the earlier sections, experiments were reviewed where human perception of orientation in the context of broadband visual noise stimuli with amplitude spectra α values ranging from 0.0 to 1.5 was found to be highly anisotropic. As shown in Figure 4b, the magnitude of the anisotropy was higher when the α value of the noise was 1.0. However, these stimuli had random phase spectra and thus contained similar amplitude spectra, but not the semantic content inherent in natural scenes. The findings also reviewed in that section, in part, addressed such an argument, but those studies did not address α directly since the focus of those studies was more on the existence of interactions between the phase-defined content biases of natural scenes and the ability to detect oriented increments of amplitude. In those experiments, stimulus image sets containing

[2] Portions of this research were presented at the third annual Vision Sciences Society meeting.

approximately equal amounts of content at all orientations were employed as control stimuli in which the results from the oriented content biased imagery could be compared. The average α value for that imagery was approximately 0.95. Thus, the question of whether or not the ability to detect oriented increments of amplitude for real-world stimuli consisting of naturalistic content would depend on α has not specifically been addressed. Here we examine this issue by utilizing sets of natural scene imagery possessing a broad range of amplitude spectrum α values to determine if the magnitude of the horizontal effect changes as a function of the α value of natural scenes. In order to test this, sets of images containing approximately equal content (amplitude) at all orientations were gathered, with each set containing images with amplitude spectra α values within one of several narrow ranges.

Methods

Apparatus. Stimuli were presented by an SGI 540 Visual Workstation on an SGI 420C monitor. To eliminate edge contours from the room and monitor bezel, a circular mask (visual angle: 59°) with a 17° circular stimulus aperture was fitted to the monitor. Resolution was set at 800 × 600, frame rate at 120 Hz, and maximum luminance was 80 cd/m^2 and was linear with camera output.

Participants. Four observers, naïve to the purpose of the experiment, participated in the current experiment. All had normal or corrected to normal vision and were screened by vision test to assure they had no residual astigmatism. The age range was 18–32 and IRB-approved informed consent was obtained.

Image selection and manipulation. Sets of imagery containing approximately equal amounts of naturalistic content at all orientations were gathered from our image database described earlier (refer to Hansen et al., 2003 for the details on how measurements of oriented content were carried out). After gathering the set of "naturally isotropic" images, each image (512 × 512 pixels) was in turn, corrected for photographic gamma and windowed with a spatial Gaussian function to eliminate edge effects incurred in the Fourier transform. All images were then Fourier transformed so that the α value could be obtained from the amplitude spectrum of each image after averaging across orientation (only 1–200 cpp were included in the calculation of α). This process resulted in 24 images that were then split into six groups of four based on α (see Figure 7a for examples). The descriptive statistics of the α values for each set are as follows: Set 1: $M = 0.659$, $SD = 0.02$; Set 2: $M = 0.725$, $SD = 0.03$; Set 3: $M = 0.854$, $SD = 0.02$; Set 4: $M = 0.949$; $SD = 0.03$; Set 5: $M = 1.061$, $SD = 0.03$; and Set 6: $M = 1.443$, $SD = 0.22$. The process of stimulus generation was as follows. First, each image was Fourier transformed and the amplitude spectrum for each image was then forced to be isotropic by averaging across all orientations at each spatial frequency (thus eliminating any small biases in amplitude; note that such a procedure effectively eliminates edge effects incurred in the Fourier transform process due to averaging across orientation; refer to Figure 5). An increment was made by weighting a given amplitude spectrum by a 45° orientation-bandwidth triangle filter with a peak centred on one of the four test orientations (see Figure 5). In addition, an isotropic version of each image was saved without an oriented increment of amplitude. This process resulted in five stimulus images created from each original image. After manipulation, each image was inverse Fourier transformed back to a spatial image, at which time they were fit with an edge-''blurred'' circle in order to eliminate the images' border edges. In order to facilitate the comparison between data obtained with these natural scene stimuli and the data from the visual noise experiments described earlier, all of the above procedures were carried out again with the same image sets, only the phase spectra were scrambled (each phase angle in each image's phase spectrum was randomly reassigned a new coordinate with the spectrum). The scrambled phase images sets served as a control condition (see Figure 7b for examples).

Procedure. Participants viewed the display from a distance of 1.325 m (stimuli subtended 10.45° visual angle) with their heads centred in a chin- and forehead rest. In this experiment a single-interval Yes/No task was employed. Observers responded by keypress to indicate whether or not an oriented increment was present. All sessions were blocked by the orientation of the test increment and took no more than 50 min to complete (with only one session allowed per day). Each block contained 240 trials, half with an oriented increment, and half without, for a total of 960 trials per session (i.e., each image

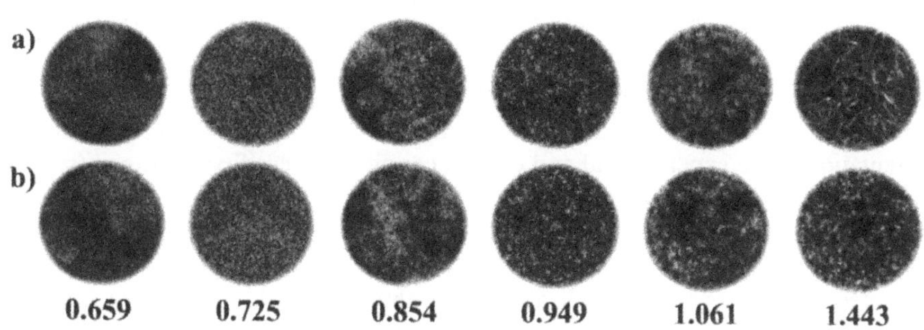

Figure 7. Stimulus examples from each α image set used in Experiment 1; the numbers at the bottom indicate the respective α value from which the examples were selected. Row (a) images that have had their amplitude spectra made isotropic, but do not contain an oriented increment of amplitude. Row (b) examples from the scrambled-phase control condition. The only difference between Rows (a) and (b) is that the phase spectra of the images in Row (b) have been scrambled.

was randomly presented 10 times, 5 with an increment and 5 without). A single trial consisted of the presentation of a fixation cross (500 ms), followed by the stimulus image (400 ms), followed by a white-noise mask (500 ms). All participants were given practice sessions with auditory feedback. An unbiased estimate of sensitivity (d′) was calculated for performance in each session (Creelman & Macmillan, 1991) and sensitivity estimates were averaged from four daily sessions for the experimental image set condition, and for the control image set condition (i.e., scrambled phase image sets), each with a different random order (total of 3840 trials per condition, per person).

Results and discussion

A 6 × 4 (Image slope set × Orientation) two-way repeated measures ANOVA was used for the initial analysis of each condition (experimental and control). For the experimental image sets, there was a significant main effect for slope, $F(5, 15) = 6.11$, $p < .05$, Cohen's $f = .671$, as well as for orientation, $F(3, 9) = 6.52$, $p < .05$, Cohen's $f = .782$; however, the interaction was not found to be significant, $F(15, 45) = 0.585$, $p > .05$, observed power = .307; and for the control image sets: there was a significant main effect for slope, $F(5, 15) = 2.93$, $p < .05$, Cohen's $f = .11$, as well as for orientation, $F(3, 9) = 8.82$, $p < .05$, Cohen's $f = 1.57$, again the interaction was not found to be significant, $F(15, 45) = 1.67$, $p > .05$, observed power = .82. As shown in Figure 8a and 8b, the two data sets are very similar which argues that the phase structure likely had little impact on performance. To examine this in more detail, the same image slope sets were compared for the experimental and the control conditions with separate 2 × 4 (Condition × Orientation) two-way repeated measures ANOVAs. As expected, there were no significant Condition × Orientation interactions: For the 0.6588 slope sets, $F(3, 9) = 0.21$, $p = .88$; for the 0.7247 slope sets, $F(3, 9) = 0.78$, $p = .54$; for the 0.8537 slope sets, $F(3, 9) = 0.02$, $p = .96$; for the 0.9494 slope sets, $F(3, 9) = 0.75$, $p = .55$; for the 1.061 slope sets, $F(3, 9) = 0.70$, $p = .57$; for the 1.443 slope sets, $F(3, 9) = 1.82$, $p = .21$. In order to determine whether or not the magnitude of the horizontal effect (i.e., worst performance for detecting horizontal increments of amplitude compared to the other three nominal orientations), the experimental and control conditions were first averaged (since the interactions were not significant) and complex contrasts were carried out for each of the different slope image sets (i.e., performance for detecting horizontal increments of amplitude compared to the average performance for detecting oriented increments of amplitude at the other three nominal orientations). Thus a 6 × 2 (slope vs. horizontal performance and the average performance at the other three nominal orientations) two-way repeated measures ANOVA was employed. There was a significant interaction between the different slope conditions and the magnitude of the horizontal effect, as determined by the complex contrasts described above, $F(3, 9) = 3.4$, $p < .05$, Cohen's $f = .154$).

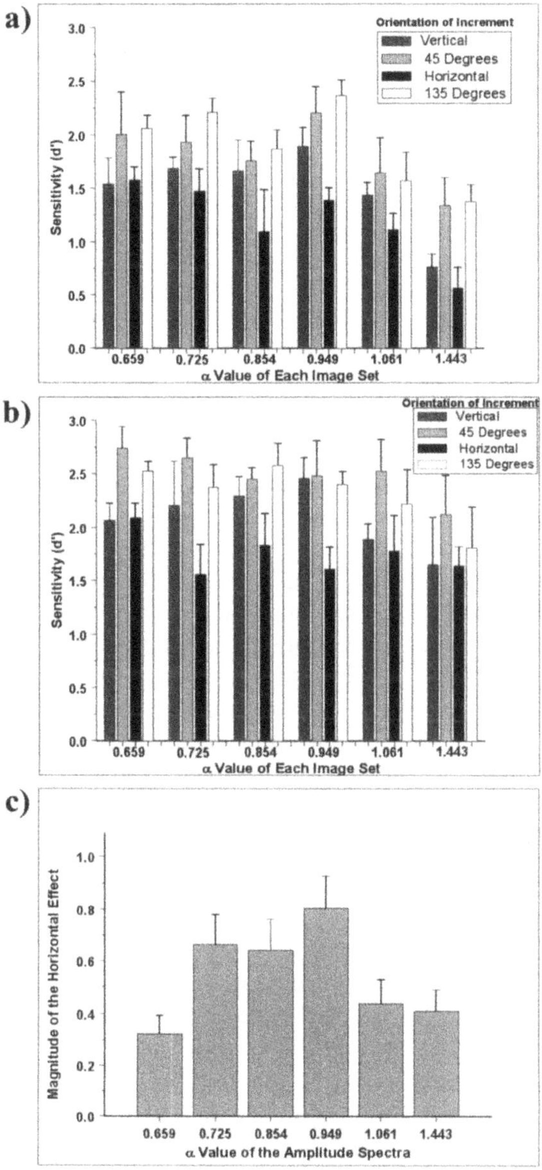

Figure 8. Data from Experiment 1. (a) Experimental condition: Average sensitivity (d′) for detecting oriented increments of amplitude (indicated by bar shading) with imagery contained in each α image set (as labelled on the abscissa). Error bars are +1 *SEM*. (b) Control condition (i.e., scrambled phase imagery): Plotted in the same manner as in (a). (c) Histogram showing the magnitude of the horizontal effect (i.e., the difference between the average ratio for vertical, 45° and 135° oblique matches and horizontal matches) for each of the α conditions (obtained by first averaging across experimental and control conditions); error bars are +1 *SEM*.

Figure 8c depicts this graphically by plotting the differences between horizontal and the average performance for the other nominal orientations.

The main effect of slope is interesting in and of itself as it argues for significant differences (averaged across orientation) in performance as a function of slope. As depicted in Figure 8a and 8b, it's not surprising that there was a significant main effect for orientation (averaged across slope conditions); however, the primary question here was whether or not performance for detecting oriented increments of amplitude at horizontal orientations (as compared to the other orientations) varied as a function of the scenes' amplitude spectra slopes (or α value). As shown in Figure 8c, the magnitude of the horizontal effect (performance for detecting horizontal increments relative to performance for detecting oriented increments at the other three nominal orientations) does indeed scale with α, with the largest magnitude observed with imagery from the $\alpha = 0.949$ image group. Interestingly, the horizontal effect magnitude histogram is quite similar in form to the distribution of α values shown in Figure 3. While the number of images in each set was small, it is worth noting that the largest horizontal effect magnitude occurred with test images that had α values most often encountered, and that this magnitude was weaker for test stimuli possessing α values less often encountered (e.g., Figure 8). The indication of the current set of results is that when the visual system is presented with imagery possessing a typical content distribution (in terms of spatial scale), the inherent gain control mechanism, mentioned earlier, operates maximally (and, in a sense, optimally) by reducing sensitivity to horizontal orientations.

EXPERIMENT 2: PERCEIVING ORIENTATION IN THE PRESENCE OF VARIABLE CONTENT BIASES[3]

In the earlier reviewed work of Essock et al. (2003), Hansen and Essock (2004) and Hansen et al. (2003), sensitivity for detecting oriented increments of amplitude in visual noise and natural scenes was discussed. The primary finding was that the ability to detect oriented increments of amplitude was worst at horizontal relative to performance at off-horizontal orientations (the horizontal effect) and that this effect remains even with natural scene stimuli containing a natural bias in content at any given orientation. As mentioned in earlier, we have suggested that this effect evolved to discount the dominant horizontal content typically present in natural scenes in order to "emphasize" content at off-horizontal orientations. Such a hypothesis would predict that sensitivity for detecting content (oriented amplitude increments) at off-horizontal orientations would be best for observers presented with imagery containing strong horizontal biases compared to imagery where such a bias is less pronounced. Here we investigate this idea with natural scene imagery containing different amounts of

[3] Portions of this study were presented at the third annual Vision Sciences Society meeting.

horizontal content biases to determine if sensitivity for detecting amplitude increments at off-horizontal orientations depends on the relative predominance of horizontal content in a given image. In order to test this, a set of natural scene imagery was compiled to serve as experimental stimuli in which the amount of horizontal content (carried by the phase spectra) was variable. Additional control image sets were also obtained that contained variable content biases at the other nominal orientations (i.e., vertical, 45°, and 135° oblique). Experiment 2 is similar in nature to those experiments conducted by Hansen and Essock (2004) and Hansen et al. (2003); however, in those experiments, the magnitude of the content biases was held constant. That is, the amount of content bias at a given orientation for the imagery of each bias set was held constant in those experiments. Here we use real-world stimuli containing variable amounts of naturalistic content at one of the four nominal orientations. The experimental condition involves the imagery with variable horizontal content biases, and it is with these images that a performance dependency with respect to detecting oriented increments of amplitude at off-horizontal orientations is expected. No such performance dependency is predicted for the control image sets with content biases at the other nominal orientations.

Methods

Apparatus. All aspects of this section are identical to the Apparatus section in Experiment 1.

Participants. Six observers, five of which were naïve to the purpose of the experiment, participated in this current experiment (one observer also participated in Experiment 1). All had normal or corrected to normal vision and were screened by vision test to assure they had no residual astigmatism. The age range was 18–32 and IRB-approved informed consent was obtained.

Image selection and manipulation. Sets of imagery containing variable amounts of naturalistic content at one of each of the four orientations were gathered from our image database described earlier. The additional criteria that had to be met for an image to be considered for use in the current experiment was that the amount of content at the other three orientations had to be approximately equal. The calculation of oriented content biases was made by Fourier transforming each image (taking all of the necessary precautions mentioned in the Methods section in the preceding experiment) and taking the ratio of summed amplitude within a 45° ''wedge'' centred at one of the nominal orientations to the amplitude contained in the rest of the spectrum for each image. This ''oriented amplitude ratio'' yields a percentage of total amplitude accounted for by a given orientation band. Since this is a proportional measurement, we were careful to make sure that the content biases at the other orientations were approximately equal. As shown in Figure 9, 15 images were selected for each set with increasing amounts of content bias at one of the nominal orientations (total of 60 images). Thus if a linear relationship exists between performance for detecting off-horizontal increments of amplitude and the relative amount of horizontal content bias present in the imagery, the experimental stimuli (i.e., images containing variable amounts of horizontal content) should reflect this. Furthermore, the control image sets should show no performance dependency with respect to content biases at the other nominal orientations. Each of the 15 images contained in the four image sets were made to possess an oriented increment of amplitude at one of four orientations by the same methods described in Experiment 1. Since the focus of the image selection process for the above-mentioned imagery was on obtaining imagery with amplitude biases at the different orientations, α was not controlled for. However, the descriptive statistics for α of the four different orientation-biased image sets indicated that the α values of the four sets were very similar, exhibiting considerable overlap. For the variable vertical content-biased image set: M = 1.05, SD = 0.21; for the variable 45° oblique content-biased image set: M = 1.16, SD = 0.27; for the horizontal variable content-biased image set: M = 1.12, SD = 0.27; and for the 135° oblique variable content-biased image set: M = 1.16, SD = 0.27. This was confirmed with a one-way repeated measures ANOVA which indicated that there were no significant differences between the α values for the four different content-biased image sets, $F(3, 42) = 0.816, p > .05$. The α values were calculated by the same method described in Experiment 1.

Procedure. All experimental conditions were identical to those described in the methods section of Experiment 1, with the following exceptions. In the interest of practicality, all 60 images were randomly assigned to one of three experiment image sets (20 per set), each containing 5 images from each of the oriented content-bias sets. Thus on any given day, a participant would be tested with one of the three experiment image sets, where trials were blocked by orientation of the amplitude increment (experimental and control images were not tested separately). Each block contained 200 trials, half with an oriented increment, and half without, for a total of 800 trials per session (i.e., each image was randomly presented 10 times, 5 with an increment and 5 without), with two sessions for each experiment image set (i.e., a total of 4800 trials), and only one session allowed per day.

Since all images were forced to be isotropic (before applying the 45° orientation bandwidth triangle filter to generate the stimuli), the oriented amplitude ratio, used to classify the original images (see above), cannot be used. Thus, in order to compare performance for detecting oriented increments of amplitude with the experimental and control image sets, a measure of the amount of phase-defined content present at the four different orientations was needed. We considered using a local oriented spatial convolution kernel to

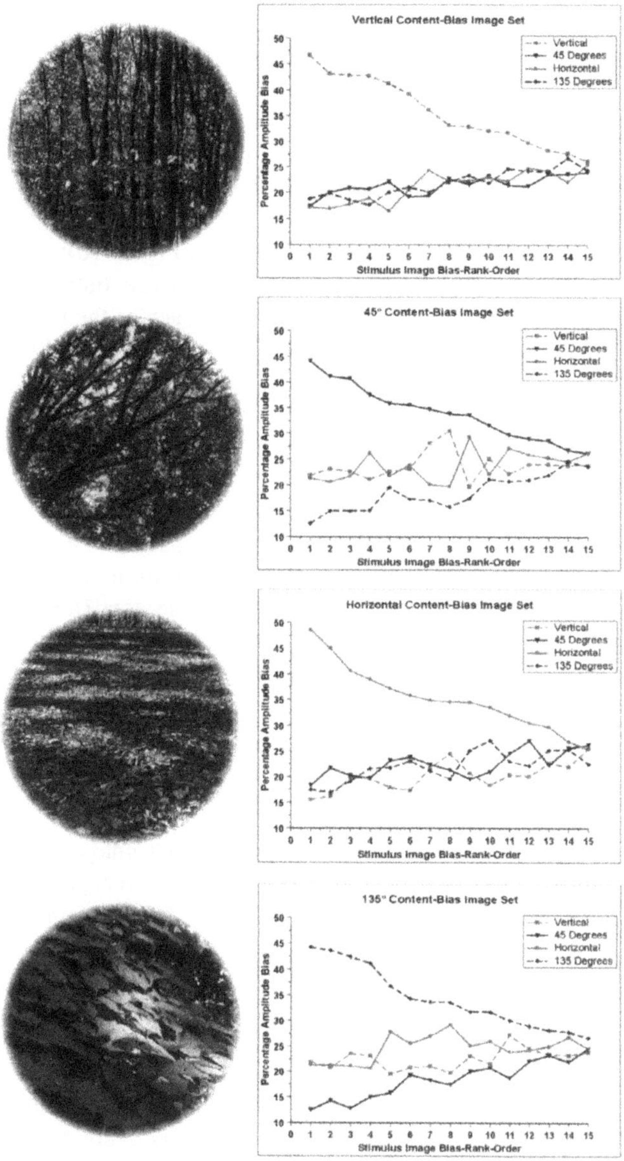

Figure 9. An example of an image (unaltered) from each of the variable orientation content-bias image sets (left column) and graphs depicting the decreasing amplitude at each orientation across the set of images (right column). All graphs are plotted in an identical fashion, with percentage of amplitude bias (i.e., ratio of the amplitude within a 45° ''wedge'' region centred at one of the four orientations to the amplitude at the other three orientations) on the ordinate. For the abscissa, images were rank ordered with respect to amplitude bias for each of the four content-biased sets in order to illustrate the decline in a given orientation's content bias across the images contained in each set.

obtain an estimate of content bias to which the performance data could be compared, where the magnitude of kernel responses was used as a measure of the oriented content biases in the natural scene stimuli. Ideally, the convolution kernel should resemble the visual receptive fields measured at the level of striate cortex. Some, (Stork & Wilson, 1990; Swanson, Wilson, & Giese, 1984; Wilson, 1991) have suggested that the sixth derivative of a Gaussian, or D6, is the function most representative of V1 receptive fields (see Figure 10); this was the kernel we decided to employ. We used the total response output from this convolution kernel as a measure of the local content at each of the four different orientations. The procedure in which this process was carried out is illustrated in Figure 10. In this procedure, the image was first Fourier transformed to obtain the amplitude spectrum. The spectrum was then "whitened" by assigning the mean amplitude value to all coordinates except for the DC coordinate, which was left at its original value. The image was then inverse Fourier transformed. Next, one of four D6 kernels (each oriented at a nominal orientation) was convolved with the image.[4] The filtered image was then passed to a "threshold" algorithm where only filter responses beyond ±1 *SD* from the mean were summed and then divided by the total potential response area (i.e., the actual number of pixels in the filtered image). The result of the process served as a measure of the percentage of filter response for each image. Responses from filters at the same orientation as each image's content bias were then correlated with performance (d′).

Results and discussion

The observers' sensitivity to the oriented content for each image in a content-biased image set was averaged across observers to yield an average performance measure for each image. However, since the focus of this experiment was on how performance for detecting oriented increments of amplitude related to the amount of horizontal content, only correlations between responses from D6 filters at the same orientation as the content bias and the sensitivity to oriented amplitude increments were calculated for each of the four test orientations. For example, the horizontal D6 filter responses to each of the 15 horizontal content bias images were correlated to performance for detecting oriented increments of amplitude at each of the four nominal orientations respectively for each horizontal content-biased image. Pearson's r calculated for horizontal D6 responses and performance for detecting oriented increments of amplitude at each of the nominal orientations show consistent positive relationships between the amount of horizontal bias and performance for detecting oriented increments of

[4] In order to obtain filter responses for the entire image, the area near the borders of the image were "mirrored" beyond the image's true edges so that filter responses affected by the image edges would not be considered as part of the response.

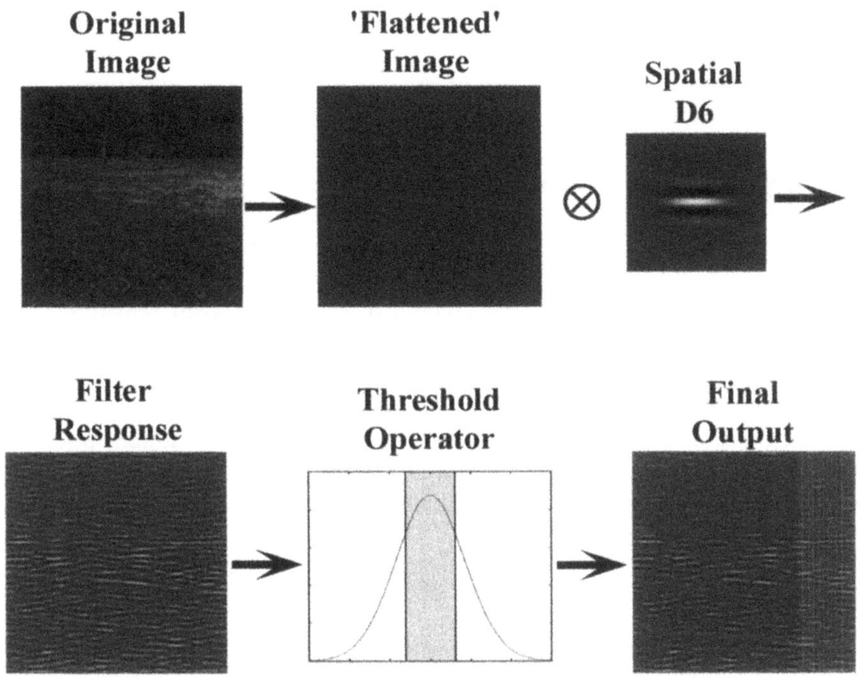

Figure 10. Illustration of the convolution process referred to in the text. From left to right, starting with an unaltered image (for this example, an image from the horizontal variable content-biased image set), which is then made to have a ''flat'' amplitude spectrum (by assigning the mean to all orientations and spatial frequencies without altering the DC component). Notice the corresponding spatial image still contains perceivable structure. The ''flattened'' image was then convolved (⊗) with the appropriate D6 filter. This process passes only local content to which the filter is tuned (orientation and spatial frequency), resulting in the filter response image. The content bias for this example image was then made by summing all pixels ±1 *SD* of the mean (threshold operator) and dividing this value by the total number of pixels in the image. This ratio served as the percentage of the total filter response signalling significant structure.

amplitude: vertical sensitivity, $r(13) = .661, p < .01$; 45° sensitivity, $r(13) = .55, p < .05$; horizontal sensitivity, $r(13) = .625, p < .05$; and 135° sensitivity, $r(13) = .439, p > .05$. The control content bias image data are as follows. For the variable vertical content bias image set: Vertical sensitivity, $r(13) = -.44, p > .05$; 45° sensitivity, $r(13) = -.39, p > .05$; horizontal sensitivity, $r(13) = -.35, p > .05$; and 135° sensitivity, $r(13) = -.31, p > .05$. For the variable 45° content biased image set: Vertical sensitivity, $r(13) = .36, p > .05$; 45° sensitivity, $r(13) = .41, p > .05$; horizontal sensitivity, $r(13) = .14, p > .05$; and 135° sensitivity, $r(13) = -.12, p > .05$. For the variable 135° bias image set: Vertical sensitivity, $r(13) = -.06, p > .05$; 45° sensitivity, $r(13) = -.06, p > .05$; horizontal sensitivity, $r(13) = -.14, p > .05$; and 135° sensitivity, $r(13) = .16, p > .05$.

Figure 11 shows a graphical depiction of the correlations listed above. The prediction made in the introduction of this experiment was that performance for detecting off-horizontal increments of amplitude will scale with stimuli that contained increasing amounts of content at horizontal orientations. The results for detecting oriented increments with the horizontal-biased image set certainly show this positive relationship, however, sensitivity for detecting oriented increments of amplitude for all four test orientations depended on the amount of horizontal structure (i.e., phase defined content) within the images. That is, the results show that the magnitude of the horizontal effect remains relatively constant with increasing horizontal content biases, and *overall* sensitivity for detecting oriented increments of amplitude increased with increasing horizontal content (see Figure 11), while no such trend is apparent with the control imagery. Since the measurement of the image-content bias was obtained from a local oriented convolution kernel, higher sensitivity to the different oriented increments of amplitude must be considered as related to the larger prevalence of globally distributed, *local* horizontal content. This strong relationship argues

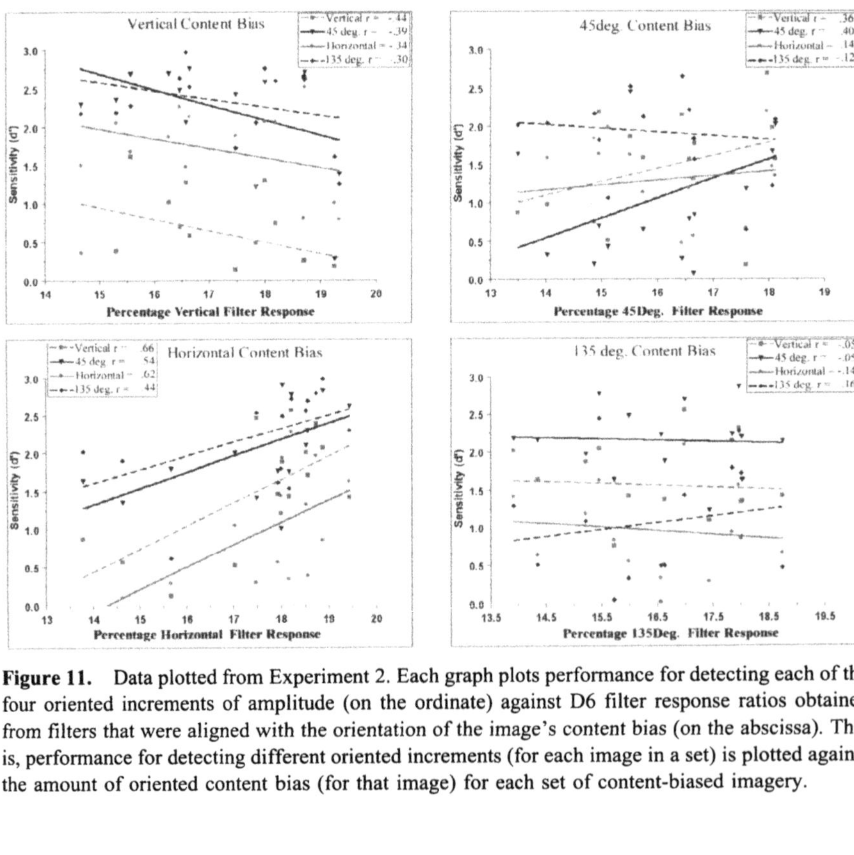

Figure 11. Data plotted from Experiment 2. Each graph plots performance for detecting each of the four oriented increments of amplitude (on the ordinate) against D6 filter response ratios obtained from filters that were aligned with the orientation of the image's content bias (on the abscissa). That is, performance for detecting different oriented increments (for each image in a set) is plotted against the amount of oriented content bias (for that image) for each set of content-biased imagery.

for yet another factor that has an impact on sensitivity for oriented content in natural scenes, that being the relative amount of globally distributed, *local* horizontal structure within an image.

GENERAL DISCUSSION

Experiments 1 and 2 demonstrated two performance dependencies for detecting oriented content in natural scene stimuli. The first was related to the global distribution of content across spatial scale, or α, and the second was an overall performance dependency on the relative amount of oriented content at or near horizontal orientations.

It was shown in Experiment 1 that the magnitude of the horizontal effect is dependent on the particular α value associated with a given scene. Similar findings, although to a smaller extent, were shown in a suprathreshold matching paradigm that utilized visual noise patterns (i.e., random phase spectra) as stimuli (Essock et al., 2003), as shown in Figure 4b. The results from Experiment 1 indicate that stimuli containing content distributed with respect to the typical α of natural scenes serve to maximize the activity/activation of the anisotropic gain control mechanism previously mentioned. This raises a question regarding the utility of such a sensitivity adjustment. Such a relation to slope may stem from the nature by which global activation of striate neurons is pooled. Given the inherent horizontal effect and its correspondence to the typical content biases observed in natural scenes, the data from Experiment 1 argue that scenes with a typical α (i.e., ≈ 1.0) will tend to possess larger content biases at horizontal relative to off-horizontal orientations. Additionally, such oriented biases would be predicted to be less prevalent for imagery possessing very shallow/steep amplitude spectra slopes, a tendency that has yet to be fully investigated.

Experiment 2 was conducted to examine whether or not the amount of horizontal content in a natural scene stimulus could serve as the factor upon which sensitivity at off-horizontal orientations was determined. It was found however, that sensitivity for increments of amplitude at each one of the test orientations scaled (i.e., positive relationship) with the amount of horizontal content contained in the natural scene stimulus. Thus, the more horizontal content present within a natural scene, the more sensitive humans are for detecting the oriented targets used in Experiment 2, with this sensitivity maintaining its anisotropic horizontal effect pattern regardless of the prevalence of horizontal content.

CONCLUDING REMARKS

The focus of the current paper was to discuss human visual performance at different test orientations as it relates to two frequently documented spatial regularities found to occur in real-world imagery. The first regularity discussed was the typical $1/f^{\alpha}$ distribution of content of natural scene amplitude spectra. While the typical range of α values encountered in real-world imagery has been the subject of some debate, we (as well as others) have posited that when the analysis is broken down in terms of the type of environment, the range typically encountered is much more narrow. The second regularity discussed was the tendency toward a bias of Fourier amplitude at the cardinal axes relative to the oblique axes. While the issue concerning whether or not there is typically more content at/near horizontal compared to vertical orientations may never be fully resolved, we have presented reasonable arguments supporting the view that horizontal content is typically more prevalent in natural scene imagery (as well as showing a corresponding performance bias).

The work of Webster and Miyahara (1997) revealed that the typical CSF obtained with standard laboratory/clinical procedures does not convey an accurate account of visual sensitivity to different spatial scales in the context of real-world viewing. Specifically, since the $1/f^{\alpha}$ distribution of spatial content in natural scenes has much more energy concentrated near the lower spatial scales, the subsequent adaptation shifts the CSF most at these scales. Thus peak sensitivity is shifted towards higher spatial scales. Empirical studies were also reviewed where human visual performance was found to be most "tolerant" to manipulations of α when the α of a given natural stimulus was close 1.0 (the value most frequently observed in a large number of studies). When considering this idea of tolerance (e.g., Tadmor & Tolhurst, 1994) in the context of the work conducted by Thomson and Foster (1997) it seems most likely that the range of tolerance is centred on the true α value characteristic of the environment an individual may find her/himself in. It is worth pointing out that this range of optimal tolerance is quite similar to the smaller ranges depicted in Figure 3a. That is, the range of tolerance is likely to be suited (possibly having been established throughout evolutionary development) to the typical deviation of α found to occur in one of the three most general environments described in the earlier sections. Additionally, several studies were reviewed that supported the idea that optimal discrimination between two different objects occurs when the imagery has amplitude spectra with an α value close to the typical value most frequently observed in natural scene imagery. However, it is very likely that optimal discrimination will occur if the α value of a given scene does not deviate outside the range of visual tolerance to changes of α. The benefit of such a processing strategy is that it relieves striate cortex from having to encode each image's α as one examines the visual environment; rather it serves to "whiten" the input (Atick & Redlich, 1992; Barlow, 1959, 2001). Specifically, the idea here is that the adapted state of the visual system to the spatial distribution of content in a typical environment (e.g., the three general environments examined in the earlier sections) "frees" visual cortex from having to "attend" to such characteristics. Given the results reviewed in earlier sections, the state of adaptation induced by a scene's α value likely serves to force the visual system to operate in an optimal fashion within the limited tolerance range as reviewed above.

In terms of human visual sensitivity to orientation in the context of broadband stimuli (visual noise as well as natural scenes), results from empirical studies were reviewed that demonstrated a visual processing anisotropy not observed with standard visual test stimuli (e.g., lines or sine gratings). That is, just as viewing broadband natural scene stimuli alters sensitivity as a function of spatial scale (as compared to tests made with single sinewave stimuli), a different relation as a function of orientation is also observed when tested with naturalistic stimuli (as opposed to simple stimuli). In addition, the typical distribution of spatial content in natural scene imagery was also shown (via Experiment 1) to have an impact on human sensitivity at different orientations. The magnitude of the behavioural horizontal effect was shown to scale with the α associated with the different sets of natural scene imagery, thereby reducing sensitivity to horizontal content according to the distribution of content at different spatial scales (with the largest magnitude of the horizontal effect occurring with imagery possessing α values most often encountered in real-world environments). Lastly, Experiment 2 showed that general sensitivity to oriented increments of amplitude was positively related to the amount of image content at horizontal orientations across a set of images. Specifically, general performance increased with increasing amounts of horizontal image content. This indicates that, although sensitivity to oriented broadband content remains anisotropic (i.e., the horizontal effect anisotropy), overall sensitivity is optimal with imagery containing large amounts of globally distributed, local horizontal content.

It is clear from the studies reviewed, as well as the experiments reported here, that human visual sensation and perception are very much dependent upon the content regularities typically encountered in real-world environments—and perhaps much more so than traditional views based upon results from experiments with simple stimuli would predict. While this review focused on two frequently documented content regularities as measured via Fourier analysis (two-dimensional space), other methods carried out in three-dimensional space have revealed content regularities that also correspond to well known visual phenomena. Specifically, the work of Howe and Purves (2002), and Yang and Purves (2003) have shown that a classic visual illusion involving perceived changes in the length of a line as a function of its orientation can be related to content regularities present in imagery obtained with a laser range scanner.

Together, the results of the two experiments reported here fit well with the evidence that the human visual system is "tuned" to operate optimally when presented with scenes/imagery possessing the typical natural distribution of spatial content, not only with respect to scale, but also with respect to the typical biases of oriented content present in the real visual world.

REFERENCES

Annis, R. C., & Frost, B. (1973). Human visual ecology and orientation anisotropies in acuity. *Science, 182*, 729–731.

Appelle, S. (1972). Perception and discrimination as a function of stimulus orientation: The oblique effect in man and animals. *Psychological Bulletin, 78*, 266–278.

Atick, J. J., & Redlich, A. N. (1992). What does the retina know about natural scenes? *Neural Computation, 4*, 196–210.

Baddeley, R. (1997). The correlational structure of natural images and the calibration of spatial representations. *Cognitive Science, 21*, 351–372.

Baddeley, R. J., & Hancock, P. J. B. (1991). A statistical analysis of natural images matches psychophysically derived orientation tuning curves. *Proceedings of the Royal Society of London: Series B, 246*, 219–223.

Barlow, H. B. (1959). Sensory mechanisms, the reduction of redundancy, and intelligence. In *Proceedings of the National Physical Laboratory symposium* (pp. 537–559). London: Her Majesty's Stationary Office.

Barlow, H. B. (2001). Redundancy reduction revisited. *Network: Computation in Neural Systems, 12*, 241–253.

Blakemore, C., & Campbell, F. W. (1969). On the existence of neurons in the human visual system selectively sensitive to the orientation and size of retinal images. *Journal of Physiology, 203*, 237–260.

Blakemore, C., & Nachmias, J. (1971). The orientation specificity of two visual after- effects. *Journal of Physiology (London), 213*, 157–174.

Bracewell, R. N. (2000). *The Fourier transform and its applications.* Boston: McGraw-Hill.

Burton, G. J., & Moorhead, I. R. (1987). Color and spatial structure in natural scenes. *Applied Optics, 26*, 157–170.

Campbell, F. W., Kulikowski, J. J., & Levinson, J. (1966). The effect of orientation on the visual resolution of gratings. *Journal of Physiology (London), 187*, 427–436.

Chapman, B., & Bonhoeffer, T. (1998). Overrepresentation of horizontal and vertical orientation preferences in developing ferret area 17. *Neurobiology, 95*, 2609–2614.

Chapman, B., Stryker, M. P., & Bonhoeffer, T. (1996). Development of orientation preference maps in ferret primary visual cortex. *Journal of Neuroscience, 16*, 6443–6453.

Coppola, D. M., Purves, H. R., McCoy, A. N., & Purves, D. (1998a). The distribution of oriented contours in the real world. *Proceedings of the National Academy of Sciences USA, 95*, 4002–4006.

Coppola, D. M., White, L. E., Fitzpatrick, D., & Purves, D. (1998b). Unequal representation of cardinal and oblique contours in ferret visual cortex. *Proceedings of the National Academy of Sciences USA, 95*, 2621–2623.

Craven, B. J. (1993). Orientation dependence of human line-length judgments matches statistical structure in real-world scenes. *Proceedings of the Royal Society of London, 253*, 101–106.

Creelman, C. D., & Macmillan, N. A. (1991). *Detection theory: A user's guide.* Cambridge, UK: Cambridge University Press.

De Valois, R. L., Yund, E. W., & Hepler, N. (1982). The orientation and direction selectivity of cells in macaque visual cortex. *Vision Research, 22*, 531–544.

Essock, E. A. (1980). The oblique effect of stimulus identification considered with respect to two classes of oblique effects. *Perception, 9*, 37–46.

Essock, E. A., DeFord, J. K., Hansen, B. C., & Sinai, M. J. (2003). Oblique stimuli are seen best (not worst!) in naturalistic broad-band stimuli: A horizontal effect. *Vision Research, 43*, 1329–1335.

Field, D. J., & Brady, N. (1997). Visual sensitivity, blur and the sources of variability in the amplitude spectra of natural scenes. *Vision Research, 37*, 3367–3383.

Furmanski, C. S., & Engel, S. A. (2000). An oblique effect in human primary visual cortex. *Nature Neuroscience, 3*, 535–536.

Hancock, P. J. B., Baddeley, R. J., & Smith, L. S. (1992). The principle components of natural images. *Network: Computation in Neural Systems, 3*, 61–70.

Hansen, B. C., & Essock, E. A. (2004). A horizontal bias in human visual processing of orientation and its correspondence to the structural components of natural scenes. *Journal of Vision, 4*, 1044–1060.

Hansen, B. C., Essock, E. A., Zheng, Y., & DeFord, J. K. (2003). Perceptual anisotropies in visual processing and their relation to natural image statistics. *Network: Computation in Neural Systems, 14*, 501–526.

Howe, C. Q., & Purves, D. (2002). Range statistics can explain the anomalous perception of length. *Proceedings of the National Academy of Sciences, 99*, 13184–13188.

Jastrow, J. (1893). On the judgement of angles and positions of lines. *American Journal of Psychology, 5*, 214–248.

Keil, M. S., & Cristóbal, G. (2000). Separating the chaff from the wheat: Possible origins of the oblique effect. *Journal of the Optical Society of America A, 17*, 697–710.

Knill, D. C., Field, D. J., & Kersten, D. (1990). Human discrimination of fractal images. *Journal of the Optical Society of America, 7*, 1113–1123.

Lashley, K. S. (1938). The mechanisms of vision: Xv. preliminary studies of the rat's capacity for detail vision. *Journal of General Psychology, 18*, 123–193.

Li, B. W., Peterson, M. R., & Freeman, R. D. (2003). The oblique effect: A neural basis in the visual cortex. *Journal of Neurophysiology, 90*, 204–217.

Maffei, L., & Campbell, F. W. (1970). Neurophysiological localization of the vertical and horizontal visual coordinates in man. *Science, 187*, 386–387.

Mansfield, R. J. W., & Ronner, S. P. (1978). Orientation anisotropy in monkey visual cortex. *Brian Research, 149*, 229–234.

Oliva, A., & Torralba, A. (2001). Modeling the shape of the scene: A holistic representation of the spatial envelope. *International Journal of Computer Vision, 42*, 145–175.

Orban, G. A., & Kennedy, H. (1980). Evidence for meridional anisotropies in orientation selectivity of visual cortical neurons. *Archives Internationales de Physiologie et de Biochimie, 88*, 13–14.

Pantle, A., & Sekuler, R. (1968). Size-detecting mechanisms in human vision. *Science, 162*, 1146–1148.

Párraga, C. A., & Tolhurst, D. J. (2000). The effect of contrast randomization on the discrimination of changes in the slopes of the amplitude spectra of natural scenes. *Perception, 29*, 1101–1116.

Párraga, C. A., Troscianko, T., & Tolhurst, D. J. (2000). Spatiochromatic properties of natural images and human vision. *Current Biology, 12*, 483–487.

Ross, H. E., & Woodhouse, J. M. (1979). Genetic and environmental factors in orientation anisotropy-field-study in the British-isles. *Perception, 8*, 507–521.

Shapley, R., & Lennie, P. (1985). Spatial frequency analysis in the visual system. *Annual Review of Neuroscience, 8*, 547–583.

Simoncelli, E. P., & Olshausen, B. A. (2001). Natural image statistics and neural representation. *Annual Review of Neuroscience, 24*, 1193–1216.

Sokol, S., Moskowitz, A., & Hansen, V. (1989). Evoked potential and preferential looking correlates of the oblique effect in 3-month-old infants. *Documenta Opthalmologica, 71*, 321–328.

Stork, D. G., & Wilson, H. R. (1990). Do gabor functions provide appropriate descriptions of visual cortical receptive fields? *Journal of the Optical Society of America A, 7*, 1362–1373.

Swanson, W. H., Wilson, H. R., & Giese, S. C. (1984). Contrast matching data predicted from contrast increment thresholds. *Vision Research, 24*, 63–75.

Switkes, E., Mayer, M. J., & Sloan, J. A. (1978). Spatial frequency analysis of the visual environment: Anisotropy and the carpentered environment hypothesis. *Vision Research, 18*, 1393–1399.

Tadmor, Y., & Tolhurst, D. J. (1994). Discrimination of changes in the second-order statistics of natural and synthetic images. *Vision Research*, *34*, 541–554.

Thomson, M. G. A. (1999). Higher-order structure in natural scenes. *Journal of the Optical Society of America A*, *16*, 1549–1553.

Thomson, M. G. A. (2001a). Beats, kurtosis and visual coding. *Network: Computation in Neural Systems*, *12*, 271–287.

Thomson, M. G. A. (2001b). Sensory coding and the second spectra of natural signals. *Physical Review Letters*, *86*, 2901–2904.

Thomson, M. G. A., & Foster, D. H. (1997). Role of second- and third-order statistics in the discriminability of natural images. *Journal of the Optical Society of America A*, *14*, 2081–2090.

Timney, B. N., & Muir, D. W. (1976). Orientation anisotropy: Incidence and magnitude in Caucasian and Chinese subjects. *Science*, *193*, 699–701.

Tolhurst, D. J., & Tadmor, Y. (1997). Band-limited contrast in natural images explains the detectability of changes in the amplitude spectra. *Vision Research*, *37*, 3203–3215.

Tolhurst, D. J., & Tadmor, Y. (2000). Discrimination of spectrally blended natural images: Optimization of the human visual system for encoding natural images. *Perception*, *29*, 1087–1100.

Tolhurst, D. J., Tadmor, Y., & Chao, T. (1992). Amplitude spectra of natural images. *Ophthalmic and Physiological Optics*, *12*, 229–232.

Torralba, A., & Oliva, A. (2003). Statistics of natural image categories. *Network: Computation in Neural Systems*, *14*, 391–412.

Van der Schaaf, A., & van Hateren, J. H. (1996). Modeling the power spectra of natural images: Statistics and information. *Vision Research*, *36*, 2759–2770.

Van Hateren, J. H., & van der Schaaf, A. (1998). Independent component filters of natural images compared with simple cells in primary visual cortex. *Proceedings of the Royal Society of London: Series B*, *265*, 359–366.

Webster, M. A., & Miyahara, E. (1997). Contrast adaptation on the spatial structure of natural images. *Journal of the Optical Society of America A*, *14*, 2355–2366.

Wilson, H. R. (1991). Psychophysical models of spatial vision and hyperacuity. *Spatial Vision*, *10*, 64–86.

Yang, Z., & Purves, D. (2003). Image/source statistics of surfaces in natural scenes. *Network: Computation in Neural Systems*, *14*, 371–390.

VISUAL COGNITION, 2005, 12 (6), 1235–1254

Distance perception from motion parallax and ground contact

Rui Ni and Myron L. Braunstein
University of California, Irvine, CA, USA
George J. Andersen
University of California, Riverside, CA, USA

Meng and Sedgwick (2001, 2002) found that the perceived distance of an object in a stationary scene was determined by the position at which it contacted the ground in the image, or by nested contact relations among intermediate surfaces. Three experiments investigated whether motion parallax would allow observers to determine the distance of a floating object without intermediate contact relations. The displays consisted of one or more computer-generated textured cylinders inserted into a motion picture or still image of an actual 3-D scene. In the motion displays, both the cylinders and the scene translated horizontally. Judged distance for a single cylinder floating above the ground was determined primarily by the location at which the object contacted the ground in the projected image (''optical contact''), but was altered in the direction indicated by motion parallax. When more than one cylinder was present and observers were asked to judge the distance of the top cylinder, judged distance moved closer to that indicated by motion parallax, almost matching that value with three cylinders. These results indicate that judged distance in a dynamic scene is affected both by optical contact and motion parallax, with motion parallax more effective when multiple objects are present.

The importance of the ground surface in specifying the layout of a 3-D scene was clearly described by Gibson (1947/1958, 1950). Gibson proposed that the perception of the visual world is not based on perceiving distances of objects from the eye across empty space but is based on perceiving relationships between objects and a background surface. The ground surface, he argued, was the most important background surface for distance perception because it is

Please address all correspondence to: Myron L. Braunstein, 3151 Social Science Plaza, University of California, Irvine, CA 92697-5100, USA. Email: mlbrauns@uci.edu

Portions of this research were presented at the 2003 meeting of the Vision Sciences Society, Sarasota, Florida. This research was supported by NIH Grant 1R01EY12437. We thank Z. Bian, C. Feria, and H. Zhong for helpful discussions.

DOI:10.1080/13506280444000724

common to all terrestrial environments and is the surface that supports objects and the locomotion of most animals. Gibson (1946/1958) provided an analysis of the relevant information available in a retinal projection of a ground surface. Specifically, he described a set of retinal gradients that provide the slant of the ground surface and distance along the ground surface. For example, a retinal gradient of texture results from the change in projected texture density in the retinal image with increasing 3-D distance, a retinal gradient of size-of-similar objects results from the change in the projected extents of similar-size objects with increasing distance, and a retinal gradient of velocity (also referred to as motion parallax) results from the inverse relationship between the projected velocities of stationary objects and their distance from the observer, during movement of the observer.

A considerable body of literature now exists on retinal gradients and the perception of surface slant. (See Howard & Rogers, 2002, for a comprehensive review of depth perception research.) Until recently, however, only a few studies have directly addressed the role of the ground surface in the perception of layout in 3-D scenes (see, for example, Sedgwick, 1983, 1987; Sedgwick & Levy, 1985; for a recent review, see Sedgwick, 2001). Some studies have considered judged distance along the ground in outdoor scenes (Levin & Haber, 1993; Toye, 1986). Loomis (e.g., Loomis, da Silva, Fujita, & Fukusima, 1992; Loomis, da Silva, Philbeck, & Fukusima, 1996; Loomis, Philbeck, & Zahorik, 2002) has examined the perception of distance, location, and shape along a ground surface in outdoor scenes, using both open-loop walking and perceptual judgements. In the open-loop walking tasks, observers viewed a target and then attempted to walk blindfolded to the location of the target, attempted to point to the target while walking along a perpendicular path, or attempted to walk towards the target after walking along an oblique path. In the perceptual judgement tasks, observers constructed depth intervals on the ground to match frontal-parallel intervals. These studies involved the perception of where objects contacted the ground, as well as information for distance from the observer to points along the ground surface.

Effects of characteristics of the ground surface on judged distance were studied by Sinai, Ooi, and He (1998). Using both a blindfolded walking task in which observers turned 90° and attempted to walk a distance equivalent to the perceived egocentric distance of a target and a perceptual judgement task in which observers turned 90° and set the distance of an object to match the perceived distance of the target, they found that a change in texture (from grass to concrete) across the distance being judged resulted in an underestimation of the distance, but a gap in the texture resulted in an overestimation. Feria, Braunstein, and Andersen (2003) replicated the effect of a change in texture on judged distance, finding that judged distance is reduced in the presence of any texture discontinuity, whether produced by a change in texture patterns, a contrast reversal of the same pattern, or a misalignment of identical regular textures, either along a ground surface in a 3-D scene or in a frontal plane.

Gibson (1950, p. 178) pointed out that the perceived position of an object along a ground surface is determined by the position at which the object contacts the ground in the optical projection—the "optical contact" position—in the absence of relative motion or stereoscopic depth. He demonstrated the effect of optical contact with a photograph (p. 179) in which the object that appeared more distant was physically closer in the scene, but was suspended above the ground by an unseen support (see also Sedgwick, 1983, 1989). More recently, Meng and Sedgwick (2001, 2002) examined the role of mediated ground contact in determining the perceived positions of objects in a stationary scene. They placed an object on a platform that was either resting on the ground or floating above the ground. The optical contact position of the floating platform was more distant from the observer than the simulated distance of the platform. Observers moved a marker along a track extended in depth in the scene to indicate the perceived position of the object. The judged location of an object on a platform depended on where the object contacted the platform in the image and on where the platform contacted the ground in the image. They referred to this combined effect as a nested contact relation.

Information about optical contact can be expected to combine with other information about the relative distances of objects in determining perceived layout. Cutting and Vishton (1995) presented an analysis of the information available from different sources, for different distances from an observer. Some of these sources, such as motion parallax, provide quantitative information about layout; other sources, such as occlusion, provide qualitative information. Although both optical contact and motion parallax provide quantitative information about layout, the use of these cues in judging relative depth relies on very different constraints. Optical contact specifies layout under a constraint that objects do not float above the ground. Motion parallax specifies layout in depth under a constraint of rigid motion, that is, no relative motion between the objects and the ground or among the objects. Both of these constraints can be violated, of course. Some objects do float above the ground. Objects and surfaces may move independently of each other.

In the present series of experiments we examined the interaction between optical contact information and motion parallax in specifying layout in a 3-D scene. If optical contact specifies a distance that is different from the simulated distance of an object (as when the object is floating above the ground), will the addition of motion parallax result in a perception of the object at its simulated distance? In a single image of a stationary scene, it is not possible to distinguish a floating object from an object on the ground unless additional information is provided. We can think of motion parallax as providing additional information that disambiguates the information provided by optical contact. But there is another interpretation of the situation: Both optical contact and motion parallax, like all depth cues, provide ambiguous information. With optical contact and motion parallax providing different indications of an object's position in the scene, the object may be perceived as lying on the ground and sliding back and forth relative to the scene (an optical contact interpretation) or as floating above the scene and moving rigidly with the scene (a motion parallax interpretation). Which interpretation is accepted by an observer may be a reflection of which underlying constraint can be relaxed or ignored, in this case, a gravity-based constraint for optical contact and a rigid motion constraint for motion parallax.

In the first experiment the optical contact position of the object was varied while the simulated distance, indicated by motion parallax, was held constant. In the second experiment, in contrast, the optical contact position was held constant while the simulated distance was varied. The third experiment examined the effect of the number of cylinders that were stacked above the ground, both for stationary scenes and for scenes shown in motion.

Although most research on the perception of 3-D objects in 3-D scenes has used either computer-generated scenes or directly observed real-world scenes, both of these methods have limitations (Sauer, Braunstein, Andersen, & Bian, 2001). Simulated scenes provide good control over stimulus characteristics but may lack the abundance of information sources available in natural environments. Real-world scenes provide this abundance of information but it is difficult to control all attributes of objects in such scenes. In the present research, we balance these two methods by adding computer-generated objects (textured cylinders and a track with a marker) to a movie of a real scene.

EXPERIMENT 1: VARIATIONS IN OPTICAL CONTACT WITH CONSTANT MOTION PARALLAX

The purpose of Experiment 1 was to investigate the effect of motion parallax information on perceived distance in the presence of optical contact information. While the distance of the object simulated by motion parallax and the simulated height of the top surface above the ground were kept constant, the optical contact position was varied by varying the vertical extent of the object, and thus the separation of the bottom of the object from the ground. This was analogous to the variation in platform heights in Meng and Sedgwick's (2001) first experiment. If the motion parallax information was sufficient to determine judged distance, a constant distance should be perceived regardless of the variations in the optical contact location.

Method

Observers. The observers were 14 students from the University of California. All observers had normal or corrected-to-normal visual acuity and were naive about the purpose of the experiment. The observers received extra credit in a psychology course for participating.

Apparatus. A Pentium 4 2 GHz computer displayed the stimuli on a 21 inch (53 cm) Dell monitor, with a resolution of 1024 (horizontal) × 768 (vertical) and a refresh rate of 85 Hz. The experiment was carried out in a darkened room. Observers viewed the displays binocularly through a 19 cm diameter collimating lens with a focal length of 75 cm, with their heads stabilized by a chin- and headrest. The distance between the observer and the display screen was 85 cm. A black viewing hood was placed between the collimating lens and the monitor, limiting the field of view to the display area, and black cloth separated the observer from the apparatus to assure that the observer would not see the location of the monitor. Responses were made using a Microsoft SideWinder joystick.

Stimuli. The displays consisted of computer-generated cylinders and a computer-generated track with a red marker, superimposed on a background movie of an actual 3-D scene. The background movie consisted of 50 photographs taken with a Kodak DC260 digital camera with a 35 mm equivalent lens. The photographs were taken at fixed positions along a horizontal track that were spaced according to a sinusoidal function, with the larger separations in the centre of the track and the smaller separations at the ends. This was intended to approximate the motion of an observer moving his or her head back and forth horizontally. The photographs were cropped from the original size of 1536 × 1024 pixels to 1024 × 768 pixels, with the original pixel aspect ratio kept unchanged. An example of a cropped photograph with a cylinder, track, and marker superimposed is shown in Figure 1.

In the 3-D simulation, the cylinder was located either 13 m (front position) or 19 m (back position) from the observer. To keep the projected sizes of the cylinders the same at these two positions, the simulated cylinder diameters were set at 36 cm and 52 cm in the front and back positions, respectively. The height (vertical dimension) of the cylinder was 9 cm, 27 cm, or 45 cm (designated as short, medium, and tall) in the front position and 13 cm, 39 cm, or 65 cm in the back position. Figure 2 shows the three cylinder heights with the cylinder located at the front position. The top surface of the cylinder was kept at a constant height above the ground at each position. For the tall cylinder (Figure 2a), the relative velocities of the cylinder and scene were consistent with a cylinder resting on the ground and translating rigidly with the scene. For the short and medium cylinders (Figures 2b and 2c), the relative speeds of the cylinder and the ground surface were consistent with one of two possibilities: (1) The object was floating above the ground but moving rigidly with the ground, or (2) the object was located on the ground but was moving relative to the ground (i.e., sliding across the ground). The displays were also consistent with intermediate possibilities combining floating and nonrigidity.

Figure 1. A frame from the stimulus display showing a cylinder in the back position and the track with the adjustable red marker.

Design. The independent variables were the height of the cylinder (tall, medium, and short) and the simulated position of the cylinder in the scene (front and back). Each observer responded to 10 replications of each of the six conditions. The 60 trials were presented in a different random order for each observer in two blocks of 30 trials, preceded by a practice block that was the same as the first block of trials.

a b c

Figure 2. Cylinders in the front position: (a) Tall, (b) medium, and (c) short. (The displays showed the full scene as illustrated in Figure 1.)

Procedure. The observers' task was to adjust the red marker on the track on the right side of the scene, as shown in Figure 1, until it matched the distance of the cylinder. This was similar to the task used by Meng and Sedgwick (2001, 2002). The marker was adjusted with a joystick and the observers pressed the trigger button on the joystick when they were satisfied with their response.

Results and discussion

Figure 3 shows judged distance as a function of cylinder height for the two scene positions, averaged across the 14 observers. For 12 of the 14 observers, judged distance increased with decreasing cylinder height (that is, increased with increasing space between the cylinder and the ground plane in the 3-D simulation) at both scene positions. A three (cylinder height) by two (scene position) repeated measures ANOVA showed significant main effects for projected height of the cylinder, $F(2, 26) = 18.17$, $p < .01$, and scene position, $F(1, 13) = 220.70$, $p < .01$, and a significant interaction, $F(2, 26) = 10.11$, $p < .01$.

If the perceived distance was determined only by the optical contact location, the results should match the dashed lines in Figure 3. If, on the other hand, the results were consistent with the motion parallax information (under an assumption of rigid motion of the cylinder with the ground surface), the cylinder

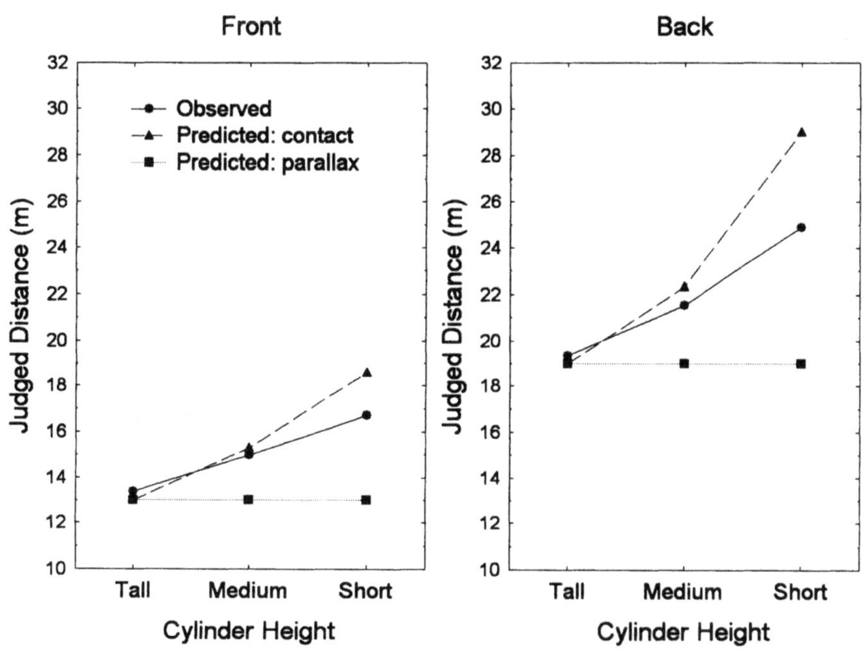

Figure 3. Judged distance as a function of cylinder height for two scene positions in Experiment 1.

should be perceived at a constant distance, regardless of its height. This is indicated by the dotted lines in Figure 3. The results indicate that motion parallax altered the judged position of the object in the scene relative to that predicted from optical contact, but that the judged position was closer to the optical contact position than to the position indicated by rigid motion. In other words, there appears to have been a compromise between these two sources of information, with greater weight given to optical contact.

Control experiment. Both stationary and motion versions of each of the displays, except those with the middle cylinder height, were included in a control experiment with 10 naive observers, using the same procedure. Results for the motion conditions were similar to those in the main experiment. In the stationary conditions the cylinder position was judged according to its optical contact position, as would be expected from Meng and Sedgwick's (2001) results.

EXPERIMENT 2: VARIATIONS IN MOTION PARALLAX WITH CONSTANT OPTICAL CONTACT

In Experiment 1, motion parallax indicated a constant cylinder position as the optical contact position was varied across trials by varying the distance of the cylinder above the ground. The motion of the cylinder was thus constant at each of the two scene positions. If motion parallax affects the perceived position of the cylinder, as indicated in Experiment 1, then variations in cylinder speed corresponding to variations in distance from the observer should result in variations in judged distance. The manipulations in Experiment 2 were complementary to those in Experiment 1 in that we varied the cylinder position indicated by motion parallax across trials, while keeping the optical contact position constant at one of two levels. The cylinder's projected size and height in the image was kept constant across all conditions. This was done by varying the cylinder's simulated height and distance above the ground, as shown in Figure 4.

Method

Observers. The observers were 16 students from the University of California, Irvine. All observers had normal or corrected-to-normal visual acuity, and all were naive about the purpose of the experiment. None had participated in Experiment 1. The observers received extra credit for participating.

Apparatus and procedure. The apparatus and procedure were the same as in Experiment 1.

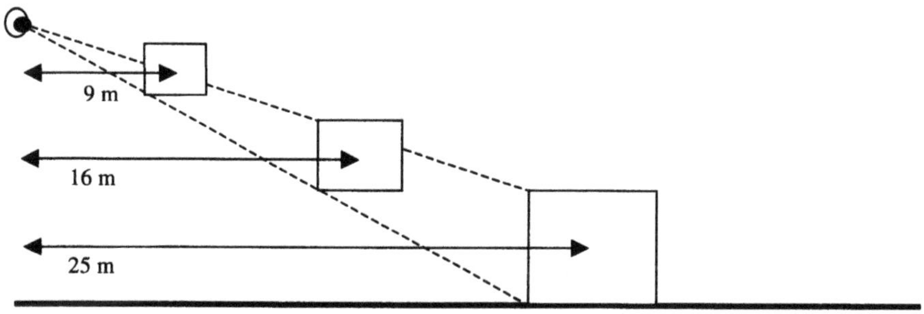

Figure 4. The three simulated cylinder positions in the front of the scene. The cylinders were adjusted in size to project the same image. The distances in the back of the scene were 16 m, 25 m, and 34 m, with corresponding adjustments in the sizes.

Stimuli. The background scene was the same as in Experiment 1. All cylinders were the same projected height as the short cylinder in Experiment 1. The projected cylinder was given one of two optical contact positions, 25 m or 34 m from the observer in the simulated scene. At each of these two image positions the cylinder was assigned a translation velocity corresponding to one of three simulated distances. At the 25 m optical contact position, the distances simulated by motion parallax were 7 m, 16 m, and 25 m. The 7 m and 16 m motion parallax conditions thus represented cylinders floating above the ground, whereas the 25 m motion parallax condition represented a cylinder lying on the ground. Similarly, at the 34 m optical contact position, the distances simulated by motion parallax were 16 m, 25 m, and 34 m, with the first two conditions representing floating cylinders and the third representing a cylinder lying on the ground.

Design. The independent variables were the optical contact position (25 m or 34 m) and the distance simulated by motion parallax (7 m, 16 m, and 25 m or 16 m, 25 m, and 34 m). Each observer responded to 10 repetitions of each of the six conditions. The 60 trials for each observer were randomly arranged in two blocks, preceded by a practice block of 30 trials.

Results and discussion

Judged cylinder position as a function of projected scene position and motion parallax, averaged across the 16 observers, is shown in Figure 5. A three (simulated distance) by two (optical contact position) repeated measures ANOVA showed significant main effects for simulated distance, $F(2, 30) = 11.25$, $p < .01$ and optical contact position, $F(1, 15) = 91.16$, $p < .01$. The interaction was not significant, $F(2, 30) = 2.83$, $p > .05$. If the perceived distance was determined only by optical contact, a constant distance should have been

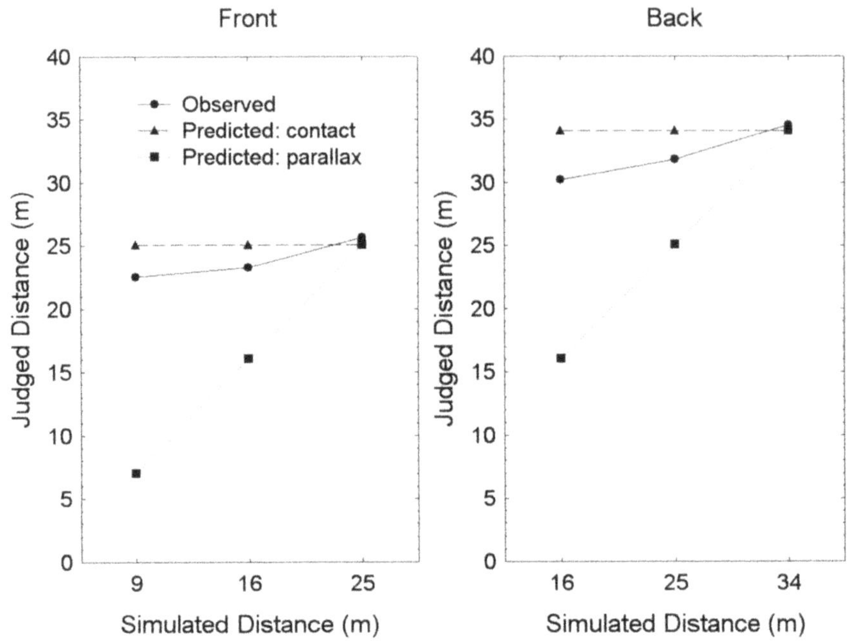

Figure 5. Judged distance as a function of simulated distance for two scene positions in Experiment 2.

judged for each of the two optical contact positions, as shown by the dashed lines. If, on the other hand, the results were consistent with the motion parallax information (under an assumption of rigid motion), the cylinders should be perceived at different distances, as indicated by the dotted lines. The results indicate that motion parallax altered the judged position of the object in the scene relative to that predicted from optical contact, but as in Experiment 1 the judged position was closer to that indicated by optical contact than to the position indicated by motion parallax. There appears to again have been a compromise between these two sources of information, with greater weight given to the optical contact information.

Control experiment. A control experiment, similar to the Experiment 1 control experiment, was conducted with 10 naive observers. Both stationary and motion versions of each of the displays, except for the intermediate positions simulated by motion parallax, were included. (This was relevant only for the motion stimuli.) Results for the motion conditions were similar to those in the main experiment, whereas in the stationary conditions the cylinder position was judged according to its optical contact position.

EXPERIMENT 3: EFFECTS OF MULTIPLE OBJECTS

In Experiments 1 and 2 there was a clear effect of motion parallax, but judged distance was determined primarily by optical contact. This may have occurred because the effectiveness of motion parallax is limited when only one object and a ground surface are displayed in motion. This is because the perception of an object moving relative to a surface is a reasonable alternative to the perception of an object moving rigidly with the surface, and appears to be preferred to a perception of an object floating above a surface. If two objects were present, however, perceiving the objects' positions according to optical contact would require perceiving the two objects and the surface each moving at a different speed. These three speeds would coincide exactly with the projected speeds that would be produced by a surface and two objects located at the same distance from the observer, all moving rigidly. With three objects, the perception of the objects moving at different speeds should be even less likely. Adding additional objects could have effects unrelated to motion, however. The objects might be grouped and thus perceived as being at the same location in the scene, that is, stacked one above the other, rather than at varying locations. For this reason we examined both motion parallax scenes and stationary scenes containing one, two or three cylinders.

Observers. The observers were 20 students from the University of California, Irvine. All observers had normal or corrected-to-normal visual acuity and were naive about the purpose of the experiment. None had participated in Experiments 1 or 2. The observers received extra credit for participating in the experiment.

Apparatus and procedure. The apparatus was the same as in Experiments 1 and 2. The procedure was the same as in Experiments 1 and 2, except that the observers' task was to judge the distance of the top cylinder, if there was more than one.

Stimuli. The background scene was the same as in Experiments 1 and 2. One, two, or three cylinders were superimposed on the scene (Figure 6). The projected size of each cylinder was the same as that of the short cylinder in Experiment 1. In the motion parallax simulation with three cylinders, the bottom cylinder was resting on the ground at 13 m or 19 m and the other two cylinders were stacked above the bottom cylinder, with simulated vertical separations of 9 cm in the near position and 13 cm in the far position. For the two-cylinder conditions only the top two were displayed and for the one-cylinder conditions only the top cylinder was displayed. The optical contact position of the top cylinder was 25 m or 34 m, regardless of the number of cylinders displayed. The scene and the cylinders were either translating horizontally or stationary.

Figure 6. Cylinder arrangements in Experiment 3. (The displays showed the full scene as illustrated in Figure 1.)

Design. There were three independent variables, the number of cylinders (1, 2, or 3), the simulated position of the cylinders (13 m or 19 m), and whether the scene was in motion or stationary. The first two variables were run within observers; the third variable was run between observers. Each observer responded to 10 repetitions of each of the six within-observer conditions. The 60 trials were divided into three blocks based on number of cylinders, with half of the observers receiving the blocks in the order 1-2-3 and half receiving the blocks in the order 3-2-1. The stimuli in each block were presented in a random order. The three experimental blocks were preceded by a practice block with the same number of cylinders as the first experimental block.

Results and discussion

Judged distance, as a function of number of cylinders, scene position, and motion, averaged across observers, is shown in Figure 7. The dashed line indicates the optical contact position of the top cylinder (or of the cylinder, if there was only one) and the dotted line indicates the position simulated by motion parallax. Both of these positions were constant across number of cylinders. A three-way ANOVA with two within-observers variables and one between-observers variable showed significant main effects for scene position, $F(1, 18) = 377.49, p < .01$, for number of cylinders, $F(2, 36) = 8.04, p < .01$ and for motion vs. stationary, $F(1, 18) = 24.33, p < .01$. As can be seen in Figure 7, the cylinder was judged as closer in each of the motion conditions than in the corresponding stationary condition, but judged distance decreased with an increase in the number of cylinders in both the stationary and the motion displays. This indicates that the effect of increasing the number of cylinders was not entirely due to motion, but may be related to grouping of the cylinders, leading to a perception of the cylinders as stacked on top of one another rather than located at different positions in depth. The effects in the stationary case are analogous to the nested contact effects reported by Meng and Sedgwick (2001, 2002) except that in the present case the objects were not in direct contact but

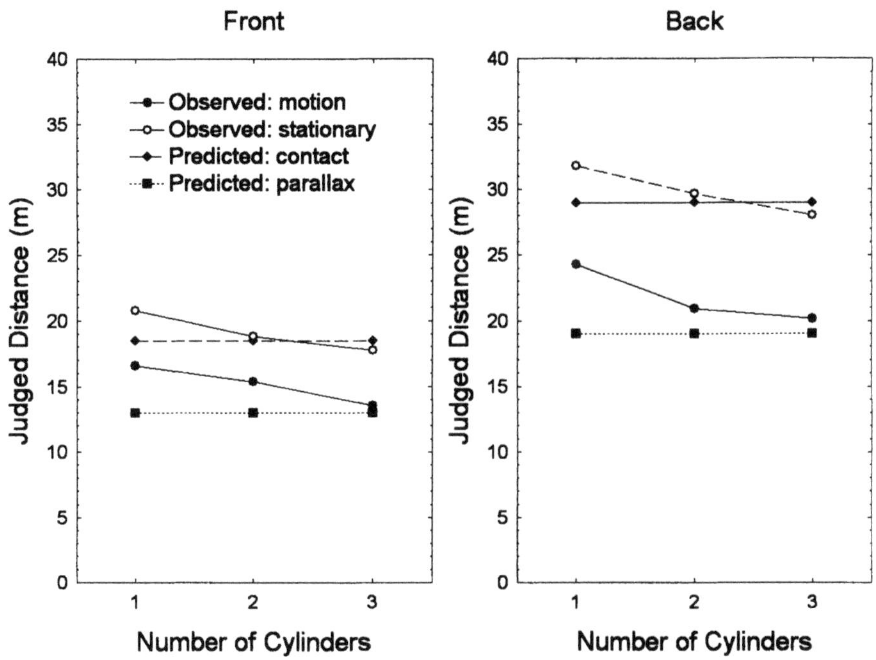

Figure 7. Judged distance as a function of number of cylinders for moving and stationary displays at two scene positions in Experiment 3.

may have been related to one another across vertical gaps. It is especially important to note that with three cylinders in motion, judged distance is very close to the distance simulated through motion parallax, demonstrating that under these particular conditions optical contact information is no longer dominant in determining judged distance.

Control studies

It is possible that the tendency to judge the distance of the top cylinder according to the optical contact position of the bottom cylinder was due to the vertical alignment of the cylinders in the 3-D simulation, rather than to their rigid motion, when more than one cylinder was present in the scene. We conducted two control experiments to examine this possibility.

Nonrigid horizontal motion. The displays were similar to the motion displays in Experiment 3, except that the cylinder speeds were altered so that the speed of each cylinder corresponded to its optical contact position. 10 naive observed participated. With this change in the cylinder speeds, the number of cylinders did not have a significant effect on judged distance.

Rigid motion without alignment. The displays were again similar to the motion displays in Experiment 3, except that the cylinders were not aligned vertically. The top cylinder was positioned to the left of the bottom cylinder's position by 27 cm in the front position and 39 cm in the back position, in the simulated 3-D scene. The middle cylinder was positioned to the left of the bottom cylinder's position by 9 cm in the front position and 13 cm in the back position. (As in the main experiment, some displays contained only the top cylinder, others the top and middle cylinders, and others all three cylinders.) The cylinders moved rigidly in accordance with their scene positions, as in the main experiment. Ten naive observers participated. The effect of number of cylinders was significant, although smaller than that found in Experiment 3. In combination, these two control experiments indicate that vertical alignment of the cylinders is neither necessary nor sufficient to account for the results of Experiment 3.

GENERAL DISCUSSION

Motion parallax clearly interacts with optical contact information in determining the perceived position of an object in a 3-D scene. With a single object moving rigidly with the ground plane, perceived location is determined primarily by optical contact but is altered when motion parallax indicates that the object is not in contact with the ground in the 3-D scene. Even when the optical contact location of an object is kept constant, varying the speed with which the object translates changes the judged distance. With two objects moving rigidly with the ground plane, judged object location was determined primarily by motion parallax. With three objects, judged location was almost equal to that indicated by motion parallax.

The effects found with multiple objects appear to have at least two components, one present in both stationary and motion parallax scenes and one specific to motion. The stationary displays of two or three cylinders could be interpreted as cylinders lying on the ground at different distances or stacked vertically at the same distance. In the first case the top cylinder would be perceived at its optical contact position. In the second case the top cylinder would be perceived at the optical contact position of the lowest cylinder. Our results indicate a tendency to perceive multiple cylinders as forming a vertical stack, rather than as each lying on the ground, with the judged distance of the top cylinder falling between its optical contact position and the optical contact position of the lowest cylinder. The cylinders are apparently grouped together, but grouping alone does not imply that they would be perceived as stacked vertically. The perception of the cylinders as stacked vertically is consistent with the equidistance tendency (Gogel, 1965), and that tendency may explain this perception.

With the addition of motion, the judged distance of the top cylinder moves still further from its optical contact position and closer to its simulated position.

This could be due to a tendency to perceive a group of objects that have projected motions compatible with rigid motion in 3-D as moving rigidly in 3-D (Ullman, 1979). If a rigidity constraint underlies the effect of number of cylinders, however, one might expect judgements to change more rapidly with an increase in the number of cylinders for the motion condition than for the stationary condition, but the slope of these two curves is about the same. Although rigidity cannot be ruled out, another explanation of the effect of motion may be that the common horizontal motion of the cylinders in 3-D enhanced the perception of the cylinders as forming a group, and thus enhanced the perception of the distance of the top cylinder being the same as that of the bottom cylinder. We have made informal observations with displays in which three cylinders were translating horizontally but were given additional, independent vertical motions. The perceived distance of the top cylinder was no different for these cylinders, which displayed nonrigid motion, than for the cylinders that moved rigidly. This suggests that it is the equal 3-D translation speed (the Gestalt principle of common fate), and not rigid motion per se, that is responsible for the effect of common horizontal motion of a group of cylinders on perceived distance.

The results of the two control experiments related to Experiment 3 suggest that the perception of multiple cylinders as located at the same position in the scene in motion parallax displays is based on common motion in 3-D and not vertical alignment of the edges of the cylinders. In the first control experiment the amplitude of the horizontal motions of the upper cylinders was altered to match their optical contact positions. With the horizontal motions no longer consistent with rigid motion or a common horizontal translation, the upper cylinders were judged as resting on the ground rather than stacked above the lower cylinder. In the second control study the cylinders were shifted horizontally in the simulated scene so they were no longer aligned vertically. This reduced but did not eliminate the tendency to judge the upper cylinders as moving rigidly with the lower cylinder. It might be expected that with further separation between the cylinders, either horizontally or vertically, the effect of multiple cylinders on perceived distance would diminish and eventually disappear.

A possible explanation for the dominance of the optical contact information in determining the perceived position of the cylinder in Experiments 1 and 2, and of the single cylinder in Experiment 3, is that the response would match the optical contact position if the observer merely aligned the marker and the cylinder horizontally in the image. We have several reasons for discounting this possibility. First, it would mean that the observers were not following instructions to respond according to a 3-D perception. Of course this is possible, but the systematic deviations from the optical contact response when motion parallax or stacking of the cylinders provided conflicting 3-D information suggests that the observers were responding to 3-D relationships and not directly to height in the image. Meng and Sedgwick's (2001, 2002) results provide further evidence that observers making similar judgements were not merely aligning the marker with the object in the image. They observed variability across different conditions that would not be expected if observers had just aligned the marker and the object horizontally in the image.

The interaction of optical contact and motion parallax in determining the judged distance of an object in a 3-D scene may be regarded as an instance of cue combination. Our data does not distinguish between models in which depth is calculated separately for each cue and a weighted average determines judged depth, and models in which information from the two sources is combined to compute a single depth estimate. We examined the distributions of responses on individual trials for conditions that suggest a compromise between the distance indicated by optical contact and the distance indicated by motion parallax. These distributions are unimodal, suggesting that the compromise occurs on individual trials and not through averaging of trials in which the responses are based on one cue or the other.

On the assumption that judged depth can be modelled as a weighted function of the distance indicated by optical contact and the distance indicated by motion parallax, we calculated the weights that best fit our data. In Experiment 1 and Experiment 2 the equation used to fit the judgement results was judgement = w_{motion} * simulated position + $w_{optical}$ * optical contact position, where $w_{optical} = 1 - w_{motion}$. The value assigned to w_{motion} for Experiment 1 and Experiment 2 was .20, using a least squares difference criterion for determining the best fit. The estimated judgements and the observed judgements for Experiments 1 and 2 are shown in Figures 8 and 9. The model accounted for 98.5% of the variance in the 12 data points in the two figures.

Two additional variables had to be considered in applying a weighted combination model to Experiment 3: The number of cylinders and whether the scene was stationary or moving. In the stationary conditions we used a weighted combination of the optical contact position of the cylinder being judged and the optical contact position of the lowest cylinder in the stack. With one cylinder present, these were of course the same. With three cylinders present the optical contact position of the lowest cylinder was the same as the simulated (physical) position in the scene of the top cylinder. With two cylinders present, the optical contact position of the lowest cylinder was intermediate between the optical contact position of the top cylinder and the simulated position of the top cylinder. We used the equation judgement = $w_{optical}$ * optical contact + w_{lowest} * lowest contact, where lowest contact is the optical contact position of the lowest cylinder and $w_{lowest} = 1 - w_{optical}$. The value used for $w_{optical}$ was .94. In the motion case we used the simulated (motion parallax) position of the top cylinder and the optical contact position of the lowest cylinder. This was based on the assumption that when multiple cylinders are present, and these cylinders have a common motion, it is the optical contact position of the lowest cylinder, not that

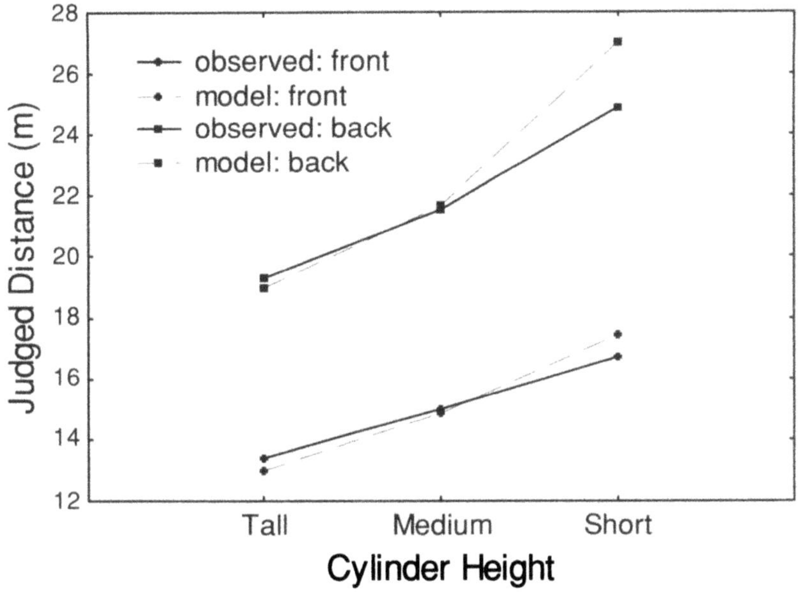

Figure 8. Comparison of model estimates to observed data in Experiment 1.

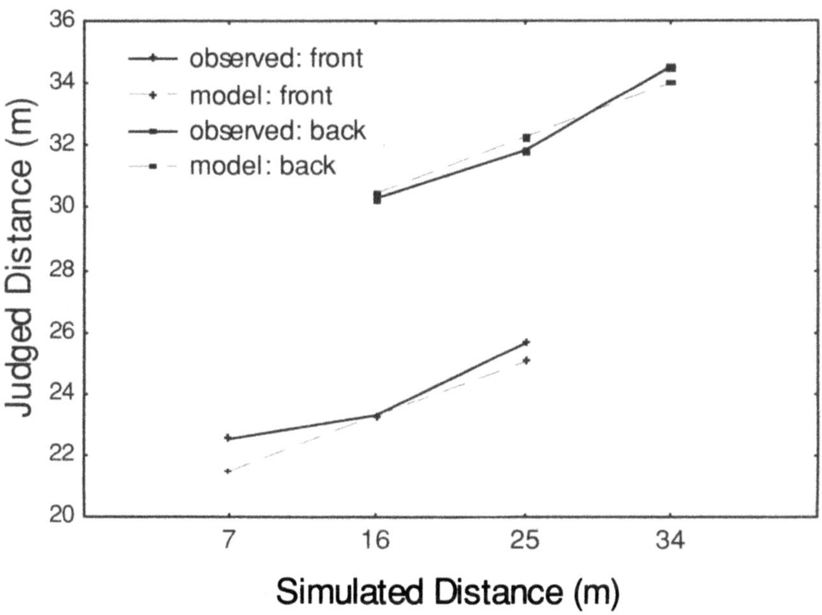

Figure 9. Comparison of model estimates to observed data in Experiment 2.

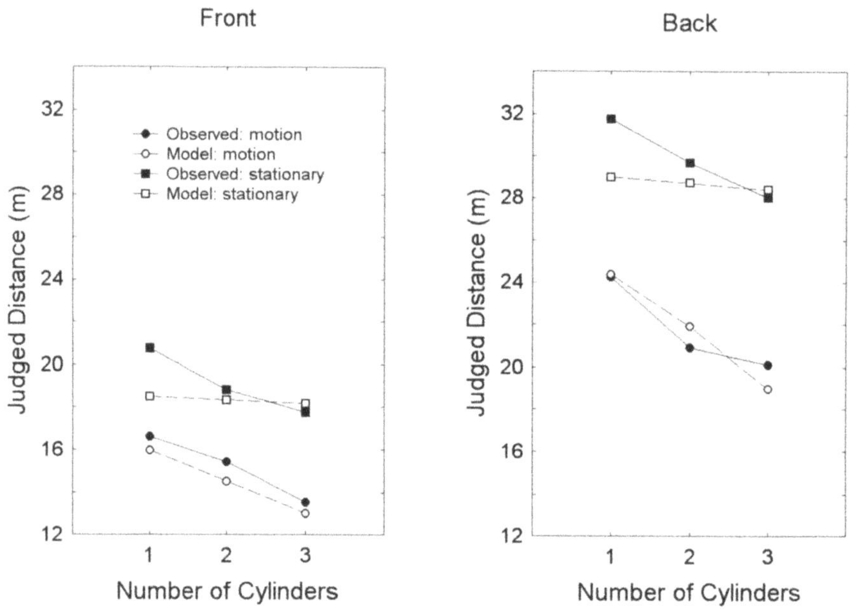

Figure 10. Comparison of model estimates to observed data in Experiment 3.

of the top cylinder, that affects the judged position of the top cylinder. The equation used was judgement = w_{motion} * simulated position + w_{lowest} * lowest contact. The value used for w_{motion} was .46. The estimated and observed judgements are compared in Figure 10. The model accounted for 95.3% of the variance in the stationary data and 97.6% of the variance in the motion data. The error in the predictions cannot be reduced for the stationary condition for one or two cylinders, without adding a constant to the prediction equation, because the judged values do not fall between the two values used in the prediction. Although these simple weighting functions generally fit the observed data, at least for the motion conditions, this does not rule out alternative possibilities for accounting for the results. The model estimates are presented only to show that the results are generally consistent with a compromise solution based on a weighted function of two variables in each experiment.

The present results provide further support for Gibson's (1946/1958, 1950) theoretical position that the perception of the distances of objects in a 3-D scene depends on how these objects are related to the ground surface. In addition, our results provide indirect support for Meng and Sedgwick's (2001, 2002) conclusion that ground contact relations can be mediated through intermediate surfaces. With more than one cylinder in a scene, whether the scene was stationary or moving, the judged position of the top cylinder appears to have been influenced by the optical contact position of the lowest cylinder, although the cylinders were not in contact with each other in the projection or in the simulated scene. This suggests that nested contact relations may propagate across gaps between objects, especially when the objects are grouped perceptually. In the present case, common 3-D motion of the cylinders appears to have been especially effective in enhancing this grouping.

REFERENCES

Cutting, J. E., & Vishton, P. M. (1995). Perceiving layout and knowing distances: The integration, relative potency, and contextual use of different information about depth. In W. Epstein & S. Rogers (Eds.), *Handbook of perception and cognition: Vol. 5. Perception of space and motion.* New York: Academic Press.

Feria, C. S., Braunstein, M. L., & Andersen, G. J. (2003). Judging distance across texture discontinuities. *Perception, 32*, 1423–1440.

Gibson, J. J. (1958). Perception of distance and space in the open air. In D. C. Beardslee & M. Wertheimer (Eds.), *Readings in perception.* Princeton, NJ: D. Van Nostrand. (Reprinted from *Motion picture testing and research*, AAF program Report no. 7, by J. J. Gibson, Ed., 1946, Washington, DC: US Government Printing Office, 1947).

Gibson, J. J. (1950). *The perception of the visual world.* Boston: Houghton Mifflin.

Gogel, W. C. (1965). Equidistance tendency and its consequences *Psychological Bulletin, 64*, 153–163.

Howard, I. P., & Rogers, B. J. (2002). *Depth perception.* Thornhill, Ontario, Canada: I. Porteous.

Levin, C., & Haber, R. N. (1993). Visual angle as a determinant of perceived interobject distance. *Perception and Psychophysics, 54*, 250–259.

Loomis, J. M., da Silva, J. A., Fujita, N., & Fukusima, S. S. (1992). Visual space perception and visually directed action. *Journal of Experimental Psychology: Human Perception and Performance, 18*, 906–921.

Loomis, J. M., da Silva, J. A., Philbeck, J. W., & Fukusima, S. S. (1996). Visual perception of location and distance. *Current Directions in Psychological Science, 5*, 72–77.

Loomis, J. M., Philbeck, J. W., & Zahorik, P. (2002). Dissociation between location and shape in visual space. *Journal of Experimental Psychology: Human Perception and Performance, 28*, 1202–1212.

Meng, J. C., & Sedgwick, H. A. (2001). Distance perception mediated through nested contact relations among surfaces. *Perception and Psychophysics, 63*, 1–15.

Meng, J. C., & Sedgwick, H. A. (2002). Distance perception across spatial discontinuities. *Perception and Psychophysics, 64*, 1–14.

Sauer, C. W., Braunstein, M. L., Andersen, G. J., & Bian, Z. (2001). Judged shape of ground plane regions in realistic 3-D scenes [Abstract]. *Journal of Vision, 1*, 41.

Sedgwick, H. A. (1983). Environment-centered representation of spatial layout: Available visual information from texture and perspective. In J. Beck, B. Hope, & A. Rosenfeld (Eds.), *Human and machine vision* (pp. 425–458). New York: Academic Press.

Sedgwick, H. A. (1987). Layout2: A production system modeling visual perspective information. In *Proceedings of the first international conference on Computer Vision* (pp. 662–666). Washington, DC: IEEE Computer Society Press.

Sedgwick, H. A. (1989). Combining multiple forms of visual information to specify contact relations in spatial layout. In P. S. Schenker (Ed.), *Sensor fusion II: Human and machine strategies* (SPIE proceedings Vol. 1198, pp. 447–458). Bellingham, WA: The International Society for Optical Engineering (SPIE).

Sedgwick, H. A. (2001). Visual space perception. In E. B. Goldstein (Ed.), *Blackwell handbook of perception* (pp. 128–167). Malden, MA: Blackwell.

Sedgwick, H. A., & Levy, S. (1985). Environment-centered and viewer-centered perception of surface orientation. *Computer Vision, Graphics, and Image Processing*, *31*, 248–260.

Sinai, M. J., Ooi, T. L., & He, Z. J. (1998). Terrain influences the accurate judgment of distance. *Nature*, *395*, 497–500.

Toye, R. C. (1986). The effect of viewing position on the perceived layout of space. *Perception and Psychophysics*, *40*, 85–92.

Ullman, S. (1979). *The interpretation of visual motion*. Cambridge, MA: MIT Press.

SUBJECT INDEX

For Product Safety Concerns and Information please contact our EU representative GPSR@taylorandfrancis.com
Taylor & Francis Verlag GmbH, Kaufingerstraße 24, 80331 München, Germany

www.ingramcontent.com/pod-product-compliance
Ingram Content Group UK Ltd.
Pitfield, Milton Keynes, MK11 3LW, UK
UKHW020023200926
475702UK00014B/2035
9781138873278